Sustainable Design in Fashion and Textiles

Addressing the urgent environmental issues resulting from conventional textile production, this book delivers essential knowledge and tools that will enable policymakers, industry professionals and students to implement sustainable practices and reduce the environmental impact of these sectors globally.

Providing a comprehensive guide on how to apply sustainable design principles, the book covers a broad range of topics and opens with an introduction to sustainable design concepts and the pressing need for change in the textile industry. It examines environmentally friendly processing techniques, sustainable fibres and materials, life-cycle analysis and the environmental impacts of current practices through case studies from Asia, Europe and North America. It promotes discussions on traceability and the integration of circular-economy principles, including upcycling, closed-loop systems and disassembly-friendly design, and showcases practical applications of these ideas. Each chapter concludes with summaries of the key concepts and practical takeaways. The book provides readers with an in-depth understanding of the commercial implications of sustainable design, including market trends, regulatory compliance, benefits of eco-labelling and certification, and the promotion of openness and accountability in the supply chain.

Designed as both a professional reference and an academic resource, this book catalyses systemic transformation by equipping stakeholders with the knowledge and tools necessary to reduce environmental impact, enhance transparency and build resilient circular-business models. In doing so, it supports industry professionals, such as fashion designers, textile engineers, sustainability consultants, businesses and policymakers, with everything they need to stay ahead in a rapidly developing sector.

Yasir Nawab is Vice Chancellor of the University of Kamalia, Pakistan, and a tenured professor at the National Textile University, Faisalabad, with an adjunct affiliation at North Carolina State University, USA. He has authored over 170 journal publications and edited more than ten books. His research focuses on sustainable textiles, the circular economy and advanced materials. He serves as co-chairman of the Pakistan Circular Textile Platform and contributes to global sustainability initiatives, including collaborations with WWF-Pakistan. Dr Nawab is a member of leading professional bodies such as the Materials Research Society and the Textile Institute, UK.

Muzzamal Hussain is CEO of KnowTex, Pakistan, and Assistant Professor at the National Textile University, Faisalabad. He also served as Director of the National Centre for Composite Materials. His research focuses on sustainable textile engineering, composite materials and alternative fibres. He has authored over 22 journal publications and several book chapters and has contributed to technology commercialisation, including licensed innovations using agricultural waste fibres. Dr Hussain is a member of professional bodies including the Society of Plastics Engineers, USA, the Institution of Engineers, Pakistan, and the Pakistan Engineering Council.

Ali Raza Shafqat is CEO of ReComp Solutions, Pakistan, and a lecturer in the Department of Textile Engineering & Design at Balochistan University of Information Technology, Engineering and Management Sciences (BUITEMS), Quetta. He is also a Ph.D. scholar at the National Textile University, Faisalabad. His research focuses on sustainable textile composites, the circular economy and the valorisation of textile and agricultural waste into innovative materials. He has contributed to multiple research projects, publications and international conferences. His work integrates academia and industry to develop scalable, sustainable solutions for the fashion and textile sector.

Shahid Rasul is Associate Professor of Mechanical Engineering at Northumbria University, UK. His research focuses on sustainable materials, the circular economy and energy systems. He leads innovative initiatives such as the Texonomy project, aimed at transforming the textile sector through circular and low-carbon solutions. With international academic experience spanning Japan, Saudi Arabia and the UK, he actively collaborates with industry and academia to address global sustainability challenges and promote clean energy and materials innovation.

Sustainable Design in Fashion and Textiles

Concepts, Case Studies, and Practical Tools

Edited by Yasir Nawab, Muzzamal Hussain, Ali Raza Shafqat, and Shahid Rasul

Routledge
Taylor & Francis Group
LONDON AND NEW YORK

Designed cover image: Shutterstock

First published 2027
by Routledge
4 Park Square, Milton Park, Abingdon, Oxon OX14 4RN

and by Routledge
605 Third Avenue, New York, NY 10158

Routledge is an imprint of the Taylor & Francis Group, an informa business

British Library Cataloguing-in-Publication Data
A catalogue record for this book is available from the British Library

Library of Congress Cataloging-in-Publication Data
Names: Nawab, Yasir editor | Hussain, Muzzamal, 1989– editor | Shafqat, Ali Raza editor | Rasul, Shahid editor
Title: Sustainable design in fashion and textiles : concepts, case studies, and practical tools / edited by Yasir Nawab, Muzzamal Hussain, Ali Raza Shafqat and Shahid Rasul.
Description: Abingdon, Oxon ; New York, NY : Routledge, 2027. | Includes bibliographical references and index.
Identifiers: LCCN 2026018308 (print) | LCCN 2026018309 (ebook) | ISBN 9781041069645 hardback | ISBN 9781041069614 paperback | ISBN 9781003638148 ebook
Subjects: LCSH: Clothing trade—Environmental aspects | Textile industry—Environmental aspects | Fashion design | Textile design | Sustainable design | Life cycle costing
Classification: LCC TT497 .S96 2026 (print) | LCC TT497 (ebook)
LC record available at https://lccn.loc.gov/2026018308
LC ebook record available at https://lccn.loc.gov/2026018309

ISBN: 9781041069645 (hbk)
ISBN: 9781041069614 (pbk)
ISBN: 9781003638148 (ebk)

DOI: 10.4324/9781003638148

Typeset in Times New Roman
by codeMantra

Contents

Abbreviations

AGEC	Anti-Gaspillage pour une Économie Circulaire
AHP	Analytic Hierarchy Process
BASF	Badisin Anili und Soda Fabric
BEE	Bureau of Energy Efficiency
C2C	cradle-to-cradle
CPS	cyber-physical systems
CSR	corporate social responsibility
CSRD	Corporate Sustainability Reporting Directive
DEI	diversity, equity and inclusion
DES	Discrete Event Simulation
DfD	design for disassembly
DfR	design for recycling
DPP	digital product passport
DSS	decision-support system
EIA	environmental impact assessment
EIB	European Investment Bank
EPD	Environmental Product Declaration
ESG	environmental, social and governance
GHG	greenhouse gas
GOTS	Global Organic Textile Standard
GRI	Global Reporting Initiative
IFC	international funding processes
ILO	International Labor Organization
IPCC	Intergovernmental Panel on Climate Change
IPM	integrated pest management
ITC	International Trade Centre
KPIs	key performance indicators
LCA	life-cycle assessment
MCI	material circularity indicator
MEG	monoethylene glycol
NGO	non-governmental organisation
OECD	Organisation for Economic Co-operation and Development
PaaS	product-as-a-service
PCR	product categories rules
PEF	product environmental footprint

PET polyester
PFC polyfluoroalkyl substance
PFMB Preferred Fibre and Materials Benchmark
PHA polyhydroxyalkanoates
PLA polylactic acid
PP pollution prevention
PTA pure terephthalic acid
RFID Radio Frequency Identification
PET recycled polyester
SASB Sustainability Accounting Standards Board
SBM sustainable business model
SBP State Bank of Pakistan
SCM supply chain management
SME small and medium-sized enterprise
TBL triple bottom line
TCFD Task Force on Climate-related Financial Disclosures
UNESCO United Nations Educational, Scientific and Cultural Organization
UNIDO United Nations Industrial Development Organization
VOC volatile organic compounds
WRAP Worldwide Responsible Accredited Production
WWF World Wildlife Fund

Contributors

Faheem Ahmad is an assistant professor in the Department of Textile Technology at National Textile University, Faisalabad, Pakistan. He holds a Ph.D. in Materials Science and Engineering from Koç University, Türkiye. His research interests include textile fibres, yarn manufacturing, non-wovens, fibre-reinforced non-woven composites, clothing comfort and nanomaterials, particularly aerogel-based textiles. He has published multiple research articles and has industry experience in textile production and planning.

Sheraz Ahmad has a Ph.D. in textiles with over 15 years of academic/industrial experience. He currently serves as a faculty member at the National Textile University, Pakistan. His research focuses on alternative natural fibres, gel-based textiles and valorisation of agricultural waste. Dr. Ahmad is one of the founders and the CEO of the Alternative Fiber Solutions, a faculty start-up, which is successfully commercialising sustainable and eco-friendly fibres and technologies in collaboration with leading industry partners in Pakistan.

Zeeshan Ahmed is a textile professional with more than 25 years of industrial experience in fibre and textile manufacturing, currently associated with Neela by Sapphire Fibres as General Manager, Research and Development. He holds an academic background in textile-related studies from the Sindh University, Jamshoro. His work focuses on textile production, material development and industrial processing. His interests include advancing efficiency and sustainability within textile manufacturing systems with a focus on practical, industry-driven solutions.

Farooq Azam is a lecturer at the Department of Textile Technology, National Textile University, Faisalabad, Pakistan. He is involved in teaching and academic activities within the textile engineering domain. His role focuses on contributing to undergraduate education and supporting academic development in textile technology.

Ghazia Batool is serving as a lecturer in Faisalabad Business School at the National Textile University, Faisalabad, Pakistan. She is also a Ph.D. scholar in finance at the National Textile University, Faisalabad. Her research focuses on sustainability, financial systems and the economic implications of environmental regulations in the textile sector. She has a strong academic background in business and finance and is actively engaged in interdisciplinary research linking sustainability and financial performance.

Shubhajit Dutta is a doctoral researcher at Aalto University, where his work focuses on textile recycling. He holds an M.Sc. (Technology) in creative sustainability with a specialisation in materials and chemical engineering from Aalto University. His research interests encompass sustainable textile manufacturing and recycling processes, functional material development and technical textiles.

Hüseyin Aksel Eren is a professor in the Department of Textile Engineering at Bursa Uludağ University, Türkiye. He holds a Ph.D. in textile engineering. His research focuses on decolorisation of textile materials, waterless dyeing, ozone-based textile treatments and sustainable textile processes. He has published scientific articles and book chapters in these areas and continues to work on improving environmentally friendly approaches in textile processing.

Matteo Gallidabino is Lecturer in Forensic Chemistry at King's College London, where he applies analytical science and modelling approaches to the study of trace materials. His research includes fibre analysis as well as the investigation and modelling of fibre shedding, transfer and fragmentation in textiles. He has particular expertise in combining experimental data with statistical and computational methods to better understand the behaviour of fibres in both forensic and environmental contexts. Through this work, he contributes to science-led approaches that can inform more sustainable garment design and support innovation towards reduced environmental impact in the fashion and textiles sector.

Muhammad Dawood Husain is currently working as a professor at the Textile Engineering Department, NED UET, Pakistan, specialising in smart and medical textiles. He holds a Ph.D. in textile engineering from the University of Manchester and a master's degree from Hochschule Niederrhein, Germany. His research interests include textile sensors, soft robotics, performance wear, sustainability and Industry 4.0. With over 22 years' teaching and research experience, he has secured significant research funding as PI/Co-PI and serves as Associate Editor of the *AATCC Journal of Research* while also contributing to advanced research in sustainable manufacturing and responsive textile systems.

Muzzamal Hussain is CEO of KnowTex, Pakistan, and Assistant Professor at the National Textile University, Faisalabad. He also serves as Director of the National Centre for Composite Materials. His research focuses on sustainable textile engineering, composite materials and alternative fibres. He has authored over 22 journal publications and several book chapters and has contributed to technology commercialisation, including licensed innovations using agricultural waste fibres. Dr Hussain is a member of professional bodies including the Society of Plastics Engineers (USA), the Institution of Engineers Pakistan, and the Pakistan Engineering Council.

Abigail Irving-Munro is an ethical fashion designer specialising in sustainable design and circular fashion systems. Her work focuses on developing innovative approaches to address the disposable nature of garments and their impact on the second-hand clothing economy. Her research investigates consumer behaviour, material fibre composition, supply-chain logistics and ethical values influencing fashion consumption. She aims to propose regenerative, localised systems that can drive structural change within the second-hand clothing supply chain and support more sustainable and responsible fashion practices.

Alana James is an associate professor in fashion at the School of Design, Arts and Creative Industries at Northumbria University. An experienced fashion designer, educator and researcher, Alana's academic career is dedicated to creating sustainable change for the future of the fashion industry by using design tools, methods and approaches. In 2023, she was awarded £1.9 million as principal investigator to lead IMPACT+, funded by UKRI Circular Fashion and Textile Programme: NetworkPlus. Alana's contribution to her field globally has been acknowledged through two prestigious international appointments, including Visiting Professor at BINUS University in Indonesia and Adjunct Professor at Kamalia University in Pakistan.

Chimdia KeChi-Okafor is an environmental forensics expert specialising in microfibre pollution and sustainability. She served as a research fellow at Northumbria University, where she led cross-disciplinary projects under the UKRI IMPACT+ Circular Fashion and Textiles programme. Her research focuses on identifying synthetic and natural fibres as well as the biodegradation products of textile fibres, across marine, freshwater and atmospheric systems. This work contributes to evidence-based solutions with industrial and policy relevance. Her work bridges science, industry and policy, with a strong emphasis on sustainability innovation and mentoring emerging researchers.

Danyal Rashid Khan is a lecturer in the Department of Textile Engineering at NED University of Engineering & Technology. His work focuses on the intersection of smart textiles and sustainable manufacturing, specifically exploring how advanced materials can modernise traditional industrial practices. A Master of Engineering Management in Industrial Management from NEDUET, Danyal's research bridges the gap between established engineering methods and emerging trends. His primary goal is to drive the global textile industry towards a more technologically integrated and environmentally responsible future. By focusing on eco-textiles and functional materials, he aims to develop solutions that are both technically feasible and sustainable.

Raja Muhammad Waseem Ullah Khan is a textile engineer and R&D professional with expertise in sustainable materials, textile recycling and advanced composite development. He is serving as Senior Manager, Research and Development at Artistic Milliners, Karachi. He has 11 research publications and four book chapters. His research focuses on circular-economy approaches, waste valorisation and the integration of AI in textile processes. He has contributed to different scholarly publications and industry-driven innovations, with a strong interest in bridging academia and industrial sustainability practices.

Awais Khatri is a professor of textile engineering at Mehran University of Engineering and Technology, Pakistan, and editor of the *Mehran University Research Journal of Engineering and Technology*. With around 23 years of experience in academia, research and industry collaboration, his expertise centres on sustainable textile processing, coloration and functional finishing. He has led multiple funded projects and contributed to curriculum development and accreditation systems. He has authored around 80 scholarly publications, including research and review articles, monographs, book chapters and practical workbooks. His work contributes to advancing sustainable design practices and strengthening linkages between academia and industry.

Shumaila Kiran is Professor and Chairperson in the Department of Applied Chemistry at Government College University, Faisalabad, Pakistan, specialising in chemistry. She holds a doctorate in chemistry from the University of Agriculture, Faisalabad (2012). Her research interests include wastewater management, synthesis of synthetic dyes, extraction and application of natural dyes, phytochemical analysis and pharmaceutical properties of medicinal plants, trace analytical chemistry and environmental electrochemistry. She has completed three research projects, published over 100 research articles and more than 40 book chapters, and edited a book on biomedical textiles. She has 14 years of teaching and research experience. During her academic career, she supervised seven Ph.D., 50 masters, and almost 100 bachelors' degree graduates.

Bushra Mushtaq is a laboratory engineer at the Textile Testing Laboratory, National Textile University, Faisalabad, Pakistan. She is also a Ph.D. scholar specialising in textile engineering and sustainable materials. Her research focuses on alternative natural fibres, hydrogels and bio-based composite materials for applications such as wound dressings, smart textiles and eco-friendly products. She has contributed to multiple peer-reviewed publications in areas including non-woven materials and sustainable textile innovations, with a strong emphasis on circularity, waste utilisation and environmentally responsible textile solutions.

Nkumbu Mutambo is a researcher and designer currently pursuing a Ph.D. at Northumbria University. Her research explores the role of heritage in Zambian fashion design, focusing on the intersection of history, identity and creative practice. With a background as a creative practitioner, she investigates how art and design shape societal narratives, particularly amplifying marginalised voices. She is also part of the Creative Fuse initiative, supporting internships and contributing to research on the development and impact of the creative and cultural industries in the north-east of England.

Josphat Igadwa Mwasiagi is currently an associate professor at the Department of Manufacturing, Industrial and Textile Engineering, Moi University, Eldoret, Kenya, specialising in textile engineering and sustainability. He also chairs ARSO TC 53 on textiles. He holds a Ph.D. in textile engineering from Donghua University in the industry areas. His research interests include textile manufacturing, technical textiles and sustainability in the textile and fashion value chain. His current projects include InTex (Innovative Business Practices and Economic Models in the Textile Value Chain), labour market assessment for TVET textile and fashion courses and valorisation of agricultural waste.

Jawad Naeem is an assistant professor at the Department of Textile Technology, National Textile University, Faisalabad, Pakistan. He holds an M.Sc. in textile engineering and is engaged in teaching and academic activities within the textile domain. His work focuses on supporting undergraduate education and contributing to the development of textile engineering knowledge and practice.

Anil Namdeo is Professor of Air Quality Management and Net Zero at Northumbria University and Commissioner on the Global Commission on Healthy Indoor Air. His research integrates environmental sustainability, air-quality science and health impacts, with particular focus on circular economy and net-zero transitions. He works on the environmental impacts of fashion, circularity and microfibres in the atmosphere as Co-investigator on the Impact+ Network. His international collaborative research spans India, Mexico, China, South-East Asia and Africa, addressing environmental justice and sustainability challenges. He is a fellow of the Royal Meteorological Society and the Institute of Air Quality Management.

Yasir Nawab is Vice Chancellor of the University of Kamalia, Pakistan, and a tenured professor at the National Textile University, Faisalabad, with an adjunct affiliation at North Carolina State University. He has authored over 170 journal publications and edited more than ten books. His research focuses on sustainable textiles, the circular economy and advanced materials. He serves as Co-chairman of the Pakistan Circular Textile Platform and contributes to global sustainability initiatives, including collaborations with WWF-Pakistan. Dr Nawab is a member of leading professional bodies such as the Materials Research Society and the Textile Institute, UK.

Evance Obara is currently a doctoral research associate at the Center for Civil Aviation Composites in Donghua University, China, specialising in high-performance fibres for military applications. He holds a master's degree in textile engineering with a major in composite materials from Donghua University. His research interests include the development of sustainable high-performance fibres, circular economy principles in advanced composites and next-generation material solutions for aerospace structures. Evance boasts a strong publication record and has pioneered recycling solutions for high-performance textiles and composites, supporting industry-funded aerospace projects. He also consults on sustainable materials and product innovation for Kenyan textile companies.

Anne Peirson-Smith is a Professor of Fashion at Northumbria University, within the School of Design, Arts and Creative Industries. She holds a Ph.D. and has an industry background in branding and public relations. Her research specialises in sustainable fashion management, fashion marketing and communication. She has published numerous articles and book chapters and has co-authored and edited several influential books in fashion studies. She also serves as Associate Editor for leading journals, including the *Journal of Fashion, Style and Popular Culture* and the *Journal of Global Fashion Marketing.*

Miranda Prendergast-Miller is an assistant professor in the School of Geography and Natural Sciences at Northumbria University. Her research focuses on soil ecology and nutrient cycling, and the impact of human activities on soil health. Her work addresses global sustainability challenges, including soil carbon sequestration, microplastic and microfibre pollution, and sustainable soil management.

Shahid Rasul is an associate professor of mechanical engineering at Northumbria University. His research focuses on sustainable materials, the circular economy and energy systems. He leads innovative initiatives such as the Texonomy project, aimed at transforming the textile sector through circular and low-carbon solutions. With international academic experience spanning Japan, Saudi Arabia and the UK, he actively collaborates with industry and academia to address global sustainability challenges and promote clean energy and materials innovation.

Sadia Razzaq is currently working as a research assistant at the National Textile University, Pakistan, specialising in organic chemistry. She holds an M.Phil. in chemistry from the Government College Women University Faisalabad. Her research interests include bioplastics and biodegradable polymers.

Md. Reazuddin Repon is currently a postdoctoral researcher at the Department of Bioproducts and Biosystems, School of Chemical Engineering, Aalto University, Finland. He previously worked at the Department of Production Engineering, Faculty of Mechanical Engineering and Design, Kaunas University of Technology. His research focuses on biomaterials, green coloration and finishing, smart textiles and recycling. His disciplinary expertise spans materials engineering, chemical engineering and environmental engineering.

Alberto Saccavini, consultant for sustainability in fashion. Over the past ten years, he has worked at all stages in the fashion supply chain from fibre production to consumer behaviour, collaborating with United Nations Industrial Development Organization (UNIDO), European Commission, Blumine srl and many others. He is a lecturer at the Milan Polytechnic, the Milano Fashion Institute and other institutions.

Tonmoy Saha is a doctoral researcher in the Textile Chemistry Group at Aalto University, working within the bioproducts and biosystems domain. With a background in fibres and textile engineering, his research focuses on sustainable textile processing, bio-based materials and recycling. His work addresses critical challenges in textile circularity, particularly the decolorisation of dyed cotton waste for fibre-to-fibre recycling. He has developed a chemical stripping approach demonstrating high decolorisation efficiency, improved whiteness and retention of mechanical properties, with successful pilot-scale validation in collaboration with industry partners.

Abdel-Fattah M. Seyam is the Charles A. Cannon Distinguished Professor of Textiles at North Carolina State University, Wilson College of Textiles. He is a globally engaged textile engineer, technologist and educator whose career spans academia and industry. His expertise focuses on the intersection of textile engineering (woven, non-woven, composite structures), advanced manufacturing (3D printing, robotics, melt-blown processes) and smart materials (electrotextiles, fibre sensors). He places strong emphasis on graduate mentorship and translating textile research towards functional, advanced textile systems.

Ali Raza Shafqat is CEO of ReComp Solutions, Pakistan, and a lecturer in the Department of Textile Engineering & Design at BUITEMS, Quetta. He is also a Ph.D. scholar at the National Textile University, Faisalabad. His research focuses on sustainable textile composites, the circular economy and the valorisation of textile and agricultural waste into innovative materials. He has contributed to multiple research projects, publications and international conferences. His work integrates academia and industry to develop scalable, sustainable solutions for the fashion and textile sector.

Kelly J. Sheridan is currently an associate professor in forensic and environmental science at Northumbria University and CEO of the Microfibre Consortium, specialising in fibre science, environmental impact assessment and cross-sector collaboration. She holds a Ph.D. in inorganic chemistry from the University of Liverpool. Her research interests include fibre fragmentation, textile fibre analysis and the development of mitigation strategies across the apparel value chain. Kelly has extensive experience working at the interface of academia and industry, leading global, science-led initiatives to reduce the environmental impact of textiles.

Amna Siddique is an assistant professor at the National Textile University, Pakistan. She specialises in textile and materials-related disciplines within the Department of Applied Sciences. Her role encompasses teaching and research in areas aligned with textile science and applied materials. As part of a leading textile-focused academic institution, her work contributes to advancing knowledge in textile engineering and interdisciplinary scientific applications.

Thomas Stanton is a lecturer in Geography at Loughborough University. His research and teaching explore the environmental prevalence, pathways, impacts and perceptions of pollution, with a focus on aquatic environments. Tom's work lies at the interface of human and physical geography. He collaborates with scholars across multiple disciplines to explore these environmental challenges, utilising interdisciplinary methods and partnerships.

Mehvish Sultan is a lecturer at the Faisalabad Business School, National Textile University, where she has been serving since 2019. She holds bachelor's and master's degrees from the same institution and was awarded a gold medal by former president of Pakistan Mamnoon Hussain for her academic excellence. She actively supervises curricular and co-curricular activities and serves as Deputy Director of the Alumni Office of the University. Her research and teaching interests include management and marketing, with a focus on integrating academic knowledge with industry-oriented practices. Currently, she's practising research in sustainability and innovation for organisations.

Sami Ur Rehman is a textile engineer and HSE (Health, Safety & Environment) professional, holding a gold medal in MS textile engineering. He has over six years of industry and academic experience. He has served as a lecturer and safety trainer at BUITEMS University and the National Textile University. Currently, he is working as an HSE officer at Doosan Power Systems Arabia in the Kingdom of Saudi Arabia. He is also the founder of EHS Solutions Pvt. Ltd., where he provides professional safety training and consultancy services.

Amna Warda is currently leading a brand named Scrunchies Knot, which mainly focuses on fashion and design for apparel. She has completed her graduation in textile design and technology from the National Textile University. She holds an M.Sc. degree in clothing and textiles from the University of Agriculture. Her research interests include online branding, art and design strategies for commercial products, manual painting, sketching and fabric manipulation techniques.

Syed Zameer Ul Hassan is an academic and researcher specialising in textile engineering and materials. He holds a Ph.D. in textile engineering from the Technical University of Liberec, Czech Republic. He has served as an assistant professor at BUITEMS, Pakistan. His research interests include textile composites, smart materials, cotton fibre analysis, pesticide residue assessment, biosensor applications and environmental impacts on textile materials. He has published in international journals and presented his work at leading global conferences. His work focuses on the interaction of textile materials with environmental and health-related factors.

Foreword

Marco Ricchetti

We are living in an era of transformation, situated at the crossroads of multiple and interrelated transitions – environmental, societal, cultural, geopolitical and technological. Collectively, these shifts impose new layers of uncertainty and risk across the strategic landscape for global industries, with the fashion system facing a particularly complex set of imperatives.

Sustainability is fundamentally interwoven with each of these transitions. While environmental stewardship and social equity constitute its core imperatives, achieving a truly sustainable trajectory necessitates profound cultural shifts and the strategic deployment of technological innovation. In the context of geopolitical tensions, the global drive towards decarbonisation is recalibrating strategic dependencies, shifting the locus of power away from oil-producing regions. This transition is introducing new forms of economic diplomacy, with mechanisms such as carbon tariffs and compensatory capital flows from high-emission economies to those most vulnerable to the climate crisis. Finally, effectively addressing global environmental impacts and scaling circular business models reinforces the need for international harmonisation of sustainability standards and for incentives for transnational cooperation.

As a cornerstone of the modern consumer economy, fashion ranks as the second-largest category in household consumption, trailing only food and beverages, accounting for approximately 10 per cent of total household expenditure on tangible goods. Latest data from Textile Exchange indicates that per-capita consumption of textile fibres has climbed steadily, reaching a peak of 16.2 kilograms in 2024, the most recent year of record. This represents a staggering 27 per cent surge over the 2019 pre-pandemic baseline. This consumption trend is further amplified by global demographic shifts. According to United Nations projections, the world population is expected to reach 8.5 billion by 2030, an increase of 500 million from the 2023 baseline. The compounding effect of rising per-capita consumption and steady population growth is projected to drive global fibre demand to approximately 150 million tonnes by 2030, a substantial leap from the 112 million tonnes recorded in 2022.

Given its global scale, economic significance, growth rate and structural features, the fashion industry bears a pivotal responsibility in the pursuit of global sustainability. The massive scale of garment production inevitably translates into a significant environmental impact, making the industry's transformation a requirement for a sustainable future.

However, while the industry's profound impact is indisputable, its precise quantification remains a subject of intense academic debate. This 'metric challenge' is perhaps most visible in the estimation of greenhouse gas emissions, a vital benchmark for climate action. Estimates vary significantly across sources, due to 'wicked' methodological complexities, non-standardised data and divergent research boundaries. These complexities render a single, definitive figure elusive, leaving the identification of robust impact reduction targets uncertain and the establishment of sound transition policies for the sector a significant challenge. Yet, this empirical uncertainty does not absolve the industry of accountability; rather, it underscores the urgent need for fashion to pivot from a driver of environmental degradation to a fundamental part of the solution.

Indeed, the textile and fashion industry has made substantive advances towards sustainability since the turn of this century. Reflecting on the past two decades, the progress achieved is evident; however, a significant and steep journey still lies ahead. The fashion industry's climate impact requires an honest assessment of both concerning trends and emerging signs of positive change. On the one hand, the empirical reality is stark: while the 2015 Paris Agreement, supported by the IPCC's 2023 Synthesis Report, mandates a 43 per cent reduction in global emissions by 2030 to limit warming to 1.5°C, industry's current emissions trajectory continues to rise. Under 'business-as-usual' scenarios, the gap between fashion's growth and planetary boundaries remains dangerously wide, suggesting that global climate targets may be slipping out of reach.

However, this systemic risk is met by an unprecedented surge in corporate decarbonisation commitments and strategic mobilisation. There is reason for optimism in the growing number of fashion entities and multinational groups formally committed to science-based targets. Crucially, most of these pledges are not restricted to direct operations (Scopes 1 and 2) but extend deep into the supply chain (Scope 3). This expansion represents a monumental shift in responsibility, requiring the active engagement of thousands of upstream partners. While we must acknowledge the persistent disparity between corporate public pledges and measurable performance, as well as recognise that these commitments vary significantly in maturity, they nonetheless signal a critical pivot. This shift from passive participation towards a design-led, proactive effort is essential to finally align the fashion system with a decarbonised future.

The current landscape has reached a critical inflection point, a trajectory shift where the broad consensus about the need for a more sustainable business model for the textile and fashion industry must finally yield to radical implementation. In this transitional space, there is a clear and strong demand for more rigorous research and more practical, scalable solutions for the industry. The *Sustainable Design in Fashion and Textiles* book directly addresses this need, providing essential food for thought.

The decision to place 'Design' at the forefront of this volume's title is both intentional and visionary. It signals a departure from the industry's historical tendency towards fragmented, end-of-pipe solutions and instead centres the discourse on the very genesis of industry impacts. Design is not merely an aesthetic exercise; it is the primary architectural phase that determines a product's life cycle. By emphasising a design-led approach, the book advocates for a holistic vision that transcends the 'silo' mentality often found in corporate sustainability departments. Without this integrated design perspective, sustainability efforts risk becoming a patchwork of reactive interventions and surface-level concessions, isolated tweaks to materials or logistics that fail to address systemic inefficiencies.

A second key contribution of this volume is the emphasis on how effective design is inextricably linked to innovation across materials, processes, and business models, which in turn is a core tenet of the modern approach to sustainability.

Finally, a further strength of this volume lies in its masterful management of complexity. Rather than simplifying the challenge, the wide-ranging topics covered across the chapters mirror the multilayered, systemic nature of sustainability itself. This comprehensive scope serves as a coordinated, holistic map. It offers readers the necessary navigational tools to move beyond superficial fixes and towards a profound, design-led transition to more sustainable business models for the textile and fashion industry.

Preface

Yasir Nawab

The fashion and textile sector stands at a defining crossroads. Since the last century, this sector has been a powerful engine of employment generation, economic growth, technology and innovation. Unfortunately, the fashion and textile sector has also been the most resource-intensive and environmentally degrading sector globally. Resource scarcity, climate change, policy evolution and ever-changing consumer expectations are demanding not just incremental adjustments but systematic redesign and fundamental transformation. This book is solely written to address this transformation. Throughout my academic career, industrial collaborations and interaction with sustainability initiatives and circular textile platforms, I have seen immense opportunities and challenges that exist in the fashion and textile sectors. Sustainable and circular design applications in fashion are not only a niche discourse but are shaping the future of this sector. Today, despite an increasingly produced literature, an immense need is appearing for an interdisciplinary resource that can weave the concepts of design, engineering, science, business and policy. This book aims to fill this gap. The scope of sustainable design in this sector must expand beyond mere material substitution and should involve systems thinking, life-cycle innovation, responsible production, circular business models and traceability. This expansion calls to fill in the gaps between environmental science, creative design and data-backed tools. Only through such initiative can this sector thrive within planetary boundaries, along with financial viability.

For this book, we have brought together diverse insights from both fashion and textiles, academia and industry, businesses and policy. The book consists of four parts, each forming well-structured chapters to guide scholars from foundational concepts and lead towards innovation in production processes, fibres, materials, ESG frameworks, eco-labelling and digital solutions. This book presents sustainability as an operating strategy that must be practical, measurable and scalable. The engagement of scholars, researchers, professionals and policymakers is a source of encouragement, due to which the sustainable transformation has become both an opportunity and a responsibility. Countries with strong dominance in the fashion and textile sector have enormous potential to lead the circular economy, only if sustainable design strategies are embedded from the very beginning.

This book is structured in such a way that it would help various audiences; it can help educators in shaping designers and engineers, business professionals struggling to navigate ESG benchmarks, policymakers in designing and crafting policy frameworks and researchers in advancing sustainability from science-led solutions. We hope that this book will help inspire innovation, stimulate creative thinking, embed critical thinking and accelerate the advancement of sustainability in this sector through collaborative projects. To redesign business models, reimagine supply chains and rethink processes, it takes courage to transform. This transformation would not be only in terms of aesthetics but an amalgam of transparency, resilience and responsibility. Through this book, we have tried to add our five cents to this transformation. If this book manages to have even modest improvement, we will have fulfilled our purpose.

Section 1

Foundations of sustainable design in textiles and fashion

1 Introduction to sustainable design in fashion and textiles

Ali Raza Shafqat and Yasir Nawab

1.1 Introduction, purpose and objectives of the book

Fashion and textiles are the second-largest waste-producing and polluting industries worldwide [1]. The commutative size of the fashion and textile market is approximately US$1.3 trillion [2]. The global fashion and textile industry employs approximately 300 million people across the whole supply chain [3]. This industry has an immense and considerable amount of environmental, social and economic footprints across the globe. In the past two decades, global textile production has nearly doubled from 58 million tonnes to 124 million tonnes from 2000 to 2023 [4]. The global textile growth is expected to reach 160 million tonnes by 2030 [5]. The textile sector is also the second-largest industrial sector that consumes a huge amount of water and roughly consumes 4 per cent of fresh water while producing approximately 20 per cent of the total industrial wastewater [6, 7].

The production of fashion and textile goods consumes a lot of resources and has detrimental consequences for the environment. The global textile industry contributes 10 per cent of greenhouse gases, which exceeds the marine and aviation industries combined [8]. This sector also has a carbon footprint of approximately 8 per cent of the total carbon footprint, and some sources predict that its carbon share will reach 26 per cent of the carbon budget by 2050 [9]. Approximately 85 per cent of all the textiles produced each year are either dumped or burned in landfills, leading to an approximate 500,000 tonnes of microplastics being released annually [10, 11]. These microplastics enter marine and terrestrial environments and impose a threat not only to humans but the whole ecosystem [12].

Textile and fashion sectors also have a great social impact as it is a very labour-intensive industry [13]. The fashion and textile industry accounts for low-paying jobs and often consists of remote production facilities that usually have unsafe work facilities, minimum wage violations, and even child labour or enslaved labour across countries like China, India and Bangladesh [14–16].

The increasing concerns of devastating environmental impacts, social injustice, resource depletion and economic instability demands an urgent need for sustainability and compliance in the fashion and textile industry. As fashion is becoming cheap and is changing rapidly, the market value is also increasing at a fast pace, with a projection to $150.8 billion in 2025 and to $291 billion by 2032, even though wide awareness that the market size of the fashion and textile industry is expected to increase immensely in the future too [17].

This book serves as an effort to position the fashion and textile industry around sustainable design. It discusses the importance of interdisciplinary actions, provides practical tools and shares real-world case studies to empower researchers, engineers, designers, business leaders and policymakers related to fashion and textiles to enhance circularity, material efficiency,

DOI: 10.4324/9781003638148-2

social equity and innovation in this sector, while keeping the dimensions of this sector within the boundaries of sustainable design.

1.2 Urgency and scope of this book

The fashion and textile sector needs an urgent transformation; now, the governments are coming to the forefront, including the European Union (EU) and France, among many others. They are developing and implementing new legislation, such as due-diligence requirements, extended producer responsibility and imposing several bans on unsold garment destruction and disposal [18]. Moreover, highly influential institutions such as the Global Fashion Agenda and the Global Fashion Summit are highlighting this. But without the immense change in the fashion and textile sector, the industry is projected to have more devastating and detrimental consequences in the future, and the regulatory and geopolitical challenges will keep rising while the industry will go into the closed streets of stagnation. Figure 1.1 demonstrates the purpose of this book in context to the issues and urgency of sustainability in fashion and textile sector.

The core objectives of this book include:

- Providing a solid foundation to the important pillars of sustainability, including environmental, economic, social and cultural dimensions
- Defining and elaborating on the concepts of sustainable design within the realm of fashion and the textile industry
- Providing the important theories, frameworks and tools that can enable scholars, practitioners and business professionals to design and develop equitable and resilient systems in their respective professional lives.
- Demonstrating the real-world case studies, including the policy updates, use of sustainable materials, educational interventions and circular business models in support of sustainable design in the fashion and textile industry.

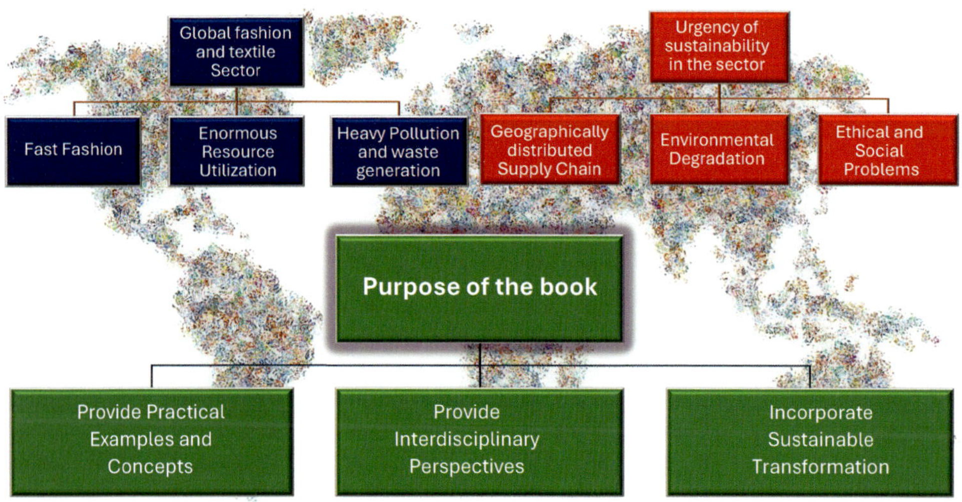

Figure 1.1 The purpose of this book in context to the issues and urgency of sustainability in fashion and textile sector.

Another important and core objective of this book is to introduce interdisciplinary learning across sustainable design as a core element of graduate studies that can enhance the core competency of scholars in design innovation, supply-chain business and engineering.

1.3 Defining sustainable design

1.3.1 The historical evolution of sustainability in fashion and textiles

In the late 20th century, the increase in broader demands of sustainability discourses emerged across the globe, which led to the introduction of sustainable design in the fashion and textile industry [19]. In 1987, the World Commission on Environment and Development published the Brundtland Report, which explained sustainable development as 'a development that meets the needs of the present generation without compromising the ability of future generations to meet their needs' [20]. This definition became the foundation for the introduction of sustainability into various industries, including the fashion and textiles industry. Figure 1.2 demonstrates the evolution of sustainability from consciousness to circular economy.

In the past few decades, sustainability in fashion and textiles has gained a lot of attention owing to environmental events and the consumer watchdogs that began drawing their attention to labour exploitation and environmental damage across the supply chains of the textile and fashion industry. Concepts and frameworks, like design for the environment, natural step and cradle to cradle proposed by McDonough and Braungart, transformed the design thinking to incorporate systems thinking, life-cycle responsibility and regenerative innovations into product and process development [21]. The fashion and textile industry began incorporating the ideas of sustainability beginning with the development of eco-friendly materials and the incorporation of corporate social responsibility strategies and life-cycle assessments (LCAs). In 2010, the sustainable design then evolved as a necessary action required to mitigate global climate change, reduce resource depletion and become an ethical and consumer-centric demand [22].

1.3.2 Sustainable design in the context of fashion and textiles

The sustainable design in context to the fashion and textile industry includes development and design of systems, processes and products with the main intention to minimise the detrimental environmental and social impacts throughout the entire life cycle of fashion and textile products. It takes into account everything, right from the fibre cultivation to the garment manufacturing, during the product's utility, leading to reuse and end-of-life disposal and remodification [23]. The sustainable design definition provides both the mindset and the methodology that balances the aesthetic value, functionality of material, ethical responsibility and economic viability alongside each other.

Figure 1.2 The evolution of sustainability from consciousness to circular economy.

The sustainable design in the fashion and textile industry integrates multiple dimensions, including environmental aspects, social dimensions, cultural, aesthetic relevance and economic resilience [24]. Environmental considerations include reducing the use of chemicals, water and energy, meanwhile encouraging the use of biodegradable and renewable materials and minimising the waste and carbon emissions [25]. The economic resilience consideration involves designing the business models in such a way that they are both environmentally responsible as well as profitable. The social dimensions include the incorporation of safe work conditions in labour practices and ensuring the well-being of workers and the community [26]. The cultural and aesthetic aspects include supporting the local economies, uplifting the heritage crafts and enhancing the long-term and emotional attachment towards the textile products. The sustainable design serves as a proactive entity rather than a reactive and it seeks to reduce the harm to the environment through innovation and foresight rather than reducing the damage after it happens.

1.3.3 The fundamental terms related to sustainability

Circular design: Circular design involves keeping the materials and end products in use through rotating them in circles, often revolving the products and materials in 9R strategies of the circular economy. This approach uses system-level design then revolves around the idea of designing the government for repair, reuse and upcycling right before initiating the manufacturing process. For example, designing for longevity, recyclability and disassembly.

Green design: This is a product-centric approach that encourages the use of environmentally benign choices. Green design offers a single stage focus at once, for example, the use of organic cotton instead of regular one or using natural fibres instead of synthetic fibres [27].

Sustainable design: Sustainable design covers the focus areas related to cultural aspects, environmental sustainability and social impacts of the product or process. It includes complete system-level thinking, for example, designing the product for disassembly, using closed-loop materials and implementing LCA strategies [28].

Eco-design: The term 'eco-design' focuses on environmental impact reduction and its scope is lifestyle, mostly process or lifestyle-focused redressal. For example, minimising the use of water, toxic dyes, or chemicals in the textile industry [29].

Ethical fashion: The term ethical fashion incorporates the key focus related to animal and human welfare into product design. It focuses on the production and sourcing ethics, and it revolves around fair trade ethics across the supply chain, for example, the use of vegan materials in textiles, ensuring reasonable or minimum work wages and ensuring a safe and healthy work environment [30].

1.4 Dimensions of sustainable design

A single focus, such as carbon-emission reduction or the use of eco-friendly materials, cannot define sustainable design in the fashion and textile industry completely. The incorporation of sustainable design in fashion and textiles spans across multidimensional considerations, as it has complex and interdependent entities linked together. The fundamental dimensions should incorporate the variables of environmental, cultural, social and economic aspects that encompass the design imperative, mitigate the risks and enhance the opportunities for systematic transformation of this sector [31].

1.4.1 Environmental dimension

The hazardous and far-reaching environmental aspects of the fashion and textile industry are renowned and well established. These environmental aspects span across immense energy

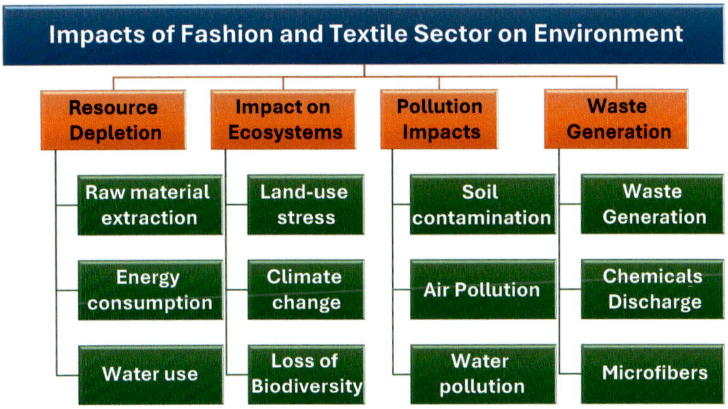

Figure 1.3 The environmental implications and impacts of fashion and textile sector.

consumption, resource depletion, land degradation, use of chemicals, enormous carbon emissions, enormous generation of waste and loss of biodiversity [3]. The environmental implications and impacts of fashion and textile sector are visually demonstrated in Figure 1.3.

1.4.1.1 Resource depletion

Textile and fashion sectors consume a huge amount of raw materials, often consuming 93 billion cubic metres of water, which is equivalent to meeting the needs of approximately 5 million people. Cotton production involves the use of a huge amount of pesticides and insecticides, approximately 11 per cent and 24 per cent respectively, making cotton the most chemically demanding crop. Textile manufacturing processes are also very energy-intensive processes and also use enormous amounts of synthetic materials. For example, polyester constitutes over 60 per cent on average in all textile products being made today, it is most widely used fibre in textiles and fashion sectors which is synthesised from petrochemicals and contributes to the dependency of fashion and textile products on fossil fuel for synthetic raw materials, widespread and extensive use of this fibre is ultimately causing resource depletion [32].

1.4.1.2 Waste generation and pollution

Approximately, the fashion and textile sector produces 92 million tonnes of solid waste annually [33]. Only the dyeing and finishing process accounts released approximately 20 per cent of industrial wastewater globally. The textile processes produce an extravagant amount of toxic chemicals, including heavy metals, formaldehyde and azo dyes in waste water, which contaminates groundwater, rivers, soil and ultimately oceans [34].

1.4.1.3 Biodiversity and climate change

Cutting of forests for human-made fibres like viscose rayon and degradation of land from the cultivation of cotton causes loss of natural ecosystem and biodiversity. The fashion and textile industry contributes to approximately 8 per cent of greenhouse gases produced by industry, and this share is projected to reach 26 per cent by 2050 without significant intervention [35, 36].

A comprehensive strategy to mitigate the environmental consequences of the textile and fashion sector is deemed necessary, and it can be accomplished by use of energy-efficient manufacturing processes, closed-loop water and chemical recycling in dyeing and finishing industry,

Figure 1.4 The environmental dimensions of fashion and textile sector.

use of low-environmental-impact materials like recycled polyester, organic cotton and hemp, and designing the products for disassembly, longevity, closed-loop recyclability and degradability at the end of their life [4, 13].

1.4.2 Social dimension

Sustainability in the textile and fashion industry does not only involve environmental aspects, but it also needs a clear explanation of ethical production and social justice aspects. The social dimension in the fashion and textile industry involves the well-being of people, providing them with proper rights with dignity, and the social dimension is involved across the complete life cycle of textile products, from the agricultural workers to the factory workers and from the consumers to waste-pickers, sorters and recyclers [37]. The social considerations for fashion and textile sector are provided in Figure 1.4.

1.4.2.1 Ethical sourcing

The global textile fashion supply chain relies immensely on labour-centric raw-material production in developing countries. These developing countries mostly have low to middle income standards, and the regulations in those countries are often weak or are not enforced. Approximately 75 million people work directly in the textile and fashion industry, and most of them are exposed to forced overtime, unsafe working conditions, job insecurity and unpaid overtime [38].

1.2.2 Fair labour practices

The fair labour practices dilemma was especially brought to light in 2013, when Rana Plaza, a local textile factory, collapsed in Bangladesh, resulting in the deaths of 1,100 factory workers [39]. This dilemma led the ethical oversight and system failure in the sourcing of materials by reputed brands. Despite a lot of efforts in fire safety in building assessment, the corporate commitments still have persistent gaps in law enforcement and supply-chain transparency.

There are a lot of trade certifications, like ethical sourcing standard SA-8000, WRAP and unionisation efforts, which are mechanisms that address social sustainability in sourcing, but only 17 per cent of the brands show proper commitment and have complete visibility of their suppliers [40].

1.4.2.2 Working conditions in the fashion and textile industry

The fashion and textile industry is one of the most challenging industry, which is labour-intensive and have deep roots in labour exploitation [41]. Many studies show that in countries like India, Bangladesh and China, the living wages in the fashion and textile industry are below those set by their respective governments. The health risks in these industries are immense, and the exposure to chemical dust and other hazardous materials is very common, especially in the informal sectors. Approximately 80 per cent of the apparel workers are female, who often complain about discrimination in salary, gender-based violence and a lack of job description.

Ethical sourcing from the suppliers should be enforced so the brands can ensure occupational health safety, decent work, and listen to the workers' voice to ensure ethical sustainability, as explained by the International Labour Organization decent work agenda.

1.4.2.4 Social justice in sustainable design

Inclusivity eco-design is one of the key elements of social sustainability [42]. It confirms that the textile product is culturally sensitive, accessible and representative of the specific community. The minority groups, including indigenous communities and people with special abilities and marginalised workers are not excluded from the design considerations. The automation of the production processes, for example, digital printing, 3D knitting, etc., often imposes a risk of increasing joblessness in the industry, especially for low-skilled labour, unless they are trained by social-protection agencies or the industry itself. To enhance social justice for sustainable design in the textile and fashion sector, the government should enforce equitable transparency and labour practices, promote community-based initiatives, especially for low-income workers and local craftsmen and people from marginalised groups. The design process should include intersectional inclusion of all these factors [43].

1.4.3 Economic dimension

Sustainability in the textiles and fashion industry includes the business models that are resilient to systemic shocks, have financial viability and are ethically grounded; it takes into account the social equity, profitability and ecological responsibility altogether.

1.4.3.1 The economic opportunities and challenges of implementing sustainability

The implementation of sustainability imposes a lot of economic challenges and opportunities in the fashion and textile sector. For example, integrating operational efficiency in the textile and fashion industry can provide benefits such as a reduction in energy and water consumption and generation of material waste, which ultimately reduces production cost. Sustainable brands are growing in differentiation and are getting huge profits just by gaining loyalty, especially from people with a sustainability-oriented mindset. The regulatory compliance for risk mitigation helps avoid financial damage and secures the reputation of the industry [44]. The economic

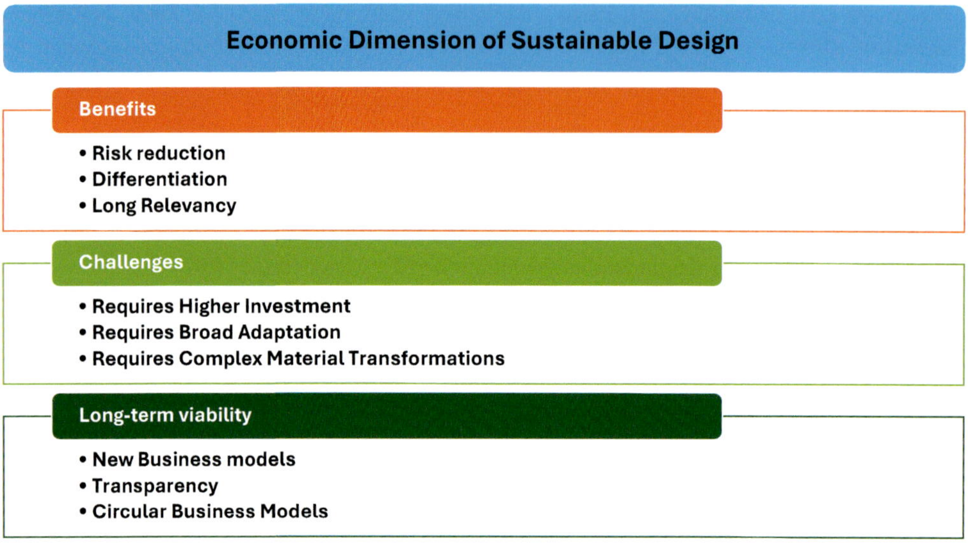

Figure 1.5 The economic considerations of implementing sustainability in fashion and textile sector.

benefits, challenges and viability of sustainability in fashion and textile sector is demonstrated in Figure 1.5.

However, the implementation of sustainability also imposes significant challenges, for example, having lower profit margins in slow fashion models while taking into account the ethical considerations, high cost of certified materials, high upfront cost of adopting cleaner technologies and ensuring fair wages. Sustainable products often incur high manufacturing costs, and consumers are often price-sensitive due to which the sales might result in reduced sales or expose sustainable products can be exposed to market volatility [45]. According to McKinsey, fashion and textile brands face increased costs by adopting sustainable practices that are ultimately paid off by life-cycle returns, or the consumer is willing to pay for sustainable materials [46].

1.4.3.2 Long-term viability and resilience

Circular business models, such as repairing, resale and reuse, and providing the product as a service, are remodelling and reshaping the textile and fashion industry. These economies are separating the material throughput from the revenue. According to a report, the global market share of second-hand garments is projected to double to $80 billion in 2027 from $40 billion in 2022 [47].

Companies are showing improved resilience to supply-chain interruptions and scarcity of materials by implementing regenerative agriculture and circular-economy practices. For example, Patagonia gained stronger customer loyalty and brand equity by increasing the use of regenerative cotton and providing repair and after-sales services [48].

1.4.4 Cultural dimension

Fashion and textile industries should take into account the identity values and intangible heritage of people by underpinning the clothing practices to incorporate sustainability. The incorporation

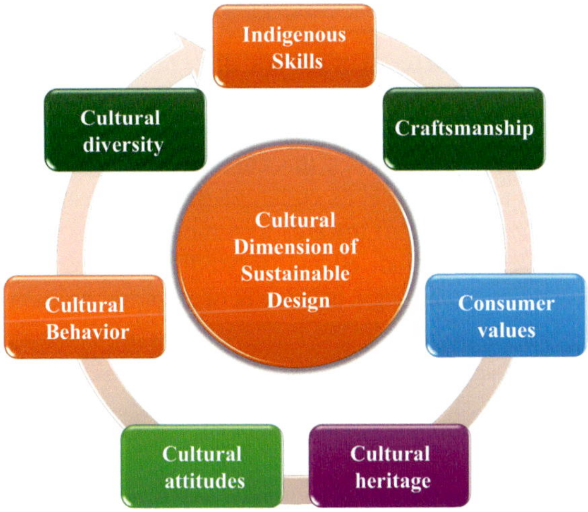

Figure 1.6 The cultural dimensions of sustainability in fashion and textile sector.

of sustainable practices requires the promotion of authentic expressions, preservation of traditional craft and enhancing the customer relationship with the government [49]. The cultural dimensions of sustainability in fashion and textile sector are provided in Figure 1.6.

Textile manufacturing is deeply rooted in almost all cultures, especially the indigenous communities in Latin America, Africa and Asia, which focus on preserving its craftsmanship for centuries-old techniques such as ajrak, batik, hand-woven fabric like khaddar and traditional natural dyeing processes. These techniques and methods are mostly environmentally friendly and often use circular resources and biodegradable materials [50].

The globalisation of the fashion and textile industry has imposed a threat to marginalised groups and has imposed risks to artisan practices that are now facing extinction due to inadequate compensation, cultural appropriation and low access to the market. Sustainable design must take into account the cultural rights and should support these craftsmen by introducing core design models that prioritise the use of local craftsmanship, integration of cultural sustainability matrices in product sustainability and provide fair labelling along with fair partnerships with local craftsmen [51].

1.4.4.1 Consumer behaviour towards sustainability

The consumption pattern of textile products is strongly influenced by cultural norms in the Global North. Disposability and consumerism are deeply entrenched in society, in which every single person purchases 60 per cent more clothing as compared to usage pattern twenty years ago [52]. This is also accompanied by wearing the garment half as long in comparison. In the Global South, the use of the circular economy practically exhibits repurposing of the products, the reuse and repair services that are often caused by resource constraints and low purchasing power are the cultural norms, for example, the India jugaar techniques, Pakistani Sindhi Ralli and the Japanese Mottoainai, etc. The efficient design while taking sustainability into account must take special care of the cultural values, aspirations and rituals of local communities. By ignoring cultural influence, products and brands could experience wider adoption or harsh customer backlash [53].

1.4.4.2 Cultural diversity and sustainability

Innovation and resilience are often a result of cultural diversity, especially in the fashion and textile industry, as diverse religions, different cultural dress codes, differences in body shapes, ethnic differences, and cultural gender identities require a strategic reshaping of practices for each group [54]. With time, brands are recognising this cultural diversity in shaping their design processes, for example, Nike's pro hijab, Levi's collaboration with local craftsmanship and Gucci hiring African designers show their shift towards a culture-sensitive product development process. The sustainable design must include ethical storytelling, community confidence, and there must be compensation while maintaining a balance between appropriation and inspiration [55].

1.5 Global challenges and trends

The fashion and textile industries are experiencing an evolving landscape in which the ecological crisis is shifting consumer expectations, technological crisis, technology disruptions, and intensified regulatory impacts are shaping the global economy. The sustainable design must take into account all these interconnected variables to maintain resilient, ethical and relevant standards [56].

1.5.1 Slow fashion versus fast fashion

Transitioning back to the slow fashion models is very important and relevant in today's world to maintain the ecological concerns of sustainability, but it definitely requires an economic shift and cultural acceptability, which is impossible without educating the customers and reconfiguration of supply chains [57].

In contrast, fast fashion relies on producing a high volume of textile and fashion products by rapid turnover and low production costs, often ignoring environmental protection and labour standards. Brands such as Zara and Shein are renowned for their fast-fashion models and are producing thousands of product designs and styles every week, which encourages customers to wear these products fewer than ten times before discarding.

1.5.2 Rise of the circular economy in the fashion industry

Circular-economy models offer alternative routes to the linear consumption model, which merely takes–make–waste. Circular-economy models, contrarily, emphasise designing the products for longevity, recycling and reuse, along with integration of regenerative systems in the fashion supply chain [57]. The circular economy in the fashion industry encourages the use of strategies, for example, using innovative materials that are circular or biodegradable, designing products for recyclability and disassembly and providing the product as a service by repair, resale and rental services. Major consortia and retailers like the Make Fashion Circular initiative by the Ellen MacArthur Foundation are encouraging circular business models, but, still, there are significant barriers to policy infrastructure and economics that impose hurdles in implementing these models [58]. The emerging terms with respect to implementation of sustainability in fashion and textile sector are highlighted in Figure 1.7.

1.5.3 Technological updates and sustainable production

In the digital age, the transformation of production processes and technological advances is reshaping sustainable design processes. For example, waterless dyeing and finishing technologies such as CO_2 dyeing are efficiently reducing the chemical and water use in textile finishing

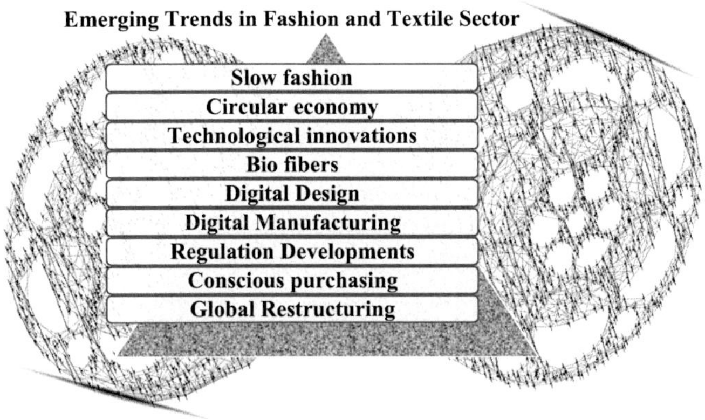

Figure 1.7 The emerging terms with respect to implementation of sustainability in fashion and textile sector are highlighted.

operations, but it incurs a lot of capital cost. The 3D knitting and other additive manufacturing techniques reduce the production of waste and enable vast product customisation. Using 3D fabrication techniques in manufacturing help anchor huge production costs, whereas use of AI integrated and blockchain technologies enable data rich sustainable design transparency and traceability. These technologies have higher operational costs due to which small industries may not be getting the proper benefits from them and may require additional training and capital to implement these technologies [59, 60].

1.5.4 Policy and regulatory landscape

Governments are taking sustainability in the fashion and textiles sector very seriously and binding the manufacturing companies with strict legislation. For example, the French AGEC (Anti-Gaspillage pour une Économie Circulaire) law imposes penalties for the destruction of pre-consumer apparel, whereas the EU has mandated launch activity and repairability while implementing the digital product passport (DPP) by 2030, the Californian Garment Worker Protection Act ensures brand accountability and transparent work wages [61, 62]. This shift from voluntary guidelines to legislation shows a paradigm shift in governments and emphasises that sustainability in product design is not optional but a regulatory necessity now [63].

1.5.5 The consumer paradigm shift

New generations are now demanding sustainability and transparency and considering these very seriously while making buying decisions. Platforms like Depop, ThredUp and Vinted have made second-hand fashion products more accessible. Whereas Export LeftOver of Pakistan has bought pre-consumer textile waste into mainstream fashion, 73 per cent of Gen Z agree to pay more for sustainable products according to a study, with the increasing awareness the scrutiny of greenwashing is rising day by day, and they are sorting certifications and verifications along with evidence-based claims from the brands [64]. Now designers must ensure and learn the implementation of ethical marketing, digital literacy and consumer expectations to meet the demands of the evolving paradigm shift and changing landscape.

1.6 Sustainability as an interdisciplinary graduate attribute

Due to the increasing demand for sustainability, it is getting recognised as one of the core competencies across all academic disciplines. Sustainability is becoming important, especially in fields such as the textile and fashion industry. The integration of sustainable design practices into graduate courses can equip future professionals with the ethical reasoning, critical thinking, systems thinking and technological education necessary for the implementation of sustainability and therefore needs a transformative change in graduate studies. Some key elements to incorporate sustainability through education institutes in design, engineering and business are provided Figure 1.8.

The significance of sustainability education is increasing day by day. According to the United Nations Educational, Scientific and Cultural Organization (UNESCO) Education for Sustainable Development Framework, learners should now be equipped to deal with complex global challenges such as biodiversity loss, inequality, climate change and unsustainable consumption patterns [65].

Now the vocational institutes and the universities must reposition their educational structures by incorporating sustainability as a learning outcome into their curriculum and implementing it across multiple disciplines, including design engineering and business programmes. The outcomes of courses should also include competence levels in ethical decision-making, application of life-cycle thinking systems, understanding of planetary limits and interdisciplinary collaborations. Sustainability is no longer a choice for education; right now, it is an essential attribute for developing leadership and innovation in textile and fashion professionals [66].

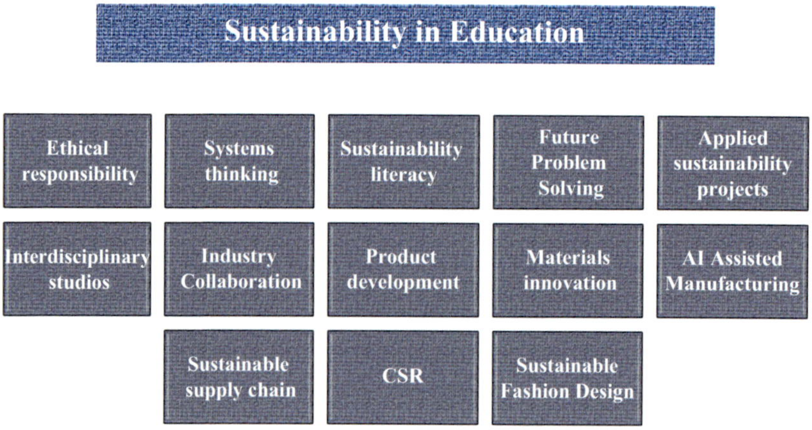

Figure 1.8 Key elements to incorporate sustainability through education institutes in graduates of design, engineering and business programmes.

1.6.1 Engineering education

Engineering disciplines should now teach topics that include innovation in recycled fibres, biodegradable and bio-based materials; LCA techniques for fibre and textile production processes; the inclusion of circular production systems; and the valorisation techniques for textile waste. The students in material science departments must be equipped with the development of biomaterials for textiles. They should be able to use textile and agricultural waste in making textile products and should be able to access the degradation profiles of these materials.

1.6.2 Business education

Business schools should include environmental, social and governance topics along with sustainable procurement transparency, traceability and green logistics and emerging technologies such as blockchain, etc. Students should be taught sustainability in the context of strategic opportunities, compliance and risk mitigation in the fashion and textile industry. Business sustainability thinking courses should also include carbon-pricing, circular business models and extended producer responsibility (EPR) compliance.

1.6.3 Design education

Design education should now teach the students to develop prototypes using upcycled materials. The students should also be able to use tools such as Adobe Substance, CLO 3D and other digital design tools to reduce waste. They should be taught the life-cycle impacts and the use of disassembly practices for textile and fashion products; and they should be able to increment MA models in ecological literacy and design activism.

Several case studies in education can be observed, including the Impact Plus project of Northumbria University in the UK. This project integrates policy design science and environmental concerns to train the students in an interdisciplinary fashion related to global problems by applications of system mapping, data visualisation and LCA. Aalto University in Finland encourages joint projects between chemists, engineers and designers to enable them to design closed-loop dyeing systems and the recyclability of multi-material products. They also encourage collaboration for the combined utilisation of business technology and aesthetic strategies [67, 68]. Hong Kong Polytechnic University encourages 3D prototyping and digital sampling to reduce the use of materials in laboratory and design studios. It encourages collaborative learning techniques by courses that include design technology ethics in fashion and sustainability in the global supply chain [69].

1.7 Conclusion

This introductory chapter established the contextual foundation for understanding the importance of sustainable design in the fashion and textile industry. The chapter began with understanding the important concepts related to the environmental concerns confronted by the sector. It discussed the key issues in the sector, ranging from labour exploitation, racial discrimination, pollution, resource depletion and overproduction of all products. The chapter then discussed sustainable design, multidisciplinary approaches and system-based techniques to address the economic inefficiencies, social injustices, ecological impacts and cultural differences through the lens of critical dimensions of sustainability, including the environmental, social, economic and cultural aspects. The chapter then highlighted strategies for waste reduction, conservation of natural resources and mitigating climate impacts, ensuring workplace safety practices and ethical labour rights implementation, preservation of traditional knowledge, diversity preservation and transformation of consumer behaviour, along with long-term profitability, long-term resilience and innovation. It should also be realised that with regulatory challenges, circular economic innovations and rapidly increasing fast fashion, everything is evolving in the context of change in consumer values. This chapter encourages the interdisciplinary application of sustainability and insists that designers, business professionals and engineers are leading their fields and should now be taught responsibly to shape the future of the textile and fashion industry and shift it towards more sustainable practices.

References

[1] S. S. Rahman, S. Siddiqua and C. Cherian, 'Sustainable Applications of Textile Waste Fiber in the Construction and Geotechnical Industries: A Retrospect', *Clean Eng Technol*, 6 (February 2022): p. 100420, doi:10.1016/J.CLET.2022.100420.

[2] T. G. Hoang, A. D. T. Binh and P. T. Nguyen, 'Fashioning a Transparent Future: Blockchain Applications in the Textile and Garment Supply Chain', in R. Nayak, H. Truong and R. Pal (eds), *Use of Digital and Advanced Technologies in the Fashion Supply Chain*, Singapore: Springer, pp. 247–66, (2025), doi:10.1007/978–81–97–7528-6_10.

[3] M. S. I. Sarker and I. Bartok, 'Global Trends of Green Manufacturing Research in the Textile Industry Using Bibliometric Analysis', *Case Studies in Chemical and Environmental Engineering*, 9 (June 2024): 100578, doi:10.1016/J.CSCEE.2023.100578.

[4] Textile Exchange, 'Global Fiber Production Reached an All-Time High of 124 Million Tonnes in 2023, According to Textile Exchange's Materials Market Report', (2024), https://textileexchange.org/news/textile-exchange-releases-2024-materials-market-report/.

[5] I. Papamichael, Irene Voukkali, Florentios Economou, Pantelitsa Loizia, Giorgos Demetriou, Mark Esposito, Vincenzo Naddeo, Marco Ciro Liscio, Paolo Sospiro and Antonis A. Zorpas 'Mobilisation of Textile Waste to Recover High Added Value Products and Energy for the Transition to Circular Economy', *Environ Res*, 242 (February 2024): 117716, doi: 10.1016/J.ENVRES.2023.117716.

[6] European Parliament, 'Fast Fashion: EU Laws for Sustainable Textile Consumption', (2020), www.europarl.europa.eu/topics/en/article/20201208STO93327/fast-fashion-eu-laws-for-sustainable-textile-consumption.

[7] K. Bailey, A. Basu and S. Sharma, 'The Environmental Impacts of Fast Fashion on Water Quality: A Systematic Review', *Water*, 14 (7) (March 2022), 1073, doi:10.3390/W14071073.

[8] P. Gazzola, E. Pavione, R. Pezzetti and D. Grechi, 'Trends in the Fashion Industry: The Perception of Sustainability and Circular Economy – A Gender/Generation Quantitative Approach', *Sustainability*, 12 (7) (April 2020): 2809, doi:10.3390/SU12072809.

[9] Decathlon, 'Why Does It Take So Much Water to Make a Cloth?', https://sustainability.decathlon.com/why-does-it-take-so-much-water-to-make-a-cloth.

[10] R. Bick, E. Halsey and C. C. Ekenga, 'The Global Environmental Injustice of Fast Fashion', *Environ Health*, 17 (1) (December 2018), doi: 10.1186/S12940-018-0433–7.

[11] M. Shamsuzzaman, Mazed Islam, Md. Abdullah Al. Mamun, Rishad Rayyaan, Kazi Sowrov, Saniyat Islam and Abu Sadat Muhammed Sayem, 'Fashion and Textile Waste Management in the Circular Economy: A Systematic Review', *Cleaner Waste Systems*, 11 (June 2025): 100268, doi: 10.1016/J.CLWAS.2025.100268.

[12] A. P. Periyasamy and A. Tehrani-Bagha, 'A Review on Microplastic Emission from Textile Materials and Its Reduction Techniques', *Polym Degrad Stab*, 199 (May 2022), doi: 10.1016/j.polymdegradstab.2022.109901.

[13] A. R. Shafqat and A. Saccavini, 'Eco-Design of Textiles', in S. Rubab Batool, S. Ahmad, Y. Nawab and M. Hussain (eds), *Circularity in Textiles*, Cham: Springer Nature, (2023), pp. 39–58, doi: 10.1007/978-3-031-49479-6_3.

[14] G. M. Mitta, 'Labor Related Challenges of Textile and Garment Industries in the Ethiopian Industrial Parks: The Perspective of Stakeholders', in R. Gronemeyer and M. Fink (eds), *Ethiopia: Awakening Crisis Outlooks* (2023), pp. 129–58, doi: 10.1007/978-3-658-41794-9_5.

[15] M. K. Kamila, S. S. Jasrotia and P. S. Kushwaha, 'Sweatshops at Work? An Exploration through Bibliometric Analysis', *Sustainable Futures*, 9 (June 2025): 100573, doi: 10.1016/J.SFTR.2025.100573.

[16] M. Russell, 'Textile Workers in Developing Countries and the European Fashion Industry', (2020), www.europarl.europa.eu/thinktank/en/document/EPRS_BRI(2020)652025.

[17] J. E. DeVoy, E. Congiusta, D. J. Lundberg, S. Findeisen and S. Bhattacharya, 'Post-consumer Textile Waste and Disposal: Differences by Socioeconomic, Demographic, and Retail Factors', *Waste Management*, 136 (December 2021): 303–9, doi: 10.1016/j.wasman.2021.10.009.

[18] S. Monseau, K. Sorsa and H. Salokangas, 'Cleaner Clothes: The EU Textile Strategy and Sustainability in the Textile Supply Chain', *Georgia Journal of International and Comparative Law*, 52 (2) (2024).

[19] A. L. Vassalo, C. G. Marques, J. T. Simões, M. M. Fernandes and S. Domingos, 'Sustainability in the Fashion Industry in Relation to Consumption in a Digital Age', *Sustainability*, 16 (13) (June 2024): 5303, doi: 10.3390/SU16135303.

[20] World Commission on Environment, 'Report of the World Commission on Environment and Development: Our Common Future Towards Sustainable Development 2. Part II. Common Challenges Population and Human Resources 4'.

[21] N. A. Ankrah, E. Manu, M. Fullen, J. Bentrar, A. Cousin, M. Mess and O. Lewald, 'Implementation of Cradle to Cradle Diversity Principles in Business Site Development Schemes', *International Journal of Urban Sustainable Development*, 10 (1) (January 2018): 92–108, doi: 10.1080/19463138.2018.1443937.

[22] B. Baldassarre, D. Keskin, J. C. Diehl, N. Bocken and G. Calabretta, 'Implementing Sustainable Design Theory in Business Practice: A Call to Action', *J Clean Prod*, 273 (November 2020): 123113, doi: 10.1016/J.JCLEPRO.2020.123113.

[23] G. Gornostaeva, 'Design and Sustainability in the Fashion Industry: The Example of Independent Labels in London', *Cleaner and Responsible Consumption*, 15 (December 2024): 100221, doi: 10.1016/J.CLRC.2024.100221.

[24] M. T. Rahaman, A. D. Pranta, M. R. Repon, M. S. Ahmed and T. Islam, 'Green Production and Consumption of Textiles and Apparel: Importance, Fabrication, Challenges and Future Prospects', *Journal of Open Innovation: Technology, Market, and Complexity*, 10 (2) (June 2024): 100280, doi: 10.1016/J.JOITMC.2024.100280.

[25] N. F. Islam, B. Gogoi, R. Saikia, B. Yousaf, M. Narayan and H. Sarma, 'Encouraging Circular Economy and Sustainable Environmental Practices by Addressing Waste Management and Biomass Energy Production', *Regional Sustainability*, 5 (4) (December 2024): 100174, doi: 10.1016/J.REGSUS.2024.100174.

[26] S. Dumitriu, C. G. Bocean, A. A. Vărzaru, A. T. Al-Floarei, N. M. Sperdea, F. L. Popescu and I.-C. Băloi, 'The Role of the Workplace Environment in Shaping Employees' Well-Being', *Sustainability*, 17 (6) (March 2025): 2613, doi: 10.3390/SU17062613.

[27] A. Batwara, V. Sharma, M. Makkar and A. Giallanza, 'An Empirical Investigation of Green Product Design and Development Strategies for Eco Industries Using Kano Model and Fuzzy AHP', *Sustainability*, 14 (July 2022): 8735, doi: 10.3390/SU14148735.

[28] C. A. Mattson, A. T. Pack, V. Lofthouse and T. Bhamra, 'Using a Product's Sustainability Space as a Design Exploration Tool', *Design Science*, 5 (2019): e1, doi: 10.1017/DSJ.2018.6.

[29] Fashion Sustainability, 'Ecodesign Principles → Term', (2026), https://fashion.sustainability-directory.com/term/ecodesign-principles/.

[30] D. K. Plannthin, 'Animal Ethics and Welfare in the Fashion and Lifestyle Industries', in S. Muthu and M. Gardetti (eds), *Green Fashion. Environmental Footprints and Eco-design of Products and Processes*, Singapore: Springer, (2016), pp. 49–122, doi: 10.1007/978–81-10–0245–8_3.

[31] L. G. L. M. Edirisinghe, A. A. P. de Alwis and M. Wijayasundara, 'Sustainable Circular Practices in the Textile Product Life Cycle: A Comprehensive Approach to Environmental Impact Mitigation', *Environmental Challenges*, 16 (August 2024): 100985, doi: 10.1016/J.ENVC.2024.100985.

[32] S. Abrishami, A. Shirali, N. Sharples, G. E. Kartal, L. Macintyre and O. Doustdar, 'Textile Recycling and Recovery: An Eco-friendly Perspective on Textile and Garment Industries Challenges', *Textile Research Journal*, 94 (23–4) (December 2024): 2815–34, doi: 10.1177/00405175241247806.
[33] A. R. Shafqat, M. Hussain, Y. Nawab, M. Ashraf, S. Ahmad and G. Batool, 'Circularity in Materials: A Review on Polymer Composites Made from Agriculture and Textile Waste', *Int J Polym Sci*, 1 (2023): 5872605, doi: 10.1155/2023/5872605.
[34] G. S. Vitale, N. Iacuzzi, S. Zingale, S. Lombardo, T. Tuttolomondo and P. Guarnaccia, 'Environmental Sustainability of Cotton: A Systematic Literature Review of Life Cycle Assessments', *J Agric Food Res*, 22 (August 2025): 102069, doi: 10.1016/J.JAFR.2025.102069.
[35] R. M. Frazier, Keren A. Vivas, Ivana Azuaje, Ramon Vera, Alonzo Pifano, Naycari Forfora, Hasan Jameel, Ericka Ford, Joel J. Pawlak, Richard Venditti and Ronalds Gonzalez, 'Beyond Cotton and Polyester: An Evaluation of Emerging Feedstocks and Conversion Methods for the Future of Fashion Industry', *Journal of Bioresources and Bioproducts*, 9 (2) (May 2024): 130–59, doi: 10.1016/J.JOBAB.2024.01.001.
[36] A. M. I. Kallio, 'Wood-Based Textile Fibre Market as Part of the Global Forest-Based Bioeconomy', *For Policy Econ*, 123 (February 2021), doi: 10.1016/j.forpol.2020.102364.
[37] V. Daukantienė, 'Analysis of the Sustainability Aspects of Fashion: A Literature Review', *Textile Research Journal*, 93 (3–4) (February 2023): 991–1002, doi: 10.1177/00405175221124971.
[38] F. Jensen and L. Whitfield, 'Leveraging Participation in Apparel Global Supply Chains through Green Industrialization Strategies: Implications for Low-Income Countries', *Ecological Economics*, 194 (April 2022): 107331, doi: 10.1016/J.ECOLECON.2021.107331.
[39] Foreign & Commonwealth Office and Department for International Development, 'The Rana Plaza Disaster: Case Study', (2014), www.gov.uk/government/case-studies/the-rana-plaza-disaster.
[40] M. Sartor, G. Orzes, C. Di Mauro, M. Ebrahimpour and G. Nassimbeni, 'The SA8000 Social Certification Standard: Literature Review and Theory-Based Research Agenda', *Int J Prod Econ*, 175 (May 2016): pp. 164–81, doi: 10.1016/J.IJPE.2016.02.018.
[41] M. Hussain and A. R. Shafqat, 'Sustainability and Materials', in K. Shaker and Y. Nawab (eds), *Engineering Materials*, Cham: Springer, (2025), pp. 257–74, doi: 10.1007/978-3-031-72263-9_11.
[42] G. Batool, M. Bashir, S. A. Baig and A. R. Shafqat, 'Ecotourism Taking Modernity to the Deprived Regions of the World', in M. Abrar, M. Asim Nawaz, F. Kiran and F. Asmi (eds), *Supporting Environmental Stability Through Ecotourism*, pp. 143–57, (2024), ICI Global, doi: 10.4018/979-8-3693-1030-4.CH008.
[43] M. Glogar, S. Petrak and M. Mahnić Naglić, 'Digital Technologies in the Sustainable Design and Development of Textiles and Clothing: A Literature Review', *Sustainability*, 17 (4) (February 2025): 1371, doi: 10.3390/SU17041371.
[44] M. A. H. Shibly, M. M. U. Hoque, M. A. H. Shibly and M. M. U. Hoque, 'Sustainability Challenges of the Textile Industry', in C.-L. Hui (ed), *Textile Industry and the Environment: Challenges, Recent Development and Future* (April 2025), IntechOpen, doi: 10.5772/INTECHOPEN.1009414.
[45] Y. J. Cai and T. M. Choi, 'A United Nations Sustainable Development Goals Perspective for Sustainable Textile and Apparel Supply Chain Management', *Transp Res E Logist Transp Rev*, 141 (September 2020): 102010, doi: 10.1016/J.TRE.2020.102010.
[46] Imran Amed, Gemma D'Auria, Marc Bain, Anita Balchandani, David Barrelet, Brian Baskin, Colleen Baum, Hannah Crump, Patricia González Méndez, Joëlle Grunberg, Alexander Li, Rahul Malik, Felix Rölkens and Daniel Zipser, 'The State of Fashion 2025', (n.d.), www.mckinsey.com/~/media/mckinsey/industries/retail/our%20insights/state%20of%20fashion/2025/the-state-of-fashion-2025-v2.pdf.
[47] M. Dehghannejad, R. Pal and K. Dissanayake, 'Cascading of Circular Business Models in the Textile and Clothing Industry: Insights for Operationalization through a Systematic Review', *Resources, Conservation & Recycling Advances*, 28 (December 2025): 200293, doi: 10.1016/J.RCRADV.2025.200293.

[48] G. C. de Oliveira Neto, R. A. R. Santos, L. F. R. Pinto, F. R. Flausino, D. E. P. de Oliveira and M. N. Seri, 'Innovative Circular Practices Integrating Business Model for Textile Industry', *J Eng Fiber Fabr*, 19 (January 2024), doi: 10.1177/15589250241226481.

[49] P. Bertola and C. Colombi, 'Can Fashion Be Sustainable? Trajectories of Change in Organizational, Products and Processes, and Socio-Cultural Contexts', *Sustainability: Science, Practice, and Policy*, 20 (1) (December 2024): 1–16, doi: 10.1080/15487733.2024.2312682.

[50] Textile School, 'Exploring Indigenous Textile Techniques: Preserving Global Cultural Heritage – Textile School', (2025), www.textileschool.com/27861/exploring-indigenous-textile-techniques-preserving-global-cultural-heritage/.

[51] V. Basile, N. Capobianco and R. Vona, 'Business Model Development in the Circular Bioeconomy: A Focus on the Fashion-Textile Industry', *J Clean Prod*, 494 (February 2025): 144944, doi: 10.1016/J.JCLEPRO.2025.144944.

[52] Business Waste, 'Fashion Waste Statistics & Facts | Textile & Clothing Waste Facts', (n.d.), www.businesswaste.co.uk/waste-facts/fashion-waste-facts-and-statistics/.

[53] J. Kumari, Y. Yang and J. Kumar, 'Tradition in Transition: How Cultural Elements in Product Design Influence Consumer Purchase Intention', *Asia Pacific Journal of Marketing and Logistics*, 37 (11) (November 2025): 3517–34, doi: 10.1108/APJML-12–2024-1879.

[54] Deepwear, 'The Impact of Cultural Diversity on Fashion Trends and Manufacturing: Deepwear', (n.d.), https://deepwear.info/blog/the-impact-of-cultural-diversity-on-fashion-trends-and-manufacturing/.

[55] S. Jhangra, 'The Role of Traditional Attire in Modern Fashion: A Study of Cultural Resilience and Adaptation', *Indian Journal of Fashion Technology*, 2 (1) (February 2025): 12–16.

[56] F. Mullick, V. A. Field, R. Assistants, H. Ahsan and M. Muhammad, 'Mapping the Sustainable Fashion Ecosystem in Pakistan', (2025), www.britishcouncil.pk/sites/default/files/textile_legacies_report.pdf.

[57] Subharthi Das, Santu Karmakar, Timilehin Olasoji Olubiyi and Santosh Kumar Behera, 'The Social and Environmental Impact of Fast Fashion', in T. Olasoji Olubiyi, S. Kumar Behera and T. Anh Tran (eds), *Global Impacts and Sustainable Practices in Fast Fashion*, pp. 311–29, (January 2025), IGI Global, doi: 10.4018/979-8-3693–7853-3.CH012.

[58] H. G. Ramírez-Escamilla, M. C. Martínez-Rodríguez, A. Padilla-Rivera, D. Domínguez-Solís and L. E. Campos-Villegas, 'Advancing toward Sustainability: A Systematic Review of Circular Economy Strategies in the Textile Industry', *Recycling*, 9 (5) (October 2024), 95, doi: 10.3390/RECYCLING9050095.

[59] A. Dobriyal and S. Sarkar, 'Greening the Textile Industry: An In-Depth Review of Eco-Printing Methods and Challenges', *Environmental Claims Journal*, 1 (2025): 1–22, doi: 10.1080/10406026.2025.2564398.

[60] M. T. Hossain, M. A. Shahid, M. G. M. Limon, I. Hossain and N. Mahmud, 'Techniques, Applications, and Challenges in Textiles for a Sustainable Future', *Journal of Open Innovation: Technology, Market, and Complexity*, 10 (1) (March 2024): 100230, doi: 10.1016/J.JOITMC.2024.100230.

[61] Lidia Lüttin, 'Full Overview of the EU Textile Strategy and Regulations', (2026), www.carbonfact.com/blog/policy/eu-regulations-for-textile-brands.

[62] L. Siliņa, I. Dāboliņa and E. Lapkovska, 'Sustainable Textile Industry: Wishful Thinking or the New Norm: A Review', *J Eng Fiber Fabr*, 19 (January 2024), doi: 10.1177/15589250231220359.

[63] Emma Feldner-Busztin, 'Fashion and Textiles Legislation: Laying Down the Key Laws – The Interline', (2024), www.theinterline.com/2024/10/24/fashion-and-textiles-legislation-laying-down-the-key-laws/.

[64] First Insight, 'The State of Consumer Spending: Gen Z Shoppers Demand Sustainable Retail', (n.d.), www.firstinsight.com/white-papers-posts/gen-z-shoppers-demand-sustainability.

[65] UNESCO, 'Education for Sustainable Development', (n.d.), www.unesco.org/en/sustainable-development/education.

[66] S. Ankareddy, G. Dorfleitner, L. Zhang and Y. S. Ok, 'Embedding Sustainability in Higher Education Institutions: A Review of Practices and Challenges', *Cleaner Environmental Systems*, 17 (June 2025): 100279, doi: 10.1016/J.CESYS.2025.100279.

[67] M. Zilvetti, M. Conti and R. Kotter, 'Sustainable Futures in UK Higher Education through Multi-Disciplinary Innovation and Advanced Practice: A Northumbria University Case Study', in W. Leal Filho, T. Dibbern, S. R. de Maya, MdC. Alarcón-del-Amo and L. M. Rives (eds), *The Contribution of Universities Towards Education for Sustainable Development. World Sustainability Series*, Cham: Springer. https://doi.org/10.1007/978-3-031-49853-4_21.

[68] S. Kyffin and M. Aftab, 'Achieving Excellence in Design Research: A Case Study of Research and Education Framework at Northumbria University', *International Journal of Arts and Technology*, 9 (3) (2016): 199–213, doi: 10.1504/IJART.2016.078609.

[69] OpenGov Asia, 'Hong Kong: PolyU's AI Innovation Advances Sustainable Fashion', (2025), https://opengovasia.com/hong-kong-polyus-ai-innovation-advances-sustainable-fashion/?c=pk.

2 Environmental impact of conventional textile production

Md. Reazuddin Repon, Shubhajit Dutta, Tonmoy Saha, Hüseyin Aksel Eren, and Shumaila Kiran

2.1 Introduction

The textile industry is a significant component of the world economy and production, but it is also one of the most harmful to the environment [1–3]. Making textiles the old-fashioned way takes a lot of resources and time. For example, you have to grow the fibres, spin them, weave them, knit them, scour them, bleach them, mercerise them, dye them, print them, finish them and make clothes [4, 5]. These steps are really bad for the environment. These impacts can be seen in a lot of places, such as when people pollute the air and water, release too many greenhouse gases (GHGs), use too much energy, make too much solid waste and let harmful chemicals out [6, 7]. The steps involved in conventional textile production are mentioned in Figure 2.1.

About 70–80 per cent of all water contamination from the textile industry comes from wet processes such as scouring, bleaching, dyeing and finishing [8]. Every year, the fashion sector utilises roughly 93 billion cubic metres of water around the world. This is around 4 per cent of all the water that is taken out of rivers and lakes. The same goes for air pollution and carbon

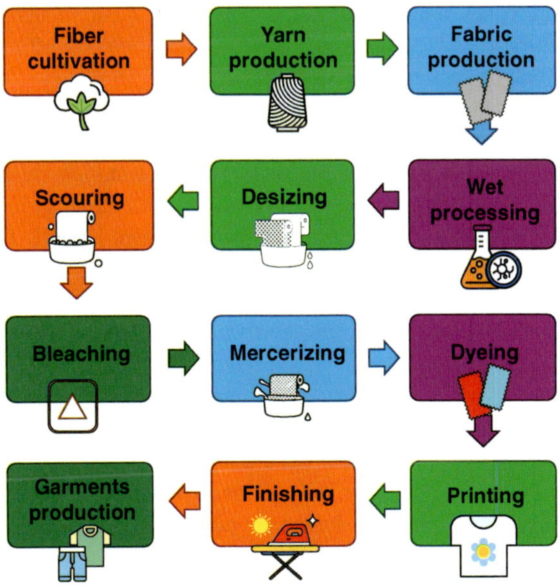

Figure 2.1 Steps involved in conventional textile production.

DOI: 10.4324/9781003638148-3

footprint. Textile production is thought to be responsible for about 10 per cent of all carbon emissions in the world [7, 9]. For instance, polyester production alone releases approximately 27 kilograms CO_2 equivalents per kilogram of fibre [10]. Solid waste is another problem for the environment, and only 14–15 per cent of textiles are recycled worldwide [11]. Because synthetic textiles don't break down, they stay in landfills and add a lot to microplastic pollution [12–14]. A wide range of dangerous chemicals, such as per- and polyfluoroalkyl substances (PFCs), flame retardants and silver nanoparticles, are always present in the textile production process and can be harmful to both people and the environment [15].

This chapter discusses the environmental effects of traditional textile production, including resource depletion, water and air pollution, high energy use, the build-up of solid waste and the use of dangerous chemicals. The need for sustainable alternatives of conventional textile production are also discussed.

2.2 Major environmental impacts of conventional textile production

2.2.1 Water pollution

The textile industry is a major environmental concern due to its intensive consumption of water and chemicals, leading to significant water pollution. Cotton farming is especially harmful during the fibre-production stage because it relies heavily on agricultural chemicals, which greatly raises the risk of polluting freshwater ecosystems with nutrients, salts and pesticides [23, 24]. Cotton only takes up 2.4 per cent of the world's arable land, but it makes up an average of 11 per cent of global pesticide sales and almost 24 per cent of global insecticide sales. This shows that it uses a lot more pesticides than other crops do. Most of the pesticides used to grow cotton are dangerous. Five of the forty-six insecticides and acaricides that make up 90 per cent of the cotton market are very dangerous, eight are dangerous, and twenty are moderately dangerous [25]. Pesticides, fertilisers and salts get into rivers, lakes and wetlands when water runs off of cotton fields and drains. These pollutants may negatively impact freshwater biodiversity directly by being toxic or indirectly by building up in the ecosystem. Also, mishandling pesticides, like washing instruments in rivers or letting chemicals spill, has the same effects: it pollutes surface waters and directly damages groundwater systems [26, 27].

During the wet processing stage, which includes scouring, bleaching, dyeing, printing and finishing, wastewater frequently has pollutants in it such heavy metals like chromium and copper, synthetic dyes, chlorine-based bleaching agents and formaldehyde resins. A lot of these chemicals are poisonous, don't break down easily and tend to build up in living things, which makes them dangerous for aquatic ecosystems in the long term [1, 28]. The conventional textile production processes and their main functions, pollutants and effluent characteristics are mentioned in Table 2.1. The viscose textile production chain is long and uses a lot of resources, including a lot of water. One tonne of cotton and viscose pulp processing requires 72 and 65 tonnes of fresh water, respectively. Moreover, the production of 1 tonne of dyed textiles requires an excessive use of water, which is nearly 120 tonnes. Following these processes, the large amounts of wastewater produced contain high levels of alkalis, organic matter, dyes, acidic compounds and harmful substances, which have a negative impact on environments [22, 29, 30]. Human-made fibres such as polyester, polyamide 6, polyamide 66, etc., release microfibers during production time and consequently during the post-consumer washing. Microplastics also pollute the water [12, 31].

Table 2.1 Conventional textile production processes and their main functions, possible pollutants and effluent characteristics [6, 16–22].

Production process	*Brief description of the process*	*Main functions*	*Possible pollutants*	*Effluent characteristics*
Fibre production	Cultivation of natural fibres or manufacturing synthetic human-made fibres	Fibre production	Pesticides, fertilisers, GHG emissions and microplastics	Low volume
Yarn production	Producing natural or synthetic yarns from different fibre sources	Yarn production	Lubricants, airborne dust particles	Low volume
Fabric production	Production of several types of fabric	Fabric production	Sizing agents,	Low volume, low BOD
Desizing	Desizing is the process of removing sizing materials from the warp yarn	Starch removal	Starch, Carboxymethyl cellulose, Polyvinyl alcohol, glucose, wax, fats, HCl	Low volume, high BOD
Scouring	Scouring is the process of removing impurities like oil, wax, fats and dirt from the fibre, yarn, or fabric	Impurities removal	NaOH, Na_2CO_3, Na_2SiO_3, wax, fragment	Low volume, strong alkalinity, colourful, high BOD
Bleaching	Bleaching is the process of removing natural or added colour from the fibre, yarn, or fabric	Whitening textile	NaOCl, NaOH, Na_2CO_3, Cl_2, H_2O_2, acids	Low volume, strong alkalinity, low BOD
Mercerisation	Mercerisation is the process of applying NaOH to the cellulosic textile to increase lustre, strength and dye uptake	Increase lustre	NaOH	Low volume, Strong alkalinity, low BOD
Dyeing	Dyeing is the process of applying dyes or pigments to fibre, yarn, or fabric to produce colourful textile materials	Colourful textile materials	Synthetic dyes, metal mordants, acids, soap, alkali, base, inorganic salt	High volume, Strong colour, high BOD
Garment manufacturing	Production of garments with cutting, sewing and assembling fabrics into finished garments		Different types of waste from cutting, sewing and packaging, microfibers released during washing.	
Printing	Printing is the process of applying colour to fabric in a specific design or pattern	Colourful textile materials	Synthetic dyes, metal mordants, formaldehyde contents, acids, soap, alkali, base, starch, gums, china clay	Very low volume, colour, high BOD
Finishing	Finishing is the special treatment (mechanical, chemical, or biological) on fabric to develop specialised functionality	Specialised functional property	Starch, tallow, finishing agents	Very low volume, low alkalinity, low BOD

Notes: GHG: greenhouse gas; BOD: biochemical oxygen demand; HCl: hydrochloric acid; NaOH: sodium hydroxide; Na2CO3: sodium carbonate; Na_2SiO_3: sodium silicate: NaOCl: sodium hypochlorite; Cl2: chlorine gas; H2O2: hydrogen peroxide.

2.2.2 Air pollution

Air pollution from conventional textile production occurs at multiple stages, from fibre production to preparing finished products. Cotton fibres can produce small particulate matter and lint dust, which can harm the air quality. On the other hand, the volatile organic compounds such as acetaldehyde, formaldehyde, and unreacted monomers can be released during the petrochemical polymerisation process of human-made fibres production [32]. This output can pollute the air. Moreover, the fibre extrusion and processing depend on fossil-fuel combustion and generate GHGs. The gases like carbon dioxide, sulphur dioxide, nitrogen oxides and other pollutants can contaminate the air [33].

Dyeing, printing and finishing processes contribute more to air pollution via emission of volatile organic compounds such as benzene, toluene and xylene from synthetic dyes and solvents [34]. Microplastics can be released to the environment during the heat-setting process (a thermal treatment) of human-made fibres [31, 35]. Toxic fumes can be released during the curing of coatings at high temperatures. Waste incineration of offcuts, defective fabrics and sludge from wastewater treatment can generate highly toxic by-products. Dioxins, polycyclic aromatic hydrocarbons and hydrogen cyanide can be produced during the burning of synthetic fibres [36, 38]. Air pollutants during conventional textile production processes are tabulated in Table 2.2.

CO: carbon monoxide; SO_2: sulfur dioxide; H_2S: hydrogen sulfide; ETP: effluent treatment plant.

2.2.3 Solid waste

The conventional textile production generates huge solid waste, and this is another major challenge for the environment. This waste generation involves fibre cultivation to post-consumer disposal. The waste contains both organic and inorganic materials. The biodegradable solid waste, such as stalks, leaves and husks from cotton, fleece from wool, is produced at the natural fibres production level [41]. On the other hand, scrap fibres and polymer sludge are generated during synthetic fibre polymerisation processes [42]. The scenario of textile waste is presented in Figure 2.2.

Cutting and sewing sections produce a large amount of solid waste (fabric cuts off). Defective products and colour-mismatched fabrics are also considered waste. Moreover, huge sludge is generated from wastewater treatment plants of dyeing, printing and finishing sections.

Table 2.2 Air pollutants during conventional textile production processes [6, 24, 39, 40].

Process involved	*Origin of pollutants*	*Pollutants*
Cotton cultivation	Production process	Small particulate matter
Spinning	Yarn preparation, carding and combing	Small particulate matter
Sizing	Emission from sizing compounds	CO, SO_2
Bleaching	Emissions from chlorine compounds	Chlorine, chlorine dioxide
Dyeing	During the coloration process with synthetic dyes	H_2S, other gases, including aniline vapours
Printing	Emission during printing and curing	Ammonia, hydrocarbons
Finishing	During functional finish application, heat-setting	Formaldehyde, lubricants
Chemical storage	From storage tanks/gallons	Volatile organic compounds
Wastewater treatment	ETP	Volatile organic compounds, toxic odour

CO: carbon monoxide; SO_2: sulfur dioxide; H_2S: hydrogen sulfide; ETP: effluent treatment plant.

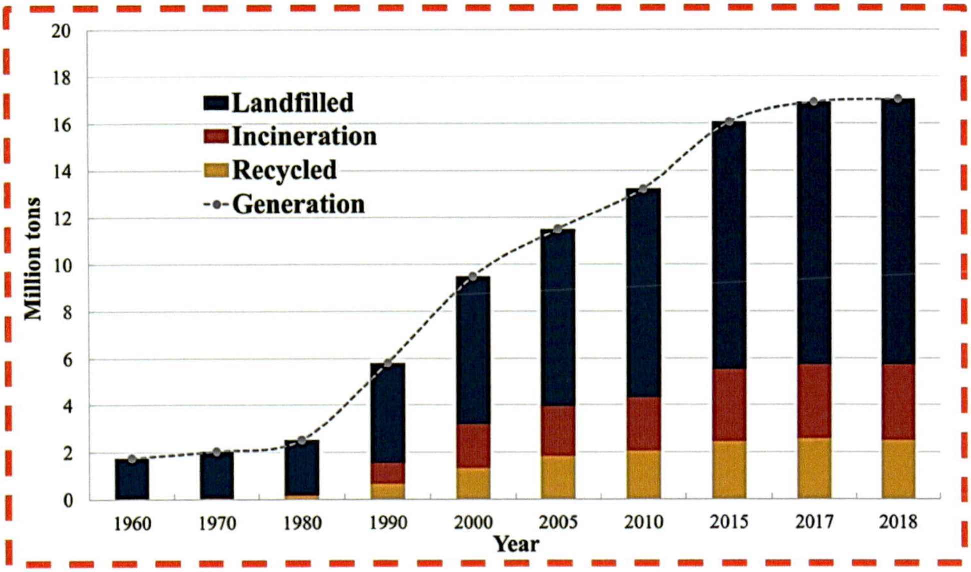

Figure 2.2 Scenario of textile waste [44, 49].

Table 2.3 Generated solid waste from conventional textile production processes [6, 24, 46–48].

Process involved	*Solid waste generated*	*Nature of the waste*
Fibre cultivation	Stalks, leaves and husk from cotton, fleece from wool and packaging materials	Stalks, leaves, husks and fleece are biodegradable, and packaging materials are non-biodegradable
Spinning process	Waste from mixing, blending, cleaning and processing waste	Biodegradable
Sizing	Lint fibres, unused starch	Biodegradable
Desizing	Packaging waste, lint from fibres, broken yarn	Lint and broken yarn are biodegradable, and packaging materials are non-biodegradable
Weaving	Oil, yarn and fabric scraps	Yarn and fabric scraps from natural fibres are biodegradable
Knitting	Oil, yarn and fabric scraps	Yarn and fabric scraps from natural fibres are biodegradable
Sewing	Stitching yarn, needle, plastic cones, trims, fabric scraps, threads	Both biodegradable and non-biodegradable
Cutting	Plastic, paper, paperboard, fabric scraps, metal clips	Both biodegradable and non-biodegradable
Dyeing	Cuts off fabrics, small dyes	Non-biodegradable
Finishing	Plastics, labels, plastic cones, trims, fabric scraps, threads	Both biodegradable and non-biodegradable
Shipping	Tape, paperboard	Both biodegradable and non- biodegradable

In a broader sense, the solid waste from conventional textile production is categorised into two types: pre-consumer textile waste (also known as industrial waste) and post-consumer textile waste (produced after use) [43–45]. Table 2.3 illustrates the solid waste generated from conventional textile production processes.

2.2.4 Deforestation

Some human-made fibres, such as viscose, lyocell, modal, acetate and cupro, are derived from wood pulp. Pulps are collected from trees such as eucalyptus, birch, beech and bamboo. The production of wood-based fibres is 7.9 million tonnes, which is approximately 6 per cent of total global fibre market [50]. If forests are managed properly and production processes are controlled (water and chemicals), then these fibres can cause less environmental impact compared to cotton and polyester. However, unsustainable sourcing such as illegal logging can lead to deforestation, biodiversity loss, soil degradation and increased GHGs [51–53].

2.2.5 Resource depletion

A large amount of resources are depleted during conventional textile production. For instance, non-renewable resources such as petroleum are needed for synthetic fibre production, especially for polyester [54]. On the other hand, huge freshwater is consumed for natural fibres cultivation and processing, especially for cotton [55]. Conversely, a large amount of textile waste is generated by the fast-fashion model because it encourages overproduction and overconsumption [56]. For instance, textile consumption increased from 17 to 19 kilograms per person between 2019 and 2022 [57]. Meanwhile, approximately 12 kilograms of textile per person are thrown away each year in the European Union.

2.2.6 The need for sustainable alternatives

A shift towards sustainable alternatives is important because of significant negative environmental impact of conventional textile production and hence it will ensure economic balance and environmental protection. Eco-friendly materials are important sustainable alternatives to conventional textile production. Traditional textile materials consume large amounts of water, chemicals and fossil fuels. Eco-friendly fibres such as hemp, linen, organic cotton and bamboo are produced and processed with less water, chemicals and energy [58]. Recycled fibres reduce the waste and decrease the need for virgin raw materials [59]. Recycled fibres also reduce pollution, carbon emission and depletion of resources.

Innovative processing techniques could be sustainable alternatives to conventional textile production. Waterless technologies such as supercritical CO_2 dyeing [60], digital printing [61], and plasma treatments [62] minimise or remove the water and consequently environmental issues.

A circular-economy approach is an important sustainable alternative to conventional textile production. The circular economy system can preserve natural resources and reduce environmental pollution and carbon emissions [45].

2.3 Conclusion

The textile industry is a major hazard to the environment throughout the whole life cycle of a textile product, from the raw material to final use and disposal. This chapter discussed the main steps in conventional textile manufacturing and pointed out the major sources of pollution at each stage and their environmental implications. The environmental footprint is mostly caused by the intensive consumption of freshwater resources, the widespread use of hazardous chemicals and considerable GHGs associated with energy demand. A significant proportion of the water used in the textile processing is discharged as wastewater. When this effluent is released without treatment, it adds dyes, salts, surfactants and other toxic chemicals to aquatic ecosystems, which puts biodiversity and ecological balance at risk. Additionally, large quantities of solid waste are generated throughout the production and consumption phases. Most of it is either thrown away in landfills or burned, which puts more stress on the environmental. The problem gets worse when synthetic fibre-based microplastics are released into the environment. These microplastics stay in the ecosystem and build up in both land and aquatic food chains. All of these problems highlight how much the textile value chain has to make technical advancements and adopt sustainable practices.

Conflicts of interest

The authors have no relevant conflicts of interest to disclose.

Availability of data and materials

The data presented herein are consistent with this article.

Acknowledgements

The authors gratefully acknowledge the technical support provided by the ZR Research Institute for Advanced Materials, Sherpur-2100, Bangladesh.

References

[1] T. Islam, M. R. Repon, T. Islam, Z. Sarwar and M. M. Rahman, 'Impact of Textile Dyes on Health and Ecosystem: A Review of Structure, Causes, and Potential Solutions', *Environ Sci Pollut R*, 30 (4) (January 2023): 9207–42, doi: 10.1007/s11356-022-24398-3.

[2] M. T. Islam, T. Islam, T. Islam and M. R. Repon, 'Synthetic Dyes for Textile Colouration: Process, Factors and Environmental Impact', *Textile & Leather Review*, 5 (2022): 327–73, doi: https://doi.org/10.31881/TLR.2022.27.

[3] F. Uddin, 'Environmental Hazard in Textile Dyeing Wastewater from Local Textile Industry', *Cellulose*, 28 (17) (November 2021): 10715–39, doi: 10.1007/s10570-021-04228-4.

[4] A. A. Sughosh Madhav, Pardeep Singh, Pradeep Kumar Mishra, 'A Review of Textile Industry: Wet Processing, Environmental Impacts, and Effluent Treatment Methods', *Environmental Quality Management*, 27 (3) (2018): 31–41, doi: https://doi.org/10.1002/tqem.21538.

[5] M. R. Repon, T. Islam, T. K. Paul, S. Jurkonienė, A. Haji, S. Shukhratov and G. F. I. Toki, 'Natural Dyes in Textile Printing: Parameters, Methods, and Performance', *Environ Sci Pollut R*, 31 (2024): 47552–83, doi: https://doi.org/10.1007/s11356-024-34424-1.

[6] M. N. Syeda Laiba Haider Naqvi, Fizzah Ayub, Abdullah Yasar, Syeda Hamayal Zahra Naqvi and Rameesha Tanveer, 'Social and Environmental Impacts in Textile Production', in P. Singh (ed.), *Dye Pollution from Textile Industry*, Singapore: Springer, (2024), pp. 423–53.

[7] S. Imran, M. A. Mujtaba, M. M. Zafar, A. Hussain, A. Mehmood, U. E. Farwa, T. Korakianitis, M. A. Kalam, H. Fayaz, C. A. Saleel, 'Assessing the Potential of Ghg Emissions for the Textile Sector: A Baseline Study', *Heliyon*, 9 (11) (November 2023), doi: 10.1016/j.heliyon.2023.e22404.

[8] V. Durand, 'Water Pollution in the Textile Industry: An Urgent Call for Sustainable Innovation', *Openaccess Government*, pp. 1–4, (4 July 2025), doi: https://doi.org/10.56367/OAG-047-11634.

[9] Z. K. Li, Y. Zhou, M. Y. Zhao, D. B. Guan and Z. F. Yang, 'The Carbon Footprint of Fast Fashion Consumption and Mitigation Strategies: A Case Study of Jeans', *Sci Total Environ*, 924 (10 May 2024), doi: 10.1016/j.scitotenv.2024.171508.

[10] M. A. Uddin, M. S. Begum, M. Ashraf, A. Azad, A. C. Adhikary and M. S. Hossain, 'Water and Chemical Consumption in the Textile Processing Industry of Bangladesh', *Plos Sustain Transf*, 2 (7) (July 2023), doi: 10.1371/journal.pstr.0000072.

[11] I. Prehu, '5 Environmental Impacts of Textile Industry, and 3 Ways to Be More Sustainable', *The Water Treatment Magazine*,(2024), https://watertreatmentmagazine.com/en/impact-textile-industry/.

[12] S. H. Akyildiz, S. Fiore, M. Bruno, H. Sezgin, I. Yalcin-Enis, B. Yalcin and R. Bellopede, 'Release of Microplastic Fibers from Synthetic Textiles During Household Washing', *Environ Pollut*, 357 (15 September 2024), doi: 10.1016/j.envpol.2024.124455.

[13] G. Vragkalis, O. Piperagkas, H. Mela and H. Karayanni, 'Microfiber Emissions through Domestic Laundry: An Estimation of Microfiber Release and Their Fate in a Medium-Sized City', *Int J Environ Sci Te*, 22 (9) (May 2025): 8025–32, doi: 10.1007/s13762-024-06186-3.

[14] A. Das, S. Dey and A. P. Das, 'Ecotoxicological Impacts of Synthetic Microfiber Pollutants and Development of Sustainable Mitigation Strategies', *Environ Chem Ecotox*, 7 (2025): 201–10, doi: 10.1016/j.enceco.2024.12.003.

[15] F. N. E. Mansoor Ahmad Bhat, Kadir Gedik and Eftade O. Gaga 'Impact of Textile Product Emissions: Toxicological Considerations in Assessing Indoor Air Quality and Human Health', in J. A. Malik and S. Marathe (eds), *Ecological and Health Effects of Building Materials*, Cham: Springer, (2021), pp. 505–41.

[16] A. Ahsan, F. Jamil, M. A. Rashad, M. Hussain, A. Inayat, P. Akhter, A. A. H. Al-Muhtaseb, K. Y. A. Lin and Y. Park, 'Wastewater from the Textile Industry: Review of the Technologies for Wastewater Treatment and Reuse', *Korean J Chem Eng*, 40 (9) (September 2023): 2060–81, doi: 10.1007/s11814-023-1475-2.

[17] M. L. Parisi, E. Fatarella, D. Spinelli, R. Pogni and R. Basosi, 'Environmental Impact Assessment of an Eco-Efficient Production for Coloured Textiles', *J Clean Prod*, 108 (1 December 2015): 514–24, doi: 10.1016/j.jclepro.2015.06.032.

[18] F. A. Esteve-Turrillas and M. de la Guardia, 'Environmental Impact of Recover Cotton in Textile Industry', *Resour Conserv Recy*, 116 (January 2017): 107–15, doi: 10.1016/j.resconrec.2016.09.034.

[19] A. S. M. R. A. Arputharaj and Sujata Saxena 'Developments in Sustainable Chemical Processing of Textiles', in S. Muthu and M. Gardetti (eds), *Green Fashion*, vol. I, Singapore: Springer, (2015), pp. 217–52.

[20] M. S. Rony Mia, Al Mojnun Shamim, Miraduzzaman Chowdhury, Suraiya Sultana, Manijah Armin, Musfic Hossain, Rozina Akter, Shuvo Dey and Hasnun Naznin, 'Review on Various Types of Pollution Problem in Textile Dyeing & Printing Industries of Bangladesh and Recommendation for Mitigation', *Journal of Textile Engineering & Fashion Technology*, 5 (4) (2019): 220–6, doi: https://doi.org/10.15406/jteft.2019.05.00205.

[21] Y. Luo, X. Y. Wu and X. M. Ding, 'Environmental Impacts of Textiles in the Use Stage: A Systematic Review', *Sustain Prod Consump*, 36 (March 2023): 233–45, doi: 10.1016/j.spc.2023.01.006.

[22] D. Mikucioniene, D. Mínguez-García, M.R. Repon, R. Milašius, G. Priniotakis, I. Chronis, K. Kiskira, R. Hogeboom, R. Belda-Anaya and P. Díaz-García, 'Understanding and Addressing the Water Footprint in the Textile Sector: A Review', *Autex Res J*, 24 (1) (28 October 2024), doi: 10.1515/aut-2024-0004.

[23] R. Kant, 'Textile Dyeing Industry: An Environmental Hazard', *Natural Science*, 4 (1) (2012): 22–6, doi: http://dx.doi.org/10.4236/ns.2012.41004.

[24] M. P. Lata Samant, Alka Goel and Manpreet Kaur, 'Impact of the Textile Industry on Global Climate Change', in K. Sadhna, R. Kumar and S. Greeshma (eds), *Climate Action through Eco-Friendly Textiles*, Singapore: Springer, (2024), pp. 11–26.

[25] J. Soth, C. Grasser and R Salerno, 'The Impact of Cotton on Fresh Water Resources and Ecosystems', World Wildlife (Background paper), (1999), https://awsassets.panda.org/downloads/impact_long.pdf.

[26] E. Karaş, 'Sustainable and Effective Management Strategies in Cotton Cultivation', in I. Y. Abdurakhmonov (ed.), *Cotton*, London: IntechOpen, (2022), pp. 1–30.

[27] C. O. Josiah Wilson Tachie-Menson, Frederick Kwame Yeboah, Flavio Odoi-Yorke, Agnes Abeley Abbey, Frank Kwekucher Ackah and Ransford Opoku Darko, 'Environmental Impact of Pesticide Runoff in Irrigated Agriculture: Evidence from Gomoa West Field, Ghana', *Discover Environment*, 3 (105) (2025): 1–19, doi: https://doi.org/10.1007/s44274-025-00268-w.

[28] C. R. Holkar, A. J. Jadhav, D. V. Pinjari, N. M. Mahamuni and A. B. Pandit, 'A Critical Review on Textile Wastewater Treatments: Possible Approaches', *J Environ Manage*, 182 (1 November 2016): 351–66, doi: 10.1016/j.jenvman.2016.07.090.

[29] J. X. Zhu, Y. D. Yang, Y. Li, P. H. Xu and L. L. Wang, 'Water Footprint Calculation and Assessment of Viscose Textile', *Ind Textila*, 71 (1) (2020): 33–40, doi: 10.35530/It.071.01.1642.

[30] D. C. Ma, C. F. Xian, J. Zhang, R. C. Zhang and Z. Y. Ouyang, 'The Evaluation of Water Footprints and Sustainable Water Utilization in Beijing', *Sustainability-Basel*, 7 (10) (October 2015): 13206–21, doi: 10.3390/su71013206.

[31] I. E. Napper and R. C. Thompson, 'Release of Synthetic Microplastic Plastic Fibres from Domestic Washing Machines: Effects of Fabric Type and Washing Conditions', *Mar Pollut Bull*, 112 (1–2) (15 November 2016): 39–45, doi: 10.1016/j.marpolbul.2016.09.025.

[32] L. Shen, E. Worrell and M. K. Patel, 'Environmental Impact Assessment of Man-Made Cellulose Fibres', *Resour Conserv Recy*, 55 (2) (December 2010): 260–74, doi: 10.1016/j.resconrec.2010.10.001.

[33] J. N. Wang and W. Azam, 'Natural Resource Scarcity, Fossil Fuel Energy Consumption, and Total Greenhouse Gas Emissions in Top Emitting Countries', *Geosci Front*, 15 (2) (March 2024): 260–74, doi: 10.1016/j.gsf.2023.101757.

[34] M. D. N. Ramos, J. P. P. Lima, S. F. de Aquino and A. Aguiar, 'A Critical Analysis of the Alternative Treatments Applied to Effluents from Brazilian Textile Industries', *J Water Process Eng*, 43 (October 2021), doi: 10.1016/j.jwpe.2021.102273.

[35] A. Rathour, A. Dhar and S. Pathania, 'Impact of Microplastics Pollution on Human Health and Aquatic Life: A Review', *Polym Int* (31 May 2025), doi: 10.1002/pi.6782.

[36] R. B. Baloyi, O. J. Gbadeyan, B. Sithole and V. Chunilall, 'Recent Advances in Recycling Technologies for Waste Textile Fabrics: A review', *Text Res J*, 94 (3–4) (February 2024): 508–29, doi: 10.1177/00405175231210239.

[37] L. A. Castillo-Suárez, A. G. Sierra-Sánchez, I. Linares-Hernández, V. Martínez-Miranda and E. A. Teutli-Sequeira, 'A Critical Review of Textile Industry Wastewater: Green Technologies for the Removal of Indigo Dyes', *Int J Environ Sci Te*, 20 (9) (September 2023): 10553–90, doi: 10.1007/s13762-023-04810-2.

[38] M. Ahmed, M. O. Mavukkandy, A. Giwa, M. Elektorowicz, E. Katsou, O. Khelifi, V. Naddeo and S. W. Hasan, 'Recent Developments in Hazardous Pollutants Removal from Wastewater and Water Reuse Within a Circular Economy', *npj Clean Water*, 5 (1) (12 April 2022), doi: 10.1038/s41545-022-00154-5.

[39] W. Jutidamrongphan, M. A. Rahman, T. Hossain, S. A. Khatun and W. D. Lamas, 'Eco-Fashion Designing to Ensure Corporate Social Responsibility within the Supply Chain in Fashion Industry', *Autex Res J*, 21 (4) (December 2021): 467–81, doi: 10.2478/aut-2020–0064.

[40] M. T. Meenaxi Tiwari and S. B. Sudha Babel, 'Air Pollution in Textile Industry', *Asian Journal of Environmental Science*, 8 (1) (2013): 64–6, https://researchjournal.co.in/online/AJES/AJES%20 8(1)/8_64-66_A.pdf.

[41] D. Marczak, L. Krzysztof and J. Misiewicz, 'Characteristics of Biodegradable Textiles Used in Environmental Engineering: A Comprehensive Review', *J Clean Prod*, 268 (20 September 2020): 1–17, doi: 10.1016/j.jclepro.2020.122129.

[42] O. Onuaguluchi and N. Banthia, 'Value-Added Reuse of Scrap Tire Polymeric Fibers in Cement-Based Structural Applications', *J Clean Prod*, 231 (10 September 2019): 543–55, doi: 10.1016/j.jclepro.2019.05.225.

[43] U. N. Haq and S. M. R. Alam, 'Implementing Circular Economy Principles in the Apparel Production Process: Reusing Pre-Consumer Waste for Sustainability of Environment and Economy', *Clean Waste Syst*, 6 (December 2023), doi: 10.1016/j.clwas.2023.100108.

[44] M. S. Abbas-Abadi, B. Tomme, B. Goshayeshi, O. Mynko, Y. Wang, S. Roy, R. Kumar, B. Baruah, K. D. Clerck, S. D. Meester and D.R. D'hooge, 'Advancing Textile Waste Recycling: Challenges and Opportunities across Polymer and Non-Polymer Fiber Types', *Polymers*, 17 (5) (March 2025), doi: 10.3390/polym17050628.

[45] S. Biyada and J. Urbonavicius, 'Circularity in Textile Waste: Challenges and Pathways to Sustainability', *Clean Eng Technol*, 24 (February 2025), doi: 10.1016/j.clet.2025.100905.

[46] K. T. Gopalakrishnan, 'Impact of Textiles on Environmental Issues: Part II', *Asian Dyer*, 9 (6) (2012): 45–51.

[47] J. Nyika and M. Dinka, 'Sustainable Management of Textile Solid Waste Materials: The Progress and Prospects', *Mater Today-Proc*, 62 (2022): 3320–4, doi: 10.1016/j.matpr.2022.04.241.

[48] Y. Wang, 'Fiber and Textile Waste Utilization', *Waste and Biomass Valorization*, 1 (2010): 135–43, doi: https://doi.org/10.1007/s12649-009-9005-y.

[49] United States Environmental Protection Agency, 'Textiles: Material-Specific Data', (2026), www.epa.gov/facts-and-figures-about-materials-waste-and-recycling/textiles-material-specific-data.

[50] T. Exchange, 'Materials Market Report 2024', *Textile Exchange*, (2024).

[51] United Nations Convention to Combat Desertification, 'Unravelling the Environmental Impact of Fibres', (2024) www.unccd.int/sites/default/files/2024-12/UNCCD-Fashion%26LAND-FINAL%20online%20publication_final_single%20pages.pdf.

[52] A. M. I. Kallio, 'Wood-Based Textile Fibre Market as Part of the Global Forest-Based Bioeconomy', *Forest Policy Econ*, 123 (February 2021), doi: 10.1016/j.forpol.2020.102364.

[53] V. Wallius and A. Näyhä, 'Wood-Based Textile Innovations in the Finnish Media: A Critical Discourse Analysis', *Forest Policy Econ*, 174 (May 2025), doi: 10.1016/j.forpol.2025.103492.

[54] K. Farhana, K. Kadirgama, A. S. F. Mahamude and M. T. Mica, 'Energy Consumption, Environmental Impact, and Implementation of Renewable Energy Resources in Global Textile Industries: An Overview Towards Circularity and Sustainability', *Materials Circular Economy*, 4 (1) (2022), doi: 10.1007/s42824-022-00059-1.

[55] A. D. La Rosa and S. A. Grammatikos, 'Comparative Life Cycle Assessment of Cotton and Other Natural Fibers for Textile Applications', *Fibers*, 7 (12) (2019): 1–8, doi: 10.3390/fib7120101.

[56] M. Shamsuzzaman, M. Islam, M. A. A. Mamun, R. Rayyaan, K. Sowrov, S. Islam, A. S. M. Sayem, 'Fashion and Textile Waste Management in the Circular Economy: A Systematic Review', *Clean Waste Syst*, 11 (2025): 1–19, doi: 10.1016/j.clwas.2025.100268.

[57] European Parliament, 'Fast Fashion: EU Laws for Sustainable Textile Consumption', (2020), www.europarl.europa.eu/topics/en/article/20201208STO93327/fast-fashion-eu-laws-for-sustainable-textile-consumption.

[58] J. Sawant, R. Guru, D. Grewal, S. C. Talekar and S. P. Kulkarni, 'Sustainability in Textiles: A Critical Review of Eco-Friendly Practices and Materials', *ShodhKosh: Journal of Visual and Performing Arts*, 5 (2) (2024): 85–97, doi: 10.29121/shodhkosh.v5.i2.2024.891.

[59] T. Wagaw and K. M. Babu, 'Textile Waste Recycling: A Need for a Stringent Paradigm Shift', *AATCC Journal of Research*, 10 (6) (2023): 376–85, doi: 10.1177/24723444231188342.

[60] G. A. Montero, C. B. Smith, W. A. Hendrix and D. L. Butcher, 'Supercritical Fluid Technology in Textile Processing: An Overview', *Industrial & Engineering Chemistry Research*, 39 (12) (2000): 4806–12, doi: 10.1021/ie0002475.

[61] D. J. Tyler, 'Textile Digital Printing Technologies', *Textile Progress*, 37 (4) (2005): 1–65, doi: 10.1533/tepr.2005.0004.

[62] A. Zille, F. R. Oliveira and A. P. Souto, 'Plasma Treatment in Textile Industry', *Plasma Processes and Polymers*, 12 (2) (2015): 98–131, doi: 10.1002/ppap.201400052.

3 Principles of sustainable design and circular design

Josphat Igadwa Mwasiagi and Evance Obara

3.1 Introduction

The basic needs for humankind include air, food and shelter. Clothing may rank second after the basic needs, and they can be considered as important needs (Figure 3.1). Other needs that are equally important include security, health and environmental safety. All needs are met by extraction of resources; therefore, a study of resource utilisation was deemed necessary, the Earth Overshoot Reports were launched in 1970, and annual records indicate a drastic increase in resource extraction. As the world population increases, there is an increase of per-capita garment needs which brings more stress to the Earth's resources and ecological regeneration [1, 2]. As shown in Figure 3.1, there is a direct correlation between consumption of textile goods (garments) and resource utilisation/extraction. Furthermore, accelerated extraction of resources is accompanied by water usage, water pollution and waste generation just to mention a few hotspots that demand urgent attention. Apart from garments, the textile and fashion industry

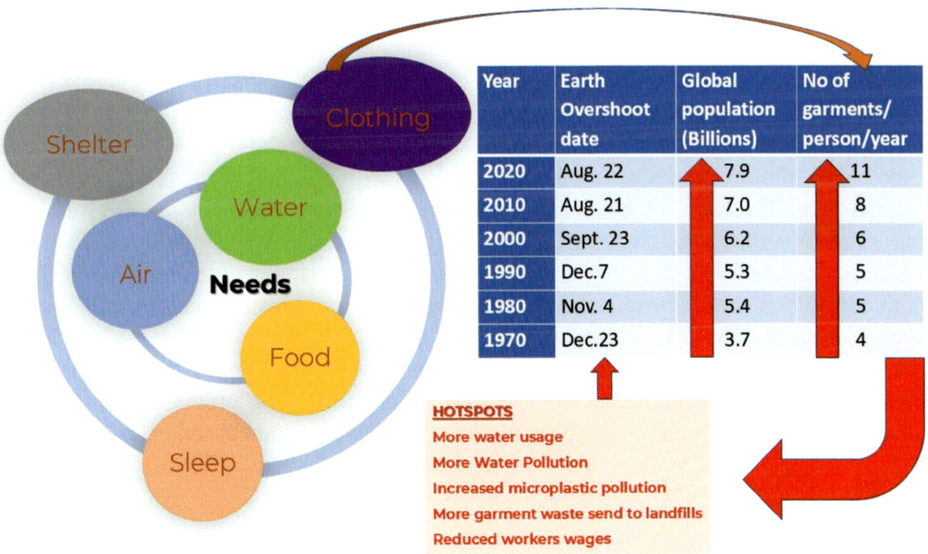

Figure 3.1 The nexus between increased consumption of garments and Earth overshoot date.

DOI: 10.4324/9781003638148-4

provides goods/products used in provision of air, food, shelter and sleep which are priority needs for human beings leading to several environmental hotspots, which include:

- usage and pollution of water and environment during dyeing and cultivation of textile fibres;
- emission of CO_2 during manufacture;
- release of 35 per cent of microplastics in water bodies from synthetic fibres [3];
- dumping of end-of-life textile goods in landfill.

Apart from environmental issues, the increase in the consumption of textile and fashion goods has also recorded some negative social hotspots such as low worker wages and inhumane working conditions [4]. Taking cognisance of the fact that the resources available on Earth are finite and that generation of waste during production and consumption causes environmental and social hotspots, the concepts of sustainability and circularity have gained prominence [5–8] as humankind endeavours to optimise the utilisation of these limited resources on Earth to meet the growing needs. This chapter will review the principles of sustainable and circular design, production of sustainable and circular fashion goods, production of sustainable textile patterns and the difference between sustainable designs and circular designs. The current challenges affecting sustainable and circular designs and future directions are also discussed, and conclusions and recommendations are given.

3.2 Sustainable design and circular economy

Traditional fashion and textiles industries have been criticised for environmental pollution and unethical practices including low wages and child labour among others. However, with the advancements in technology and the call for a more sustainable world, consumer preferences have greatly changed. According to the 2023 Eurobarometer survey on the EU ecolabel, a majority of Europeans consider the environmental impact of products rather than the purchasing decisions [9]. This has subsequently led to global competition among top brands as the local businesses remain in despair; adoption of sustainable practices comes at a cost [10]. In this regard, sustainable design and circular design have addressed these shortcomings in the textile value chain by modifying the production, use and disposal mechanisms. Sustainable design is a holistic approach that creates products, services and systems with minimal negative environmental and social impacts. It focuses on efficiency, reduction and responsibility to reduce resource depletion, improve well-being and lower pollution without compromising the future. On the other hand, circular design is a transformative framework that focuses on eliminating waste, prioritising closed-loop material flows, innovation and regeneration. This fosters the circular economy by ensuring restoration of natural capital, economic systems and waste reduction. These approaches aim at reducing environmental footprints through proper material selection, resource efficiency, durable and timeless design, low-impact manufacturing processes, waste reduction, ethical and transparent supply chains, consumer engagements and education among others.

3.2.1 Sustainable design principles

Sustainable design is a comprehensive approach that creates products and systems with the aim of minimising environmental, social and economic impacts; to promote the health and well-being of humankind. This approach challenges the conventional linear model of production which for a long time has ignored environmental stewardship and social equity but rather focused on the economic viability of products. Focusing on textile and fashion industries, sustainable design not only advocates resource efficiency (water, energy), innovation and technology, but goes further and incorporates social responsibility, aesthetic and cultural aspects. Below, we discuss these aspects and components in details.

3.2.1.1 Resource efficiency

This is a key aspect in sustainable design as it is the centre stone of major activities in production. Garment manufacturing involves raw material sourcing up to final packaging and distribution of the product. At the raw material sourcing, choosing organic materials (organic cotton, banana, sisal, bamboo, kenaf, etc.) that require minimal fertilisers and pesticides ensures efficient use of the resources. During harvesting, efficient harvesting methods that reduce waste should be adopted and the entire plant should be used (seeds, fibres, stem) [11]. At the processing facility, energy-efficient spinning methods (compact, air-jet, open-end spinning) [12] that reduce energy consumptions should be adopted. At the cutting and sewing stages, the patterns should be designed to maximise the fabric use by implementing strategies like zero-waste pattern cutting which eliminate offcuts [13]. At the quality control stage, durability tests should be conducted to ensure longer service life and reduce waste, while at the final packaging stage, recyclable materials should be adopted to reduce waste and promote sustainability. Furthermore, priority should be given to local distribution to reduce unnecessary energy consumption. Additionally, the consumer should be reliably informed about the end-of-life options such as reuse and redistribution that promote sustainability.

3.2.1.2 Waste reduction

With the increasing population, continuous consumption has led to overproduction and a subsequent increase in waste generated. Efficient usage of resources is centred at the energy and water use, material selection, simplicity and modular design. The choice of material determines its efficient usage and waste generated; for instance, the use of lightweight materials in fashion design not only generates less waste but may improve the comfort of the wearer [14]. In addition, precision-manufacturing techniques such as 3D printing reduce offcuts and scrap. The zero-waste manufacturing techniques (lean manufacturing [15], industrial symbiosis [16] and digital prototyping) ensure waste reduction by reducing excess inventory and defects and ensure that waste from one process becomes raw material for the next one. Although these methods are promising in waste reduction, the associated high costs, the need for skilled personnel and competition from the fast-fashion model among other factors hinder their full implementation. The final product should be presented to the consumer with minimal, reusable and biodegradable packaging materials to reduce waste. Material design should therefore be as simple as possible with all functionality aspects put in place to reduce waste generation. The products should be easily repaired, refurbished, or recycled by replacing only the parts rather than the entire component. Finally, end-of-life factors such as take-back programmes, design for recyclability and biodegradability should be given priority as they jointly reduce waste generation.

3.2.1.3 User-centred design

Conventional designs rely on product functionality with little attention to the user's needs; simply to say, fitting the user to the machine rather than fitting the machine to the user. Ergonomically, products and systems should prioritise the needs, preferences and emotional behaviour of the user. The user should be involved in the entire product development through interviews, observations, workshops, etc., to tailor the design to actual use, reduce waste and improve comfortability. While doing so, the design should consider the population percentiles (25th, 50th, 95th percentiles) to either reduce overproduction or underproduction that can lead to waste or lack of fit (comfortability) for the population. In addition, behavioural change is a key factor in design, for instance emotional aspects. Textile products should be designed in a manner that encourages emotional bonds with the user to encourage long-term use.

3.2.1.4 Ethical production

In the conventional linear economy, the main goal of doing business was to make profit without any consideration of ethical issues. In the circular economy, specifically in sustainable design, ethics is prioritised over profit during the production of goods. The workers operate in a safe working environment free from hazards and they receive fair wages to meet their daily needs. Moreover, policies that prohibit child labour are set and well implemented, and raw material sourcing is done in a sustainable manner (local sourcing, sustainable materials) to reduce environmental impacts [17]. The supply chain is transparent and traceable to allow consumers to make informed choices regarding where and how the products are made, any allergens present and whether the materials used are sustainable [18]. Efficient energy methods should be employed, including renewable sources to reduce carbon footprints and minimise waste. Ethical production goes beyond human and extends to animal welfare as well, ensuring that all animals involved in production (sheep for wool, goat for mohair, silkworms for silk, etc.) are treated ethically and humanely as reported by Buller et al. [19]: for example, sheep shearing with proper blade tension or reeling silk from cocoons only upon the emergence of the moth. Finally, consumers should be educated on the importance of ethical production and encouraging responsible choices that adhere to ethical practices.

3.2.1.5 Durability and timelessness

The product should remain functional while retaining its aesthetic value over time, especially in fashion and textile industry. Product durability ensures waste reduction, bypassing frequent replacements by encouraging consumers to purchase higher-quality products. In a nutshell, products should be designed for longevity while conserving resources to improve consumer value. Apparel manufacturers can enhance the product's lifespan by producing high-quality fabrics through modular designs and reinforced stitching to enhance lifespan [20]. Another good practice by the manufacturer could be to offer repair services to the consumers (take-back programmes) and encourage them to mend the garments rather than discarding them fully. Raw-material sourcing should be sustainable; therefore, it is highly encouraged to select high-quality and durable materials. Also, product versatility should be embraced by designing products that can be styled in several ways to suit several occasions while providing the consumers with clear instructions on how to manage the products.

3.3 Circular design principles

While sustainable design aims at minimising environmental, social and economic impacts, circular design goes beyond this and designs restorative products for their entire life cycle and focuses on strategies that eliminate waste (reuse, repair, recycling) or even biodegradation. This approach is anchored in the circular economy's principles and ensures that resources remain in use for a longer duration. In the textile and fashion industries, it introduces new innovations such as closed-loop systems of recycling and modular garments among others through technological advancements, regulatory considerations and consumer awareness. Below are some of the principles of circular design discussed in detail:

3.3.1 Design for longevity

The basic principle of circular design is creating durable products to extend their life span. The fast-fashion industry designs products for short-term use, therefore quality is compromised. Design for longevity opposes this approach by selecting only high-quality materials that are renewable, biodegradable and recyclable to extend the final product's lifespan [21]. In the process of extending the product's lifespan, allowances should be given for repairability and upscaling (modular designs) without complete replacement of the product. By designing for durability, waste generation is reduced due to reduction in frequent replacements. Another consideration is timeless design where designs are created to remain stylish for a longer period (classic styles) without any need for frequent replacement. The use of neutral colours with versatile colour palettes ensures easy mixing and matching with other garments. Finally, the consumer needs further education on the benefits of longevity designs and public awareness needs to be created to encourage them to prioritise the purchase of durable products to eliminate environmental impacts. This can be achieved through storytelling by sharing the materials used, being transparent on the production processes used and creating circular fashion initiatives through take-back programmes for recycling [18].

3.3.2 Design for disassembly

Circular design emphasises creating products that can be easily separated during repair, reuse, or recycling. Design for disassembly (DfD) in fashion and textiles relies on simple assembly techniques with modular construction and minimal joints to reduce the use of complex sewing methods and encourage easy replacements or upgrades [22]. The materials used should be homogeneous for easy recycling [23], and, when blending is necessary, it should be done in a way that eases disassembly and recycling. Moreover, the type of materials used should be clearly labelled for easy identification during recycling, and clear instructions should be provided to guide the disassembly process. In the assembly, the key parts should be easily accessible during repair, and the use of standardised parts should be encouraged. The products should be designed with considerations of the end-of-life management [24] and take-back programmes in place to close the product's life-cycle loops.

3.3.3 Design for recycling

Designing for recycling (DfR) ensures the recyclability of finished products without significant loss in quality. At the production stage, the selected materials should be compatible with existing recycling technologies, and blends of different fibres that complicate the recycling process should be avoided [23]. In addition, excessive use of adhesives which may hinder recycling process should be avoided, and, instead, mechanical fasteners that are easy to disassemble should be encouraged. The material should be clearly labelled to guide the recycling procedure, and, where possible, symbols should be used to indicate recyclability [25]. The consumer should be adequately informed on how to effectively recycle the product at the end of life, and proper take-back measures should be put in place by the product manufacturer. Finally, manufacturers should work closely with recycling plants for effective recycling, closing the material loops and scaling the business models through proper communication channels.

3.3.4 Design for biodegradability

This design approach creates products that can naturally decompose at the end of the life cycle without causing harm to the environment. Akter et al. [26] have studied the conversion of textile wastes to wealth and deduced that about 92 million tonnes of these wastes end up in landfill annually. As seen in DfR and DfD, material selection is a key aspect in circular design; design for biodegradability selects materials that can easily decompose at the end of their life cycle, ensuring minimal carbon footprints. For instance, biodegradable fibres such as hemp, sisal and cotton among others reduce the pressure on landfills. Another emerging technology especially in composites is the use of hybridised materials where biodegradable materials are jointly used with synthetic ones to reduce the percentage of non-biodegradable materials especially where specific strength is required. Design for biodegradability is most suited for non-recyclable materials; designers should therefore ensure that toxic finishes or dyes are avoided during the processing stage as they could contaminate soil during decomposition [27].

3.3.5 Material recovery systems

At the end of their life cycle, textiles need to be collected and repurposed to divert them from landfills. Therefore, garment manufacturers should adopt take-back programmes that collect used clothing for remanufacturing, refurbishing, or repairing and then reselling them. First, systems that collect the used products must be set in place, followed by sorting (classifying products based on their conditions and materials) and finally material recovery in which mechanical, physical, or chemical methods may be used to separate the individual fibres. This system keeps the material in circulation and provides avenues for cost savings and new value propositions that further enhance trade and foster the circular economy. Material recovery systems are gaining traction worldwide due to the rise in advocacy for sustainability; however, global textile recycling is only about 1 per cent of the total textile and apparel which calls for more efforts in this sector.

3.3.6 Product-as-a-service models

Due to the increasing population, overproduction has occurred in the past decades, which has subsequently led to the depletion of the limited natural resources; space for humankind gets scarce, and more waste is generated. Circular design attempts to solve this challenge by embracing alternative consumption models such as leasing property instead of owning. In textile fields, manufacturers can allow the consumers to rent their products for temporary use or return the purchased goods for recycling to reduce waste and conserve the ecosystem [28]. The products can also be customised to fit individual use, and businesses can adopt customised production (produced only in order) to avoid overproduction which could lead to waste. Product-as-a-service (PaaS) supports the concept of longevity design by producing high-quality products that extend their life cycle for recycling or refurbishing. The model is more suited for the younger generation or those with financial constraints who would prefer leasing to owning. The model further adopts second-hand and resale integration, allowing consumers to trade their products in a sustainable manner.

3.4 Production of sustainable and circular fashion goods

The production of sustainable fashion goods directly responds to the social, economic and environmental shortcomings, which in turn affects its ecological footprint. The modern textile and fashion sector has seen advancements due to innovations, driven by consumer demand and new governmental interventions. Herein, sustainable fashion production is detailed through discussion on materials practices, ethics and labour practices, waste management and certifications to guide the future direction in the field.

3.4.1 Materials and eco-friendly practices

Sustainable materials form the foundation of sustainable fashion, prioritizing the selection of green materials over synthetic counterparts. Organic cotton is biodegradable and does not require pesticides during production stage; Tencel is soft and has a closed-loop production channel, while recycled polyester (rPET) reduces plastic waste and pressure on landfill [29]. Recently, the use of deadstock fabrics (unused, surplus or leftover textiles) has ensured cost-effectiveness during manufacturing processes, reduced waste, lowered carbon footprints and ensured unique and limited-edition designs. While the list of eco-friendly materials has extended to next-gen materials such as plant-based leathers, algae-based fabrics and lab-grown textiles such as bio-fabricated dyes and spider silk [30], it is important to point out that not all eco-friendly materials may be fully sustainable. Take, for instance, viscose which requires millions of trees and tonnes of water to produce a single tonne of fabric; these points need further attention (in-depth life-cycle assessment [LCA]) to ensure the true definition of sustainable sourcing. In addition, sustainable sourcing demands for ethical practices, transparency and blockchain traceability as shall be discussed in detail in the subsequent subsections. On top of the fabric materials, sustainable dyes and printing materials need further considerations in sustainable designs.

3.4.2 Ethical labour and supply chain transparency

Ethical labour practices addressing tangible issues with labour exploitation have recurred historically. Fair labour practices should not only address minimum wages but living wages, safe working conditions and worker empowerment through union formation or other suitable grievances mechanisms. On the other hand, supply-chain traceability encompasses radical transparency from the supplier, blockchain technology and supplier third-party audits. Certifications for ethical and transparent sourcing such as the Global Organic Textile Standard (GOTS), OEKO-TEX, WRAP (Worldwide Responsible Accredited Production) and B Corp eliminate harmful chemicals in textile production and enhance ethics and integrity. According to the OEKO-TEX 2024 report, 'Made in Green' contributed to improved working conditions for about 2 million textile workers globally [31], while GOTS embeds labour and social requirements that prohibit discrimination and child labour and encourage safe and hygienic working conditions [32]. Moreover, to respond to labour-related risks in the workplace, GOTS currently aligns its due-diligence approach with guidance from the Organisation for Economic Co-operation and Development (OECD) [33]. With increasing consumer awareness, brands are required to disclose their raw materials source and how they treat their customers. As a way of strengthening ethical standards, the European Union has recently proposed new rules that commission companies to account for their environmental and social impacts. Moreover, techniques such as QR codes can be used to mirror the factory conditions and their carbon footprints [34].

3.4.3 Green manufacturing processes

Sustainable brands adopt resource-efficient processes to reduce water use, energy consumption and waste generation. Water-saving techniques such as waterless dyeing and closed-loop water systems have reported success towards sustainable practices in textile and fashion sectors. However, the adoption of new dyeing techniques such as digital printing [35] that apply dye precisely, air-dyeing which saves up to 95 per cent water and CO_2 dyeing which uses pressurised CO_2 instead of water have greatly impacted on the manufacturing processes [36]. Heat recovery systems such as steam recovery from incinerators and boilers ensure energy efficiency and cuts on the overall production costs. Moreover, low-temperature processes such as enzyme washing and cold pad batch dyeing (room temperature) ensure energy efficiency in the operations. The use of non-toxic and low chemical processes during manufacturing further reduces pressure on the environment through actions such as eco-friendly finishes, green chemistry and low-impact dyes. Eco-friendly finishes such as plasma treatment and ozone bleaching ensure safer alternatives to chemical coatings and excessive use of chlorine especially for denim fabrics. Enzymatic bio-polishing as an aspect of green chemistry additionally softens the textile material bypassing the use of silicones and ensures sustainability. To fully implement low-impact manufacturing, circular production models such as on-demand production, DfD, take-back and recycling mechanisms should be implemented [37].

3.4.5 Waste reduction and circular economy

The category of textile waste is diverse, ranging from pre-consumer to post-consumer waste. The former, also called industry waste, rises from the production processes while post-consumer waste results towards the product's end of life. Waste reduction is a censorious aspect in sustainable fashion production that needs utmost attention due to the sector's overproduction. Environmentally, waste generation leads to pollution which in turn depletes the natural reserves. Economically, the potential loss in valuable materials leads to a decline in circular economy practices, while social impacts include unethical labour practices such as excessive overtime which may stem from the pressure from manufacturers to produce more goods to cater for fast-fashion needs. To address it, circular economic principles have been adopted by different industries and scholars. The Ellen MacArthur Foundation [38] has documented three key principles in circular fashion: (1) waste and pollution elimination; (2) natural systems regeneration; and (3) keeping materials in use. Three key factors to be considered in sustainable design are economic, social and environmental and circular design, resource recycling, longevity and system integration and should be considered as shown in Table 3.1. Designing fashion out of waste entails designing products for longevity, recyclability and systems integration. Some of the highlighted key practices in waste reduction (which is a key feature in circular design) are now discussed at each stage of the life cycle.

3.4.5.1 Design stage

Technically, the occurrence of fashion waste should be prevented before it happens, and strategies like modular design, zero-waste pattern cutting and mono-material construction should be given priority to reduce waste or enable recycling. Modular designs produce detachable garments with interchangeable sleeves. Zero-waste pattern cutting eliminate fabric offcuts through such methods as 3D knitting and tessellation among others, while mono-material construction encourages recycling by avoiding fabric mixtures [30].

3.4.5.2 Production stage

The concept of lean manufacturing is introduced through strategies such as industrial symbiosis, waterless dyeing and make-to-order production techniques. Overproduction is a major challenge facing many industries with majority of the end products being directed at landfill. Lean production ensures only what is required is produced to align with consumer demand [39].

Table 3.1 Life-cycle thinking template.

		Impacts of sustainable design			*Impacts of circular design*		
Phase	*Activity*	*Environmental impacts*	*Social impacts*	*Economic impacts*	*Resource recycles*	*Longevity*	*Systems integration*
Fibre production	Growing of natural fibres						
	Manufacture of synthetic fibre						
Yarn and fabric production	Yarn spinning and fabric manufacture						
	Fabric coloration and finishing						
	Fabric finishing						
Garment production and packaging	Designing, cutting, sewing and garment finishing						
Distribution marketing	Transportation of products						
Consumption	Usage and cleaning						
End of life	Disposal						
	Material repurposes						
	Fibre recovery						

The zero-waste pattern making seamlessly cuts the fabric while integrating durability and high-quality garments. In addition, employee motivation at the production stage increases morale, improves efficiency and fosters innovation in sustainable practices. Natural fibre manufacture generates different forms of waste such as agricultural waste (stem, leaves, etc.), chemical residue from pesticides and fertiliser, and short fibres and dust from carding and cleaning operations. On the other hand, synthetic fibres such as polyester shed microplastics during washing together with other spin finish waste (lubricants and oil) [40]. At the yarn manufacturing stage, carding and combing waste, plastic cones, wrappers and other waste such as airborne fibre particles are potential causes of respiratory infections. Fabric production waste such as broken threads, trimmed edges and chemical sizing waste among others usually pollutes water and pose risk of injury to children (broken needles) when disposed of in open air. Dyeing and finishing waste remains to be the major source of pollution during textile manufacture: the auxiliary chemicals such as dispersants, surfactants, mordants and salt, pose serious risk to water bodies and affect aquatic life and humans altogether; wastewater with toxic dyes and heavy metals accelerate this course. The volatile organic compounds (VOC) from dyeing or finishing pollute air while the workers exposed to azo dyes may develop cancerous health risks [41]. Therefore, management of this waste is paramount, and sustainable solutions are required for waste minimisation by treating effluents, using eco-friendly dyeing and obtaining sustainability certifications previously mentioned in the earlier sections.

3.4.5.3 Consumption stage (use)

The garment lifespan should be extended at the consumption stage through consumer education programmes on care maintenance practices such as proper storage, best washing practices, repair, resale, reuse or upcycling among other strategies. Designing textiles as self-healing material systems where repeated damage delay failure and replacements can extend the garment's life and reduce its footprint [42]. Another aspect is design for repairability and longevity; fabric designs should embrace the use of high-quality and durable fibres, incorporate detachable components to the fabric structure such as zippers and cuffs and prioritise versatile styles to encourage longer usage. Technological advancements through material innovation and smart care tools can further guide consumers on the optimal usage through repair tutorials and other apps. It then goes without mention that proper governmental policies, such as tax relief or subsidies for the organisations encouraging sustainability can further impact the implementation of waste reduction measures in a more sustainable manner.

3.4.5.4 Distribution and marketing

The linear model of take–make–dispose for a long time has depended on mass production with a centralised distribution aided by complex logistics networks and priority given to profit making than environmental stewardship. Inventory management also relies on this overproduction to keep products available to consumers, which leads to discarded products and waste generation. Integrating sustainable and circular models in the distribution of textiles and fashion minimises these environmental impacts and promotes circular economy through low-impact logistics with green transportation systems (shipping and use of electric vehicles). Localised production and distribution reduce carbon footprints, save on costs, and, if shipping is done, it should be consolidated to reduce emissions and excessive packaging. On the same note, the packaging materials should be either biodegradable, recyclable, or reusable, and consumers should be encouraged to return them for reuse. New emerging trends in distribution include partnerships with carbon-neutral logistics providers for green transportation means, using AI and data analytics for the prediction of demand and upcycling of deadstock to reduce waste.

For sustainable marketing, the supply chain should be transparent; the manufacturer should honestly disclose the source and nature of the raw materials used with clear certification labels. Technological advancements through digital fashion reduce physical sampling and waste while secondary market promotions such as rental and subscription services and resale platforms, offer circular options and should be encouraged during marketing.

3.4.5.5 End-of-life stage

The conventional model of take–make–dispose upon usage (see Figure 3.2) saw textiles discarded in landfills and commercial incinerators, leading to soil or air pollution. As a result, it is important to close the loop not only to reduce landfill waste but to save on operational costs and lower the demand for virgin materials through recycling. The garments should be designed for disassembly, shredded back into fibres, and take-back initiatives should be put in place to collect old garments for recycling. In addition, upcycling should be encouraged to transform the existing garments into new unique ones, thereby reducing the need for new production.

3.5 Sustainable packaging and logistics

In the textile and fashion sector, sustainable packaging is a critical part of sustainable production and addresses the mechanisms of transportation and delivery as well. It is reported that millions of tonnes of plastic enter landfill annually, and 90 per cent of single-use plastics in fashion are not recycled significantly, leading to carbon emissions. Sustainable packaging embraces the utilisation of biodegradable and recyclable materials such as bioplastics thus supporting

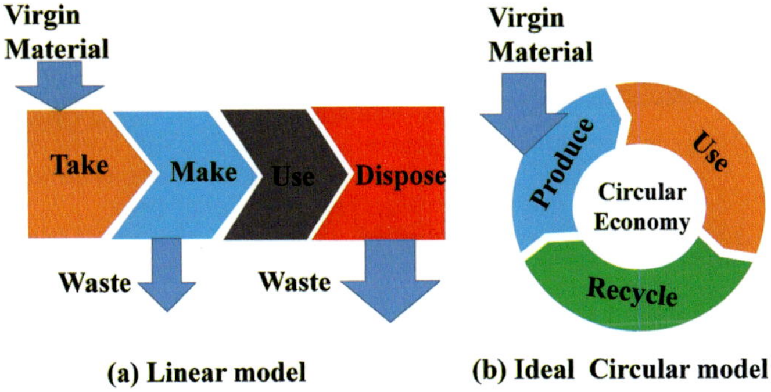

Figure 3.2 (a) Linear model. (b) Circular model.

the circular economy. For example, recycled cardboard and paper boxes lower the demand for virgin materials, reduce deforestation and support the closed-loop systems. Packaging design should allow for reusability to extend the materials' life cycle, and it is a good practice to eliminate unnecessary layers to lower emissions during shipments and reduce waste accumulation. Moreover, creating incentives for consumers who return packaging to the company is a milestone for setting reuse and recycling measures (take-back programmes). It is the responsibility of the manufacturer to ensure transparent labelling on the packaging, with clear instructions on disposal methods to further guide the consumers on sustainable packaging managements.

3.6 Production of sustainable textile patterns

The production of sustainable textile patterns is based on designs and textures that minimise environmental impact while ensuring economic viability and social equity. The sustainable textile practices are integrated into the design, production (including dyeing and printing) and finishing processes. The sustainable pattern design strategies include the following:

3.6.1 Zero-waste pattern design

Zero-waste pattern drafting supports both sustainable and circular designs by addressing important ethical and environmental setbacks in the fashion industry. This entails designing patterns that use the entire fabric without scraps (Table 3.2). Techniques such as nesting layouts and seamless designs optimise the pattern pieces and ensure perfect fit. Kimono-inspired cuts also ensure fewer seams and rectangular pieces that minimise waste. Efficient pattern layouts ensure the entire fabric usage, optimising the layout to avoid offcuts and may involve geometric shapes that interlock without gaps. While the creative use of scraps repurposes the offcuts (if any) into the design elements such as trims and patches, minimalist and versatile designs reduce complexity and fabric usage and maintain the aesthetic values desired. In addition, holistic design approaches integrate pattern drafting with garment construction while considering the fabrics dimensions from the outset [13].

3.6.2 Modular pattern designs

This technique aims at creating patterns from identical and compatible geometric modules with methods such as tessellated shapes and modular components. In tessellated shapes, patterns are

Table 3.2 Differences between zero-waste and conventional pattern cutting.

Feature	*Zero-waste method*	*Traditional cutting method*
Model	Uses geometric shapes with shared seams. Waste elimination is integrated at the onset	Uses darts, seams and flat 2D patterns with no fixed rules on waste elimination
Fabric utilisation	Aims at utilising the entire fabric without waste	The patterns are laid out linearly, leaving offcuts generates waste
Cost	Economical due to limited waste disposal and lower fabric needs	Higher fabric consumption and waste generated
Efficiency	Faster, fewer cuts are needed due to shared cutting lines	Slower, each piece is cut separately and there are minimal shared lines
Grading/sizing	There is no room for expansion as the pieces fit well, thus grading is challenging	Lengths and widths can be easily adjusted

built from repeating shapes that fit seamlessly, while modular components design fabrics from easily replaceable and standardised pieces for easier repair and disassembly. Adaptability design allows garments to be reconfigured and could include digital pattern optimisation. In full implementation of modular designs, factors such as DfD, the kit of parts approach, and reversible modules are considered. In DfD, seams are simplified, and the use of mono-materials is encouraged while the kit of parts approach ensures compatibility of the pattern modules for radical waste reduction, extended lifespan and resource efficiency [22].

3.6.3 Efficient marker planning

A marker is a technical diagram representing the arrangements of pattern pieces for different garments based on length and width prior to cutting. Therefore, marker planning and nesting optimises the number of cut pieces from a given fabric width or length, achievable through grainline flexibility in which pieces can be placed effectively without constraints; CAD/CAM software for advanced nesting algorithms, or pattern piece resizing with slight adjustments to curves that improve nesting. These aspects lower the carbon footprint by reducing the overall fabric consumption per garment. However, the adoption of software depends on the balance between human expertise, initial cost and the desired efficiency even for complex fabric designs. Recently, digital nesting (automated, computerised and optimised garment pattern pieces) has emerged with algorithms such as interactive nesting, automated nesting and multi-garment nesting for efficiency maximisation during marker planning. The multi-garment nesting combines pieces from different styles in one marker with higher utilisation than a single marker, interactive nesting is manually adjusted for further optimisation, while multi-length nesting optimises across markers of different lengths of the garment.

3.6.4 Timeless aesthetics

Structural patterns such as checks and stripes encourage longer fabric use by remaining relevant for a longer duration. These patterns must remain simple by avoiding complex elements and be versatile to increase wearability but remain neutral with the character to balance the understated tones with subtle and emphasise on quality to remain visually appealing over time [43]. Different design approaches have been used in the implementation of timeless textile patterns: herringbone, geometrics and stripes and checks. The monochromatic herringbone patterns have remained adaptable in the design with single colour patterns that create depth without overwhelming the eye. Moreover, nature-inspired motifs such as bio-mimicry and organic forms resonate universally and encourage sustainability in the industry. Other than waste reduction and product's lifespan extension, timeless aesthetics influence emotional connection and align with sustainable practices, extended to sustainable material integration and digital designs.

3.7 Innovative texturing and finishing

3.7.1 Bio-based finishes

Bio-based finishes remain biodegradable, enhance functionality and create timeless patterns. To enhance their functionality, these finishes are guided by such principles as non-toxicity, eco-friendliness, timeless integration and functional enhancements to improve fabric durability and water resistance. Plant-based coatings such as waxes and oils create water-resistant finishes that are not only renewable but lower in VOCs than synthetic finishes [44]. On a similar note, natural dye finishes such as madder root when combined with resist dyeing create biodegradable fabrics with timeless softness and organic hues. The use of microbial finishes made from bacteria and fungi is gaining traction with alternative timeless applications and sustainable benefits are potential for closed-loop systems. Finally, bio-enzyme finishes creating subtle textures and patterns which decompose naturally and reduce landfill impact [45].

3.7.2 Mechanical finishes

Techniques such as laser etching, brushing and embossing alter the structure or surface of the fabric without chemical treatments, which would have required water and chemical use otherwise polluting the environment. These methods enhance fabric strength by extending their life cycle while maintaining recyclability and upcycling processes. The patterns created should complement the diverse styles and remain relevant for different seasons with a lower environmental impact. For instance, embossed patterns of minimal geometries offer a timeless look, uses minimal water and chemicals to render the consumers with durable products. Laser etching precisely cuts into the fabric's surface with intricate designs without chemical usage, remaining sustainable and offering timeless applications [46]. Similarly, calendaring ensures improved fabric durability with glossy finishes at minimal chemical use and lower water consumption.

3.8 Difference between sustainable designs and circular designs

Although both sustainable and circular designs reduce environmental impact, their scope and approach are quite different. While sustainable design creates products and systems that minimise social, economic and environmental impacts, circular design creates systems for reuse, recycling or regeneration in a closed-loop manner. Table 3.3 highlights these differences in detail:

3.9 Challenges and future directions

Sustainable designs have reported progress in reducing environmental impacts in textile and fashion industries. However, significant challenges still hinder their progress to achieve full implementation. Due to the high cost of organic materials, small-scale tailors end up surrendering to fast fashion, which relies on synthetic fibres. Moving forward, the low living standards arising from a lack of employment opportunities further reduce consumer demand, making it harder

Table 3.3 Differences between sustainable design and circular design.

Description	*Sustainable design*	*Circular design*
Scope	Broad-includes energy efficiency, ethical sourcing and social equity	Specific focus on material loops, reuse, repair and regeneration
Aim	Minimise social, economic and environmental hotspots	Optimise material usage by feeding back production waste into the system
Life-cycle thinking	Works on product life cycle, but may also operate in a linear system	Inculcates circularity from the start
End-of-life strategy	Can include recycling or safe disposal	Prioritises reuse, remanufacture and biological or technical cycling
System thinking	May address sustainability within isolated products or processes	Requires holistic systems thinking; product, service, business model and ecosystem
Examples	Using organic cotton, using banana fibre, reducing water use, ethical labour, using natural dyes	Fibre recovery from garments and designing garments for disassembly

for big brands to be sustainable. Also, complex supply chains with multiple suppliers, manufacturers and distributors hinder the full implementation, where lack of transparency (some brands falsely market as sustainable) interferes with the verification of certification and fair labour practices. The energy-intensive production of fibres and fabrics, both for natural and recycled fabrics, further adds to the cost of goods and hinders the purchasing power of these products.

Circular designs, on the other hand, are faced with a lack of recycling facilities either due to the developing technologies or lack of awareness by the producers, which hinders waste collection. Moreover, majority of garment manufacturers rely on blended fibres which are difficult to recycle. Second-hand clothes importation, especially in developing countries, further hinders recycling efforts at the end of the life cycle. The consumer mindset has also barely changed; resale, rental or garment repair may take quite some time before being fully adopted by the consumers who still rely on buying them and discarding them at the end of the life cycle. Another hurdle is the scenario where fast fashion may conflict with the circular design principles such as DfD, producing garments that are not easily disassembled. Overproduction and lack of scalability measures remain major issues that need to be addressed through technological adoption to digitise inventory management and orders. Lack of proper logistics and distribution channels further limit the implementation of take-back programmes, which need a collaborative approach to address. Therefore, the future of sustainable and circular designs relies on collaborative approaches incorporating innovation and technological advancements in materials science, digital tools for design optimisation, shift in consumer behaviour and policy and regulations.

3.10 Conclusion and recommendations

The principles of sustainable and circular design in textile and fashion sector are discussed; production of sustainable and circular fashion goods and sustainable textile patterns are elaborated. The differences arising from the two are then elaborated based on scope and approach. Sustainable design creates products and systems which minimise social, economic and environmental impacts, while circular design creates products and systems enhancing reuse, recycling or regeneration. Combining these two approaches forms a cohesive strategy that reduces environmental harm and resource wastage during production and use.

Circular and sustainable designs differ in policies, where the former prioritises waste elimination and pollution control, while the latter integrates social and environmental aspects across supply chains. Circular design mandates renewable or recycled materials that are safe and designed for longevity, disassembly and recyclability. Different principles such as reuse, repair, reduce and recycle are introduced to minimise overreliance on virgin resources. Incorporation of digital tools further enhances transparency and traceability for circularity and ensures extended producer responsibility. Sustainable design policies promote collaboration among stakeholders to achieve resource efficiency, green chemistry and other circular aspects via LCA. Initiatives such as modular design, zero-waste patterns and robust construction jointly ensure reduction in carbon and water footprints through new recycling infrastructures and techniques.

Both strategies face challenges such as the cost of implementation, consumer behaviour, lack of recycling facilities, scalability issues and the complex global supply chains. Addressing these challenges needs multidisciplinary actions by industry players, government and other sector collaborators. The future of sustainability and circularity is promising when incorporated with innovations and collaboration in the areas of materials science, closed-loop systems and appropriate government regulations and incentives to ensure longevity design, recyclability and reuse.

References

[1] United Nations Environment Programme, *Environmental Costs and Circular Solutions for Fast Fashion*, United Nations Environment Programme, (2022), https://coilink.org/20.500.12592/s5n168.

[2] M. Wackernagel and D. Lin, 'Earth Overshoot Day', In *Handbook of the Anthropocene: Humans between Heritage and Future*, Cham: Springer International Publishing, 2023, pp. 569–72.

[3] J. C. Prata, 'Microplastics in Wastewater: State of the Knowledge on Sources, Fate and Solutions', *Mar Pollut Bull*, 129 (1) (April 2018): 262–5, doi: 10.1016/J.MARPOLBUL.2018.02.046.

[4] S. L. Naqvi, M. Nadeem, F. Ayub, A. Yasar, S. H. Naqvi and R. Tanveer, 'Social and Environmental Impacts in Textile Production', in *Dye Pollution from Textile Industry: Challenges and Opportunities for Sustainable Development*, Singapore: Springer Nature Singapore, (2024), pp. 423–53.

[5] J. I. Mwasiagi, N. S. Mpofu, E. Kariuki and K. Chepkwony, 'Eco-innovation: A Case Study of the Kenyan Textile and Apparel Industry', in *Circularity in Textiles*, Cham: Springer Nature Switzerland, (2023), pp. 19–38.

[6] M. Awais, A. Mehboob and S. Majid, 'An Exploratory Study on Circular Economy in Pakistan Textile Sector: Antecedents, Barriers and Implementation', *Research Journal for Societal Issues*, 6 (2) (June 2024): 683–95, doi: 10.56976/rjsi.v6i2.245.

[7] R. Pal, 'Sustainable Design and Business Models in Textile and Fashion Industry', in *Sustainability in the Textile Industry*, Singapore: Springer Singapore, (2016), pp. 109–38.

[8] I. Kazancoglu, Y. Kazancoglu, E. Yarimoglu and A. Kahraman, 'A Conceptual Framework for Barriers of Circular Supply Chains for Sustainability in the Textile Industry', *Sustainable Development*, 28 (5) (2020): 1477–92, doi.org/10.1002/sd.2100.

[9] European Commission, 'Flash Eurobarometer 367 (Attitudes of Europeans towards Building the Single Market for Green Products)', GESIS, (2013), doi: https://doi.org/10.4232/1.11737.

[10] J. A. Purwandani and G. Michaud, 'What Are the Drivers and Barriers for Green Business Practice Adoption for SMEs?', *Environ Syst Decis*, 41 (4) (December 2021): 577–93.

[11] S. Yousef, M. Tatariants, M. Tichonovas, Z. Sarwar, I. Jonuškienė and L. Kliucininkas, 'A New Strategy for Using Textile Waste as a Sustainable Source of Recovered Cotton', *Resour Conserv Recycl*, 145 (June 2019): 359–69, doi: 10.1016/J.RESCONREC.2019.02.031.

[12] T. Fu, Y. Zhang, N. T. Akankwasa, N. Chen and H. Lin, 'Study on the Model of Semi-Open-End Twist in Compact Spinning with Lattice Apron', *Textile Research Journal*, 91 (5–6) (March 2021): 467–79, doi: 10.1177/0040517520942545.

[13] A. M. James, B. M. Roberts and A. Kuznia, 'Transforming the Sequential Process of Fashion Production: Where Zero-Waste Pattern Cutting Takes the Lead in Creative Design', *International Journal of Fashion Design, Technology and Education*, 9 (2) (May 2016): 142–52, doi: 10.1080/17543266.2016.1167253.

[14] A. Jayakumar, S. Radoor, J. T. Kim, J. W. Rhim, J. Parameswaranpillai and S. Siengchin, 'Lightweight and Sustainable Materials: A Global Scenario', in *Lightweight and Sustainable Composite Materials: Preparation, Properties and Applications*, pp. 1–18, (January 2023), doi: 10.1016/B978-0-323-95189-0.00001-9.

[15] S. Tejada, S. Valdez, O. Yildiz, R. Salas-Castro and J.C. Alvarez, 'A Data-Driven Lean Manufacturing Framework for Enhancing Productivity in Textile Micro-Enterprises', *Sustainability*, 17 (11) (2025): 5207, doi: 10.3390/SU17115207.

[16] V. S. Yadav and A. Majumdar, 'Mitigating the Barriers of Industrial Symbiosis for Waste Management: An Integrated Decision-Making Framework for the Textile and Clothing Industry', *Waste Management & Research*, 42 (7) (July 2024): 544–55, doi: 10.1177/0734242X231197367.

[17] L. Lawson, L.M. Degenstein, B. Bates, W. Chute, D. King and P.I. Dolez, 'Cellulose Textiles from Hemp Biomass: Opportunities and Challenges', *Sustainability*, 14 (22) (2022): 15337, doi: 10.3390/SU142215337.

[18] S. Garcia-Torres, M. Rey-Garcia, J. Sáenz and S. Seuring, 'Traceability and Transparency for Sustainable Fashion-Apparel Supply Chains', *Journal of Fashion Marketing and Management*, 26 (2) (February 2022): 344–64, doi: 10.1108/JFMM-07-2020-0125.

[19] H. Buller, H. Blokhuis, P. Jensen and L. Keeling, 'Towards Farm Animal Welfare and Sustainability', *Animals*, 8 (6) (2018): 81, doi: 10.3390/ANI8060081.

[20] I. R. Chowdhury, J. Summerscales, I. R. Chowdhury and J. Summerscales, 'Woven Fabrics for Composite Reinforcement: A Review', *Journal of Composites Science*, 8 (7) (July 2024), doi: 10.3390/JCS8070280.

[21] A. R. Shafqat and A. Saccavini, 'Eco-Design of Textiles', in *Circularity in Textiles*, Cham: Springer Nature Switzerland, (2023), pp. 39–58.

[22] M. B. Ramzan, M. S. Habib, M. Omair, J. Naeem, H. Mustafa, M. W. Iqbal and A. I. Malik, 'Role of Design for Disassembly in Educating Consumers for Circular Behavior', *Sustainability*, 15 (21) (2023): 15505, doi: 10.3390/SU152115505.

[23] A. K. Azad, U. N. Haq, M. M. Khairul Akter and M. A. Uddin, 'Recycling Practices of Pre-Consumer Waste Generated from Textile Industry', in *Sustainable Manufacturing Practices in the Textiles and Fashion Sector*, Cham: Springer Nature Switzerland, (2024), pp. 301–24, doi: 10.1007/978-3-031-51362-6_12.

[24] G. Bonifazi, I. D'Adamo, C. Grosso and R. Palmieri, 'Advancing Business Strategy in End-of-Life Management for the Fashion Industry', *Bus Strategy Environ*, 34 (6) (September 2025): 6814–27, doi: 10.1002/BSE.4318.

[25] C. A. Hall, K. M. Hasling and U. Ræbild, 'Design for Recyclability: Textiles', in *Springer Handbook of Circular Plastics Economy*, Cham: Springer Nature Switzerland, (2025), pp. 583–99, doi: 10.1007/978-3-031-66209-6_29.

[26] M. Akter, H. R. Anik and S. Mahmud, 'Conversion of Textile Waste to Wealth and Their Industrial Utilization', in *From Waste to Wealth*, Singapore: Springer Nature Singapore, 2024, pp. 669–738, doi: 10.1007/978-81-99-7552-5_31.

[27] J. Jaisri and S. Balaji, 'Biodegradable Elegance: Assessing Luxury Textile Biodegradability', in *Crafting Sustainability in Luxury Textiles for a Zero-Waste Future*, Cham: Springer Nature Switzerland, (2025), pp. 43–55, doi: 10.1007/978-3-031-86773-6_5.

[28] T. Haukkala, H. Salmenperä and J. P. Ovaska, 'Multi-system Interactions among Finnish Textile Companies for Adopting Circular Economy Practices', in *Circular Economy and Sustainability*, (July 2025), pp. 1–25, doi: 10.1007/S43615-025-00622-3.

[29] A. P. Periyasamy and J. Militky, 'Sustainability in Regenerated Textile Fibers', in *Sustainability in the Textile and Apparel Industries: Sourcing Synthetic and Novel Alternative Raw Materials*, Cham: Springer International Publishing, (2020), pp. 63–95, doi: 10.1007/978-3-030-38013-7_4.

[30] A. Crawford, *Designer's Guide to Lab Practice*, Abingdon and New York: Routledge, (2023).

[31] OEKO-TEX, 'Made in Green: Social and Environmental Criteria and Certification System', (2024). www.oeko-tex.com.

[32] Global Organic Textile Standard (GOTS), 'Home page', (2020), www.global-standard.org.

[33] R. Woodward, *The Organisation for Economic Co-Operation and Development (OECD)*, Abingdon and New York: Routledge, (2009).

[34] M. Behnam and T. L. MacLean, 'Where Is the Accountability in International Accountability Standards? A Decoupling Perspective', *Business Ethics Quarterly*, 21 (1) (January 2011): 45–72, doi: 10.5840/BEQ20112113.

[35] S. Petrak, M. Mahnić Naglić, M. Glogar and A. Tomljenović, 'Assessment of Textile Material Properties and the Impact of Digital Ink-Jet Fabric Printing on 3D Simulation as a Sustainable Method for Garment Prototyping', *Sustainability*, 17 (4) (2025): 1388, doi: 10.3390/SU17041388.

[36] S. Maiti, M. Shahid and R. V. Adivarekar (eds), *Sustainable Coloration Techniques in Textiles*, Singapore: Springer Nature Singapore, (2025).

[37] M. Flanagan, 'Environmental Impact Assessment of Fabric Samples: Defining a Low Impact Design Standard for the Textile Industry', (2024), https://repository.upenn.edu/handle/20.500.14332/60585.

[38] Ellen MacArthur Foundation, 'A New Textiles Economy: Redesigning Fashion's Future', (2017), www.ellenmacarthurfoundation.org/a-new-textiles-economy.

[39] M. George, N. M. Ngoc, L. K. Nhi, V. N. Tung, L. P. Truc and R. Nayak, 'Lean Manufacturing: Case Studies from Fashion and Textile Industries', in *Lean Supply Chain Management in Fashion and Textile Industry*, Singapore: Springer Nature Singapore, (2022), pp. 269–90.

[40] A. Šaravanja, T. Pušić and T. Dekanić, 'Microplastics in Wastewater by Washing Polyester Fabrics', *Materials*, 15 (7) (April 2022): 2683, doi: 10.3390/MA15072683.

[41] W. U. Khan, S. Ahmed, Y. Dhoble and S. Madhav, 'A Critical Review of Hazardous Waste Generation from Textile Industries and Associated Ecological Impacts', *Journal of the Indian Chemical Society*, 100 (1) (January 2023): 100829, doi: 10.1016/J.JICS.2022.100829.

[42] T. Y. Lo, H. H. Su, J. H. Ho, C. W. Chang, H. R. Chen, H. H. Hsu, K. J. Chang and J. T. Chen, 'Intrinsic Repeated Self-Healing Textiles: Developing Electrospun Fabrics for Enhanced Durability and Stretchability', *ACS Omega*, 9 (52) (2024): 51623–30, doi: 10.1021/ACSOMEGA.4C09296.

[43] M. A. Casto and M. DeLong, 'Exploring Esthetic Response to Classic as a Means to Slow Fashion', *Fashion Practice*, 11 (1) (January 2019): 105–31, doi: 10.1080/17569370.2019.1565378.

[44] R. K. Seidu, B. Eghan and R. Acquaye, 'A Review of Circular Fashion and Bio-based Materials in the Fashion Industry', *Circular Economy and Sustainability*, 4 (1) (March 2024): 693–715, doi: 10.1007/S43615-023-00303-Z.

[45] P. Dey, P. Dey, M. B. Hoque, B. Baria, M. M. Rahman, S. Shovon and D. Das, 'Sustainable and Eco-Friendly Natural Dyes: A Holistic Review on Sources, Extraction, and Application Prospects', *Textile Research Journal*, 95 (19–20) (2025): 2472–99.

[46] F. Unal, A. Yavas and O. Avinc, 'Sustainability in Textile Design with Laser Technology', in *Sustainability in the Textile and Apparel Industries: Sustainable Textiles, Clothing Design and Repurposing*, Cham: Springer International Publishing, (2020), pp. 263–87.

4 Science-led design for sustainable fashion futures

Nkumbu Mutambo, Alana M. James, Chimdia keChi-Okafor, Abigail Irving-Munro, Anne Peirson-Smith, Anil Namdeo, Miranda Prendergast-Miller, Kelly J. Sheridan, Thomas Stanton, and Matteo Gallidabino

4.1 Introduction

The fashion and textile value chain is a resource-intensive, complex global network of stakeholders. From raw material extraction until the end of life, the industry is responsible for alarming levels of pollution, resource depletion and waste. Therefore, there is a critical need for stakeholders involved in both production and consumption to reframe their practice towards modes of engagement that mitigate against these negative impacts [1]. Reimagining the current take–make–waste model towards more circular business systems fundamentally necessitates a transformation of consumer behaviour and attitudes and a shift towards product development and production practices that are grounded in sustainability [2, 3]. Increased awareness of this imperative has resulted in growing consumer demand for responsibly sourced products and services [4]. The urgent need for action to catalyse such goals has seen the introduction of voluntary and mandatory legislation from policymakers like the Textiles 2030 road map from WRAP (Waste Resource Action Protocol) in the UK aimed at lowering emissions [5] and the California Responsible Textile Recovery Act in the USA which regulates the presence of harmful chemicals and the recycling and disposal of products [6]. Similarly, under the broader Green Deal, the European Union (EU) has launched their EU Strategy for Sustainable and Circular Textiles [7]. In addition to promoting transparency and traceability through the introduction of digital product passports, this strategy has set out to increase product lifespans by requiring that by 2030 textiles entering the EU market will be designed to be repairable, durable and recyclable. Although EU policy is not representative of the entire global legislative landscape, as a market accounting for 35 per cent of global imports of apparel, its policies have the potential to shape action far beyond its borders. While the practical application of the Extended Producer Responsibility (EPR) scheme remains unclear due to ongoing consultation, textile products, including apparel and footwear, are a key priority area in policy development. Moreover, the EU's introduction of eco-design requirements as part of the broader Eco-design for Sustainable Products Regulation scheme further indicates the growing importance of utilising science-led environmental data to inform product development and design practice, as well as a growing responsibility that producers and retailers will have beyond the point of sale. As such, it is critical from a market and policy perspective for stakeholders upstream within the value chain to gain the skills and knowledge needed to engage in the shift towards eco-design as an industry norm.

Eco-design has emerged as an applied framework for creating goods and services with minimal deleterious social, economic and environmental effects by encouraging a concern for issues beyond aesthetics and functionality during product development [8–10]. However, achieving this aim and engaging effectively in such transformational action cannot occur without taking a holistic account of the entire product life cycle, beyond the production and distribution stage.

DOI: 10.4324/9781003638148-5

Effective eco-design applications need to be grounded in environmental performance, as well as an awareness of how such performance is assessed and communicated. Thus, this research is foregrounded in the need for whole life-cycle thinking as well as an impact-informed approach to eco-design.

This chapter examines the role that environmental impact assessment tools can play in informing and supporting effective eco-design to drive sustainability in the fashion and textiles industry (FTI). While eco-design may be actively practised by a relatively small number of stakeholders within the product life cycle, this research acknowledges the limitations of siloed approaches in addressing the complex and networked challenges. Drawing on secondary research and a breadth of stakeholder perspectives through primary data collection, this project aimed: (1) to establish how environmental impact in the FTI is currently quantified and communicated in support of a whole life-cycle thinking approach to eco-design; (2) to identify the current challenges and potential drivers of conducting impact assessment through the adoption of a science-led approach towards sustainable innovation.

4.2 Eco-design and whole life-cycle thinking

Formulated on the definition of sustainable development being described as addressing present needs without placing the ability of future generations to do the same in jeopardy [11], eco-design aims to support the development of environmentally innocuous [8] products and services. The core aim of eco-design is to incorporate environmental concerns with other values like functionality, price and ergonomics to enable lower-impact production and consumption while still ensuring satisfaction and meeting the needs of users [12, 13]. While adopting an eco-design approach cannot strictly guarantee commercial success [13] or a shift in consumer behaviour, it does aid in creating the conditions that can support sustainable production, use and disposal of products. To achieve this, advocates argue for the incorporation of whole life-cycle thinking or a comprehensive view of potential impacts at every life-cycle stage (see Figure 4.1) from raw-material extraction, beyond the company gate and through to end of life [14, 15]. In adopting this approach, design decisions shaping material procurement, product reuse potential and recyclability are not siloed or limited in focus, resulting in unintended negative consequences (e.g., incorporating a new chemical component that helps reduce water emissions during the dyeing process while ignoring the fact that such an intervention would impede recyclability at the end of life). In essence, the key aim is to identify intervention points that would result in maximum reduction of impact across an entire life cycle. Such an approach is acknowledged to be challenging [16] and does not ensure easy solutions; however, it does support innovation and product/process development that is considered robust. For example, a study exploring e-textiles highlighted the importance of considering the end-of-life disposal fate of materials during the eco-design process, particularly considering limits in recycling infrastructure in many regions. A suggested action through the adoption of eco-design would be for textile designers to work with stakeholders in the recycling industry to develop products that could be recycled easily [17].

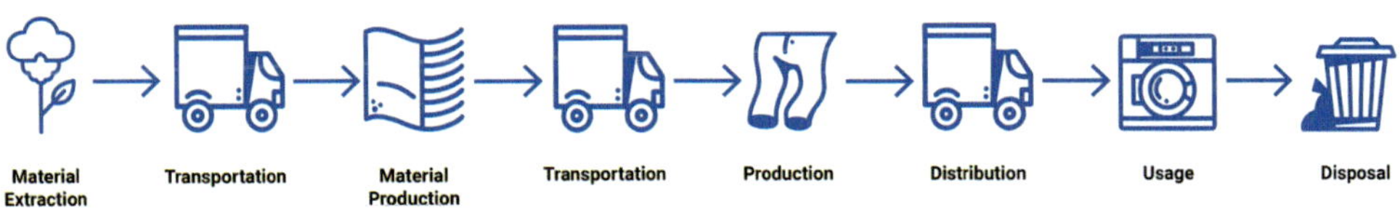

Figure 4.1 The linear product life cycle.

The primary eco-design values of reduce, reuse and recycle can be achieved through the implementation of a variety of principles or steps. Le Pochat et al. [18] distil the approach into two phases beginning with environmental assessment, which in turn informs environmental improvement. In this way, environmental assessment from a whole life-cycle perspective is a critical part of eco-design practice, irrespective of the specific process or framework used. Moreover, Baumann et al. [19] argue that the methods and frameworks used to quantify environmental consequences including emission, effects and exposures at each life-cycle phase are essential to eco-design practice. Therefore, adopting a whole life-cycle perspective requires an understanding of how environmental performance is measured and communicated. While Lindahl and Ekermann [14] have systematically categorised general eco-design tools, and Kozlowski et al. [20] have analysed a broad range of tools that can support sustainable fashion design practice, this chapter sets out to explore science-led impact assessment tools and approaches used within the FTI, which can support decision-making within an eco-design context.

Impact assessment tools have a variety of applications ranging from informing high-level policy and business decisions to supporting consumer acquisition behaviours. In addition to this, they have value in informing design and innovation practice by highlighting hotspots or stages in the life cycle with high levels of negative impact [19]. Such hotspots can be framed as critical intervention points for environmental improvement. In essence, impact-informed eco-design is centred on supporting product and material development decisions through science-based assessment and evidence. This aligns with ISO 14006: 2020, which outlines the principles and six steps of incorporating eco-design into organisational processes.

4.3 Environmental impact assessment in the fashion and textile industry

Within the FTI there are several factors influencing textile and product development, including economic considerations, product functionality and regulatory requirements. Utilising an impact-informed eco-design approach can support the integration of environmental concerns into this process through science-based assessment and evidence. Furthermore, identifying effective strategies and design solutions for addressing environmental concerns requires an understanding of how negative impacts are collated and disseminated. This section draws on an extensive review of secondary literature, including journal articles, publications from non-governmental organisations (NGOs), articles, online indexes and fashion brands, to highlight and map a selection of methods, tools and platforms used to quantify the FTI's environmental performance. Additionally, challenges and barriers limiting the efficacy of these tools are identified and discussed to provide a coherent picture of the contemporary environmental impact assessment (EIA) landscape.

The environmental impact of the FTI is calculated and articulated through a range of interconnected tools, platforms and metrics [21] including data sources, assessment and communication tools (see Figure 4.2). While there are a variety of ways in which information is shared between stakeholders and across tools, Figure 4.2 illustrates the fundamental relationship of EIA tools through the flow of data. Although their intended audiences, users and approaches may differ, the mechanisms described in this section play an integral role in shaping current understandings and practices aimed at transitioning towards more circular and sustainable approaches in the FTI.

4.3.1 Data

Adopting a science-led, evidence-based impact quantification approach requires the input of accurate, reliable and authentic data, with the nature of information needed varying depending

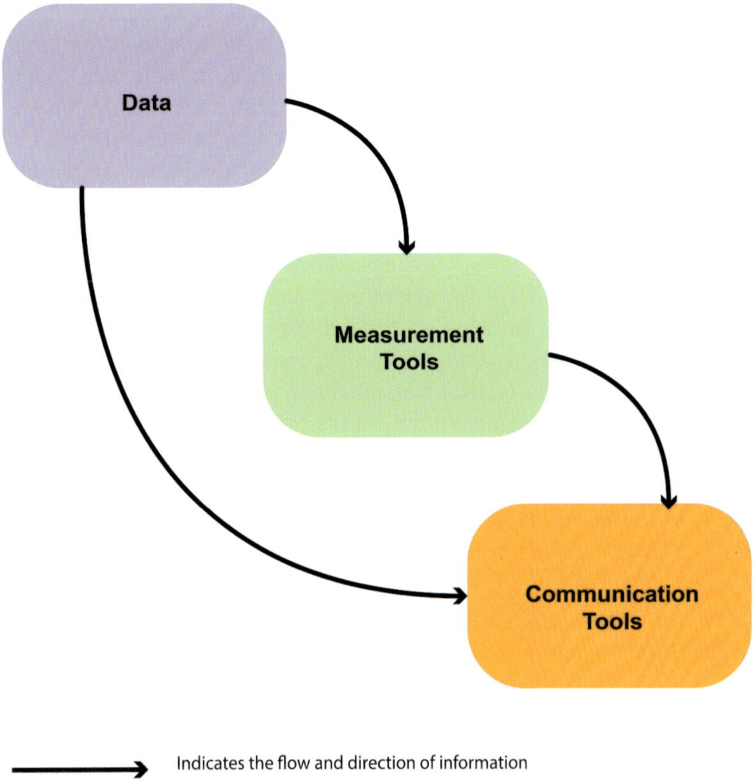

Figure 4.2 The EIA system.

on factors like the assessment method and scope. Moreover, data can range from information about the energy mix and infrastructure in a particular production location to the amount of water used to wash a garment over the course of its usable life. Gathering the data required to assess the entire product life cycle is a logistical challenge due in large, to the complex and globally dispersed nature of the fashion and textiles production networks [22, 23]. Aside from the geographic distances from involved stakeholders, traceability is further complicated by variations and inconsistencies in the operational cultures, legal requirements and political systems that stakeholders are embedded in. Moreover, the varied acquisition habits and care practices implemented during the use phase add greater intricacy and nuance when attempting to gather relevant and robust data to determine environmental performance. In addition to the challenge of collecting data, the aforementioned factors also contribute to challenges in verifying the authenticity of data, leading to the rise in what is referred to as 'Zombie data' [24]. Some examples include drastic variations in figures used to represent the impact of a cotton T-shirt [24] to the use of outdated or inaccurate data in impact assessment [25].

Primary data sourced directly from stakeholders like manufacturers, raw material producers, retailers and consumers can be supplemented by secondary data can be sourced from a range of predetermined databases which provide generic representative data. Platforms commonly behind a paywall such as Gabi, Shera and Ecoinvent are often marketed alongside proprietary life-cycle assessment (LCA) software services. Typical theses databases are utilised in scenarios where primary data is not available or is difficult to obtain. Additionally, proprietary assessment platforms draw on databases to inform impact calculations. For instance, the Higg Index uses

Sphera, while Carbon Fact uses Ecoinvent. From an eco-design perspective, it is important to consider what data was used to inform impact calculations in order to support transparency in decision-making. Additionally, the proliferation of unverified 'zombie data' increases the importance for eco-design practitioners to engage critically with data. Finally, the implications of erroneous data are further amplified by the fact that such data has the potential to inform both assessment and communication tools, potentially multiplying the impact of inaccurate data.

4.3.2 Assessment tools

There exist multiple assessment tools for determining environmental impact, each implementing a different methodology and focusing on a variety of impact points across the value chain. For example, footprint modelling describes a set of approaches used to calculate a specific impact category like greenhouse gas (GHG) emissions at a particular point in time and is used to examine impacts holistically at multiple points in the product life cycle. However, findings of this quantification tool can be used along with other data to inform alternative methodologies like LCA, depending on the study scope. LCA is a quantification tool used to calculate a range of impact categories across the stages of a product's or material life cycle through a four-step process of goal and scope definition, life-cycle inventory, LCA and interpretation [26]. LCA can incorporate the assessment of a range of environmental indicators such as land use, human toxicity, water use, GHG emission and eutrophication. Due to its utility in conducting holistic assessments across the product life cycle, LCA and LCA-based tools are commonly used to support the eco-design process [8, 27]. LCAs typically conform to ISO 14040, which outlines a broad framework to follow; however, the methodology's application varies, limiting the comparability of findings. While comparability of assessment findings is not essential, it is important to be aware of when drawing on assessment data to inform eco-design decisions to avoid erroneous comparisons. Furthermore, while the name LCA might suggest the consideration of the whole life cycle, it is important to note that not all LCAs consider the entire product life cycle in their calculations. Some have limited system boundaries, focusing, for instance, only on the first few stages of production (sometimes described as cradle-to-gate or gate-to-gate). Such limitations in scope may be a result of limited data or narrowly defined aims. Though this practice may be logical, it is important to consider the applicability of the information derived from an LCA with limited system boundaries. Lastly, it must be noted that performing a full-scale LCA, utilising on the ground primary data, requires extensive expertise and time, which makes it challenging for many small and medium enterprises with limited resources and finances to implement [10].

Alongside its suite of legislative initiatives, the EU has developed an LCA-based tool called the Product Environmental Footprint (PEF) [28]. This methodology is a quantification system covering thirteen common product categories (T-shirts, swimwear, boots, pants, etc.) to calculate impacts across the stage of the product or material life cycle. Furthermore, each PEF category has corresponding Product Categories Rules (PCRs) outlining the methodologies used to collate impact for that type of product. This level of standardisation means that findings of assessments for products in the same category can be used to make comparative evaluations. Similar to the PEF, Environmental Product Declarations (EPD) are LCA-based assessment tools used to quantify impact across life-cycle stages [29]. Like the PEF, EPDs have PCRs defining the specific method of assessing products from a variety of industries, including the FTI.

In an effort to incorporate an economic dimension to assessment data from environmentally focused methods such as the LCA and footprint modelling, eco-efficiency analysis was developed by the World Business Council for Sustainable Development [30]. The method was further advanced by BASF (Badisin Anili und Soda Fabric), a means of examining the economic cost-effectiveness of a product relative to its environmental impact. While eco-design places an emphasis on environmental dimensions, in addition to supporting decision-making during the process, eco-efficiency can also be used to support communication with different stakeholders, especially when attempting to communicate the value of environmental strategy.

Lastly, it is beyond the scope of this chapter to expansively discuss all possible assessment tools and platforms; however, it is important to note that a range of *proprietary tools* are marketed to fashion and textiles firms for conducting impact assessment. These tools are paid services with offerings including comprehensive data gathering, impact calculation and reporting tools. Some examples include Carbon Fact, the Higg Index and Vaayu, all of which offer LCA-based assessment. From an eco-design perspective, it is important to consider the methodologies that underpin these tools as well as the nature and source of data drawn upon to inform these tools.

4.3.3 Communication

The deepening efforts of FTI stakeholders to communicate about their strides towards greater sustainability have been met with increased scrutiny and regulatory pressure [31, 32]. Therefore, it is important to understand the mechanisms as well as the challenges of environmental impact communication. While public disclosures from stakeholders, reports by NGOs and brand rating systems such as the Transparency Index, Ethical Index and Good on You, are valuable sources of information that can be used to inform eco-design decision-making, eco-labels arguably have the most relevance within the scope of this discussion. In addition to their function as mediums for displaying information about the environmental performance of products and business, eco-labels are also categorised as voluntary eco-credentials, which can indicate a stakeholder's engagement in certifying a product or service to a specific standard. For instance, the Global Organic Textile Standard certifies that garments contain 70 per cent organic fibres [33], while Oeko-Tex Standard 100 indicates specific material characteristics such as the absence of harmful chemicals [34]. With over 100 fashion and textiles related eco-labels covering standards like GHG emissions, labour practices and animal welfare, much has been discussed regarding the potential challenges. Aside from the large quantity of labels, the overload of information and similarity in design can confuse [35], especially for consumers and other stakeholders who are unfamiliar with the nuances of the EIA and the sustainability landscape as a whole [36]. Although eco-labels can provide useful information in communicating science-based information regarding critical stages in a product life cycle, it is important to consider the varied nature and features of eco-labels when using them to inform an eco-design process.

There are three primary types of eco-labels recognised across the FTI: the first aligns to ISO 14024 standard and denote claims that allow for comparison, with data based on assessment requiring third party verification; the second aligns to ISO 14021 and features unverified claims; the third aligns to ISO 14025 and denotes claims based on LCA data that is third-party verified, but, importantly, is not comparable between products. In addition to challenges related to their profusion, the efficacy of eco-labels has been called into question due to a lack of clear or harmonised methodologies. For instance, a study by Diekel et al. [37] points to the lack of comparability of prominent FTI eco-labels due to differences in life-cycle stages considered. Moreover, the study noted a lack of consideration of critical life-cycle stages such as the distribution and use phase in some LCA-based eco-label methodologies. In addition to the need for the greater refinement of eco-labelling systems, the challenges presented above point to the imperative for eco-design practitioners to approach impact communication tools with thoughtful and critical consideration. The type of label, the nature of the claims being made, as well as the methodologies utilised, must be examined to support a whole life-cycle approach and efficiently incorporate eco-design practices.

4.4 Methodology

As noted above, several gaps in knowledge were identified following an extensive review of literature. These primarily centred around the lived experience of implementation of these assessment tools from an industry stakeholder perspective. Therefore, this chapter aims to present the importance of an impact-informed approach to eco-design that draws on science-led data to accurately direct future action in the FTI. Additionally, this chapter draws on a range of stakeholders to expound upon some of the drivers and challenges involved in operationalising such an approach within the current industry context. The insights generated highlight key areas that must be considered when embedding impact-informed perspectives into design and production approaches. Adopting a holistic approach to the FTI value chain, in the study participants included industry professionals involved in manufacture, design and environmental and social governance, business owners, local policymakers, academics, consumers and representatives from multi-stakeholder initiatives.

During a day-long symposium, eighty-five participants were divided into eight focus groups, each examining both current practice and opportunities for the future of the EIA landscape. Groups of eight to twelve people engaged in discussions facilitated by a research team member, with data recorded via note-taking and audio recording tools. Informed by an extensive review of secondary literature, the workshops centred on five emerging themes: (1) transparency, (2) policy and governance, (3) environmental measures, (4) economics and (5) cross-disciplinary collaboration. These discussions were divided into two parts, each running for forty-five to sixty minutes. The first centred on understanding the nature and issues of the sector's EIA landscape, with the following lines of inquiry being explored:

1. Current practice: challenges and barriers to the implementation of and engagement with, assessment methodologies
2. Key gaps in knowledge: during the second part of the focus group discussions participants concentrated on collective action to co-create possible solutions and explored the following key factors:
    - key drivers that will enable change in the future
    - possible outcomes or successes of these actions
    - the generation of potential positive impact.

This extensive gathering of qualitative primary research documenting rich focus group discussions was followed by data transcription. The research team then analysed the transcripts and notes taken by participants using the general inductive method [38]. Using this qualitative method, data were read and reread to identify emergent themes. These were then organised to form superordinate or overarching themes.

4.5 Findings

Several key insights emerged from these two-part discussions carried out across all focus-group discussions, with findings pointing to aspects of the impact assessment landscape that need to be addressed in order to improve the efficiency of and implementation. As eco-design practice should be informed by environmental assessment [8, 18, 19] the efficacy of the practice is dependent in part on the accuracy and authenticity of the metrics and measures used. Therefore, these findings can be read as critical issues that need to be considered and tackled by stakeholders to ensure a science-led approach to the future of EIA. Headline insights include the need for greater education across the industry. Participant accounts also emphasised the need for improvements in current assessment tools and data-gathering mechanisms. Lastly, the role of cross-disciplinary collaboration as a driver for innovation was also expounded upon, highlighting the necessity of science and the creative industries to collaboratively work together.

4.5.1 The need for more robust assessment and measurement tools

The importance of a holistic approach to impact assessment was emphasised across all groups. However, participants expressed concerns over perceived shortcomings in both the methods and measures used. For instance, participants pointed to the need for greater comparability of LCA-based findings. To achieve this aim, they called for globally coordinated policy initiatives and standardised/harmonised methodologies. An example of this would be the adoption of a system akin to the EU's PEF with its defined assessment protocols or PCRs. Moreover, the problematic practice of over-extrapolation of small data sets and reliance on secondary data, including aggregated regional or global averages in LCA-based calculations, was noted as limiting the veracity of claims that could be made. Participants also discussed the need for increased use of robust data (i.e., primary and verifiable) and for more research to address gaps in knowledge and understanding. For instance, they pointed to the dearth of data about product life-cycle stages beyond the production and distribution stages. Alongside this need for more data about the use and disposal phases of the life cycle, the imperative to build a more robust understanding of the impact related to a wider range of textile fibres was raised. Areas highlighted include more research about impacts related to natural fibres, as well as the need for a consideration of emerging areas of concern, like microfibre pollution. Lastly, participants felt that much of the current impact assessment practice in the sector has a limited focus on impact categories (i.e., reporting only on GHG emissions) as well as material types (i.e., biased focus on synthetic fibres alone).

4.5.2 The role of data quality and authenticity

Data featured as a key point of discussion across all thematic groupings. The dynamic nature of products and value chains in the FTI (including the varied conditions shaping production and consumption contexts) was acknowledged as a major contributor to challenges related to data-gathering. Therefore, improvements in data quality were urgently emphasised. Some key issues highlighted include the high financial cost of contracting/hiring experts to analyse and interpret data. Moreover, the difficulties of gathering and managing data from a globally dispersed supply chain, with multiple tiers of subcontractors, were also noted as a barrier to traceability and verifiability of data. Participants also identified the cost and time associated with collecting data were also identified as a barrier limiting the engagement of small and large firms alike in gathering primary information. In light of these issues, policy initiatives were discussed as a means of supporting the harmonisation of data management systems by reducing confusion and the burden (particularly on suppliers) of gathering and sharing data. Additionally, the role that data plays in accurate assessment was extensively discussed, highlighting a need for continuous improvements in data quality and authenticity to better support effective eco-design.

4.5.3 Cross-disciplinary collaboration as a driver for innovation

Cross-disciplinary collaboration was identified as an important driver for innovation and change in the impact assessment of the fashion and textiles sector. This aligns with the notion that robust and effective eco-design must include a range of stakeholders and draw upon a variety of expertise [27]. While instances of good practice led by industry and academics alike were acknowledged by participants, there were still several barriers identified. For instance, there is a need for more funding that supports cross-disciplinary ways of working within academia and industry. Participants also noted the necessity for improvements in operational structures within institutions and businesses that would allow for more capacity and greater visibility of good practice. Moreover, although industry/private firms were identified as having flexible internal structures which enable collaboration and rapid change, it was noted that such agility can be hampered by a lack of shared knowledge and language. For instance, in many design departments, teams often lack technical knowledge about the production process and EIA. This leads to collaborators within the same organisation being unable to communicate effectively across departmental lines. Therefore, enabling truly collaborative ways of working will require increased training and education, as well as investment (financial and human capital) from businesses and organisations. In essence, participants emphasised the need for cross-disciplinarity to be embedded in educational and business contexts alike. Not as an afterthought or bolt-on, but as an integral part of organisational culture and curricula. Finally, the inclusion of knowledge and perspectives from multiple stakeholders was discussed as a means of ensuring that actions like target setting, the development of test methods and eco-innovation in the sector promote a just transition. This will help to ensure that eco-design measures are not inadvertently reinforcing or creating greater social and economic inequalities through their implementation.

4.6 Conclusion

As the imperative to adopt eco-design practice increases in the FTI, this chapter has highlighted the critical role that environmental assessment plays in supporting the transition towards greater circularity and sustainability. The analysis of literature and stakeholder perspectives has revealed the critical issues that impact the efficacy of current eco-design measures and the potential value in greater levels of science-led approaches being integrated. Furthermore, the chapter advocates for a holistic approach to be adopted, working collaboratively with a range of stakeholders to ensure collective action and a collective effort towards a sustainable future for the FTI. Consequently, this chapter will conclude with recommendations to inform this future action from an industry, academia and policy-maker perspective.

4.6.1 Industry stakeholders

The primary engagement with stakeholders during the focus group discussions evidenced a clear need for improved training and education regarding the EIA tools across the FTI. Although it is not feasible or necessary for all industry actors to be experts in conducting impact assessment methodologies like LCA, it is essential for the industry professionals involved in various stages of product, process and service development to have a robust understanding of how assessment is conducted, as well as some identified current limitations. Moreover, industry stakeholders have a role to play in improving the data collation practices as well as promoting the use of primary robust data. There is also a need for greater promotion of collaboration with other stakeholders, including those in academia and government, to address gaps and improve assessment methodologies.

4.6.2 Academic stakeholders

Researchers located in the higher education space are actively encouraged to engage in activities that broaden understanding of impacts across the entire project life cycle and, critically, how design decision-making can significantly reduce these impacts. Further engagement in under-researched stages such as the environmental consequences of behavioural activities during the consumer use and disposal phases of the product life cycle. As evidenced through this research, the development of interdisciplinary action and integration of science-led data to inform and direct future innovation in the development of products, processes and behaviours across the FTI value chain is essential. The academic community need to harness the agility they possess to enable this collaboration, fostering new communities of practice to evidence the value of this collective effort. There is also a need to develop curricula that respond to identified skill and knowledge gaps and support the creation of shared languages and hybrid methodologies. This will help to ensure that future FTI professionals are equipped with the tools they need to engage meaningfully in collective action. (For an example of a life cycle informed eco-design training approach trailed in the food and beverage industry see Favi et al. [4].)

4.6.3 Policymakers

As the push towards eco-design as a norm across the FTI is driven in large part by policy initiatives, global policymakers must consider ways in which legislation can be leveraged to improve the practical implementation of eco-design strategies to ensure that they are effective in supporting the transition towards greater levels of sustainability. The globally dispersed nature of the FTI presents a range of challenges fundamentally rooted in the profusion and misalignment of initiatives, tools and standards. Legislation can be used to harmonise impact assessment approaches at a global scale. This could encourage greater comparability of findings and potentially reduce the costs associated with assessment. Furthermore, in the development of existing and emerging legislation, the critical role of science-led data must be integrated to ensure decision-making, funding allocation and future policy development are rooted in accurate, reliable and authentic EIA.

References

[1] K. Niinimäki and M. Durrani, 'Repairing Fashion Cultures: From Disposable to Repairable', *Responsible Consumption and Production*, 12 (1) (2020): 153–67.

[2] K. Niinimäki, 'Fashion in a Circular Economy', in *Sustainability in Fashion*, Cham: Palgrave Macmillan, (2017), pp. 151–69.

[3] O. D. Elisha, 'Moving Beyond Take-Make-Dispose to Take-Make-Use for Sustainable Economy', *Int J Sci Res Educ*, 13 (3) (2020): 497–516.

[4] C. Favi and M. Marconi, 'Product Eco-design in the Era of the Circular Economy', *Sustainability*, 17 (1) (2024): 213, doi: 10.3390/su17010213.

[5] WRAP, 'Textiles 2030 Circularity Pathway', (2021), www.wrap.ngo/sites/default/files/2021-04Textiles%202030%20Circularity%20Pathway.pdf.

[6] UK Fashion & Textile Association, 'California Signs Bill for Textiles EPR Scheme', (n.d.), https://ukft.org/california-textiles-epr-oct24/.

[7] European Union, *EU Strategy for Sustainable and Circular Textiles*, Brussels: European Commission.

[8] H. Brezet, 'Ecodesign: A Promising Approach to Sustainable Production and Consumption', United Nations Environmental Programme (UNEP), (1997).

[9] K. Niinimäki, 'Ecodesign and Textiles', *Research Journal of Textile and Apparel*, 10 (3) (2006): 67–75, doi: 10.1108/rjta-10–03–2006-b009.

[10] J. Henry, 'Life Cycle Thinking and Eco-Design: An Overview', *Journal of Policy Options*, 5 (3) (2022): 30–7.

[11] United Nations, 'Our Common Future', (1987). www.are.admin.ch/are/en/home/media/publications/sustainable-development/brundtland-report.html.

[12] M. Crul and J. C. Diehl, 'Design for Sustainability: Moving from Incremental Towards Radical Design Approaches', in Transitions to Sustainability, NZSSES Conference, Auckland, New Zealand, (2010).

[13] L. Maccioni, Y. Borgianni and D. C. A. Pigosso, 'Can the Choice of Eco-design Principles Affect Products' Success?', *Design Science*, 5 (2019), e25, doi: 10.1017/dsj.2019.24.

[14] M. Lindahl and S. Ekermann, 'Structure for Categorization of EcoDesign Methods and Tools', in *Re-engineering Manufacturing for Sustainability*, Singapore: Springer, (2013), pp. 117–22.

[15] K. Joachimiak-Lechman, 'Life Cycle Perspective in Design and Product development', *Engineering Management in Production and Services*, 16 (3) (2024): 143–56, doi: 10.2478/emj-2024–0029.

[16] C. Bakker, F. Wang, J. Huisman and M. Den Hollander, 'Products that Go Round: Exploring Product Life Extension through Design', *Journal of Cleaner Production*, 69 (2014): 10–16.

[17] A. R. Köhler, 'Challenges for Eco-design of Emerging Technologies: The Case of Electronic Textiles', *Materials and Design*, 51 (2013): 51–60.

[18] S. Le Pochat, G. Bertoluci and D. Froelich, 'Integrating Ecodesign by Conducting Changes in SMEs', *Journal of Cleaner Production*, 15 (7) (2007): 671–80, doi: 10.1016/j.jclepro.2006.01.004.

[19] H. Baumann, F. Boons and A. Bragd, 'Mapping the Green Product Development Field: Engineering, Policy and Business Perspectives', *Journal of Cleaner Production*, 10 (5) (2002): 409–25.

[20] A. Kozlowski, M. Bardecki and C. Searcy, 'Tools for Sustainable Fashion Design: An Analysis of Their Fitness for Purpose', *Sustainability*, 11 (13) (2019): 3581.

[21] N. Mutambo, A. Peirson-Smith, C. KeChi-Okafor, A. Irving-Munro, K. J. Sheridan, M. T. Prendergast-Miller, A. Namdeo, T. Stanton, M. D. Gallidabino, and A. James, 'Mapping the Environmental Impact Assessment Landscape in the Fashion and Textile Industries: Critical Gaps and Challenges', *Sustainability*, 16 (19) (2024): 8377, doi: 10.3390/su16198377.

[22] Y. Luo, K. Song, X. Ding and X. Wu, 'Environmental Sustainability of Textiles and Apparel: A Review of Evaluation Methods', *Environmental Impact Assessment Review*, 86 (2021): 106497.

[23] T. K. Agrawal and R. Pal, 'Traceability in Textile and Clothing Supply Chains: Classifying Implementation Factors and Information Sets Via Delphi Study', *Sustainability*, 11 (6) (2019): 1698.

[24] Planet Tracker, 'Zombie Data: Fashionably Fake Facts', (February 2022), https://planet-tracker.org/zombie-data-fashionably-fake-facts/.

[25] M. C. Lanfranchi and Elizabeth L. Cline, 'Cotton: A Case Study in Misinformation', *Transformers Foundation*, (2021), https://static1.squarespace.com/static/5efdeb17898fb81c1491fb04/t/61de9a24d5a36752adcbf737/1641978418846/CottonPaper_120122_TransformersFoundation_.pdf.

[26] Joint Research Centre and Institute for Environment and Sustainability, 'International Reference Life Cycle Data System (ILCD) Handbook', (2006), doi: https://doi.org/10.2788/38479

[27] F. Vallet, B. Eynard, D. Millet, S. G. Mahut, B. Tyl and G. Bertoluci, 'Using Eco-design Tools: An Overview of Experts' Practices', *Design Studies*, 34 (3) (2013): 345–77, doi: 10.1016/j.destud.2012.10.001.

[28] PEF Apparel Footwear, 'What Is the PEF Methodology?', (n.d.), https://pefapparelandfootwear.eu/.

[29] TIE System, 'Environmental Product Declarations', (n.d.), www.environdec.com/all-about-epds/the-epd.

[30] S. Shiwanthi, E. Lokupitiya and S. Peiris, 'Evaluation of the Environmental and Economic Performances of Three Selected Textile Factories in Biyagama Export Processing Zone Sri Lanka', *Environmental Development*, 27 (2018): 70–82.

[31] Competition and Markets Authority, 'ASOS, Boohoo and Asda: Greenwashing Investigation', (2023), www.gov.uk/cma-cases/asos-boohoo-and-asda-greenwashing-investigation.

[32] Centrum tot Bevordering van de Import uit ontwikkelingslanden, 'What Is the Demand for Apparel on the European Market?', (March 2026), www.cbi.eu/market-information/apparel/what-demand.

[33] Global Standard, 'The Standard', (n.d.), https://global-standard.org/the-standard.

[34] Oeko-Tex Service, 'Oeko-Tex® Standard 100: The Original Safety Standard', (n.d.), www.oeko-tex.com/en/our-standards/oeko-tex-standard-100.

[35] P. Ziyeh and M. Cinelli, 'A Framework to Navigate Eco-Labels in the Textile and Clothing Industry', *Sustainability*, 15 (19) (2023): 14170, www.mdpi.com/2071-1050/15/19/14170.

[36] D. Brécard, 'Consumer Misperception of Eco-Labels, Green Market Structure and Welfare', *Journal of Regulatory Economics*, 51 (3) (2017): 340–64, doi: 10.1007/s11149-017-9328-8.

[37] F. Diekel, N. Mikosch, V. Bach and M. Finkbeiner, 'Life Cycle Based Comparison of Textile Ecolabels', *Sustainability*, 13 (4) (2021): 1751.

[38] D. R. Thomas, 'A General Inductive Approach for Analyzing Qualitative Evaluation Data', *American Journal of Evaluation*, 27 (2) (2006): 237–46.

Section 2

Materials, processes and innovations

5 Innovation in sustainable fibres and materials

Bushra Mushtaq, Sheraz Ahmad, Faheem Ahmad, and Farooq Azam

5.1 Introduction

The fashion and clothing industries are at a crossroads, with the urgent need to transition to sustainability. For many years, the industry has been regarded as one of the most intensive and detrimental to the environment. Excessive irrigation in cotton production, chemical pollution from dyeing and finishing, microplastic pollution from synthetic fibres and the massive amounts of post-consumer textile residues that end up in landfills or are burned all contribute to water scarcity [1]. The growing global consumption patterns make these problems worse, and temporary measures are not enough. Still, materials such as organic cotton, recycled polyester and regular linen are often mixed up with the term 'sustainable' solutions, pointing out the questions of their scalability, the environmental benefits in the long run and their impact on the root causes of the non-eco-friendly practices [2, 3]. Recycled polyester, for instance, keeps releasing microplastics into rivers, and organic cotton is still a huge consumer of land and water. These limits emphasise the need for a more fundamental and far-reaching switch in the industry materials [4, 5].

The correlation of material science and eco-friendly techniques has resulted in a new generation of fibres and textiles that are sustainable throughout their manufacturing process. Besides gradual improvements being introduced to traditional fibres, scientists and designers, with an ever-growing number of businessmen and women, are exploring new raw material sources, not just to lessen the impact on the environment but also to help strengthen the development of regeneration systems. Normally, such changes come from biomass methods, agriculture and industry waste recycling and the latest circular recycling technologies, which are aimed at the closed-loop system [6, 7]. Mycelium fabrics and bacterial cellulose textiles are the future of lab-grown, eco-friendly materials that have the least demand for land and water. To illustrate, using farm discards such as pineapple leaves, banana pseudo-stem and grape skins to produce fibres is a creative approach to recycling already discarded biomass [8, 9]. Similarly, continuous improvements in chemical recycling have made it possible to molecularly decompose and breakdown down polyester and nylon fibres, leading to luxurious fabrics produced without using new petroleum resources.

This shift towards material innovation not only concerns the resilience and preservation of the environment but also considers the latter's direct improvement. Modern fibres can turn resource conservation into a sustainable chain, carbon capture and ecosystem regeneration, instead of just reducing their consumption [9, 10]. In the case of such a transition, industry leaders point to the close relationship between design thinking, scientific research and industrial scale-up as the main factors necessary in this change. It is the material considered the future of fashion by the international sustainable-textile expo that the events have in common with the industry, in the acceptance of the fact that the future of fashion is not through incremental improvements

DOI: 10.4324/9781003638148-7

but through revolutionary materials that are designed to be eco-friendly. This chapter provides a summary of the most innovative trends in sustainable fibres by combining information from academic institutions, material science labs and creative fashion brands. These new materials have the power to change not just the fashion and textile industry but also more general economic and cultural discussions about innovation, consumption and environmental preservation.

5.2 Overview of sustainable fibre innovation

Innovation in sustainable fibre is no longer limited to replacing a hazardous input with a less hazardous one. Inspired by natural systems, it signifies a fundamental change in the paradigm from a linear 'take–make–dispose' model to a circular and restorative one. Several important strategies define this new wave of innovation.

5.2.1 Bio waste and bio-fabricated materials

The search for sustainability in the fashion and textiles sector has promoted the use of bio-based and bio-fabricated materials, which have living organisms as active material producers. In contrast to conventionally produced fibres from petrochemical sources or from heavily cultivated plant crops, bio-fabricated materials are produced from microorganisms such as bacteria, yeast, algae and mycelium. Thus, the place of production is transferred from the industrial factory floor into controlled laboratory or farm environments [11, 12]. This technology is not only an interesting technical achievement but also a change in the way we think about and treat living systems. By recognising living systems as co-producers of textiles, this example is a practical representation of the principles of circularity, resource minimisation and ecology integration [13].

5.2.2 Bacterial cellulose-based materials

Among the most important producers of the first bio-fabricated products are the microorganisms, such as *Komagataeibacter xylinus*, that are used to create bacterial cellulose-based products [14]. These microbes, during the fermentation process, synthesise cellulose fibrils of nanometre size (one billionth of a metre), which eventually result in pellicles that are not only very pure but also very strong and biodegradable. Early inventors, such as Suzanne Lee's BioCouture project, demonstrated the adaptability of bacterial cellulose in prototype garments, whereas current businesses are commercialising the technology on a bigger scale [15]. For example, Modern Meadow has resulted in the development of 'bio-leather' materials that eliminate the need for animal hides and chemical tanning. These bacterial cellulose-based textiles may be grown into precise forms, coloured during fermentation and tailored at the molecular level, considerably reducing downstream chemical processing and textile waste [16, 17].

5.2.3 Yeast-based bio-fabrication

In parallel, bio-fabrication using yeast has expanded the possibilities for sustainable textiles, particularly tailored protein manufacturing. Synthetic biology enables yeast strains to be developed to create proteins that mimic natural fibres such as spider silk [18]. Bolt Threads, a California-based biomaterials company, pioneered the use of Microsilk fibre, a bioengineered alternative to silk [19]. The collaboration between the corporation and designer Stella McCartney is the most important case study. Their cooperation gave rise to a prototype garment made of Microsilk, which merges high-fashion design with modern biotechnology. The McCartney–Bolt

Threads partnership is a sign of bioengineered fibres' power to enter the high-fashion market and at the same time cut down the resource usage from sericulture to a minimum [20].

5.2.4 Algae-based materials

Algae products have been very successful from the start, especially in the case of fibres, polymers and pigments. Seaweed-derived alginate has been transformed into films and fibres, thus offering a biodegradable substitute for synthetic fabrics [21]. The case study of AlgiKnit (previously Keel Labs) exemplifies this, as it produces yards from kelp that are not only biodegradable but also renewable and fit for use in the fashion and footwear industries [22]. In addition to using less freshwater and land than cotton or polyester, their biomaterials sequester carbon during cultivation. AlgiKnit's pilot projects with fashion brands demonstrate how algae-based textiles have the potential to progress from lab proofs of concept to commercially viable goods [23]. Similar to this, companies such as Vollebak have experimented with replacing synthetic dyes with pigments derived from algae, focusing on the sustainable dual use of algae as a structural material and a source of pigment [24, 25].

5.2.5 Mycelium-based materials

Mycelium-based materials, which replicate the appearance and functionality of leather without the environmental and animal welfare costs associated with animal husbandry, are arguably the most well-known example of bio-fabrication. Mycelium 'leathers' cultivated on crop waste substrates have been developed by companies such as MycoWorks and Bolt Threads (Mylo) [26, 27]. Mycelium grows quickly, requires little energy and can be made into sturdy, elastic sheets that are used in clothing, purses and shoes. This novel approach in the field of technology has led to significant collaborations in fashion: the Adidas Stan Smith Mylo shoe, made from mycelium leather, was introduced in 2021, and it is one of the most recognisable fashion items in the world made with mycelium [28]. Stella McCartney revealed two garments along with a Mylo bag during Paris Fashion Week, thus indicating the use of fungi in the luxury sector [29]. In addition, the Lululemon brand, which specialises in outdoor activewear, collaborated with Bolt Threads to show the variety of applications of mycelium textiles by making yoga mats and accessories of Mylo. Research indicates that eco-friendly leathers are no longer being tested only in niche markets but are gradually gaining acceptance in emerging ones [19].

5.3 Shift of bio-based and bio-fabricated materials

The trend towards using bio-based and bio-fabricated materials is paving the way for a variety of new opportunities in design, production and consumer interaction. The fact that these products are 'grown' rather than produced through extraction methods allows the designers to take part in the process right from the beginning to set the textures, strengths and colours that they want, which results in a very personalised product that uses fewer resources in post-production. The power of design is illustrated through case studies, such as the luxury of bioengineered Stella McCartney's Microsilk pieces or the sneaker aimed at the sportswear mass market done by Adidas using Mylo leather. The latter case of Stan Smith, where no performance loss was incurred in the process, is another good example of the versatility of design. The projects are helping to educate the public, making them more aware of the sustainable options available and at the same time, strengthening the biomaterials along with the circular economy.

Table 5.1 Comparative overview of bio-based and bio-fabricated textile innovation.

Organism used	*Company*	*Product/material*	*Brand collaboration*	*Sustainability benefit*
Bacteria (*komagataeibacter xylinus*)	Modern Meadow (Biocouture)	Bacterial cellulose, Bio-leather	Experimental garments	Biodegradable, avoid animal leather tanning, low energy and chemical use
Yeast (engineered strains)	Bolt Threads	MicroSilk Bioengineered spider silk	Stella McCartney	Mimic spider silk, reduce land, water and ethical issues
Algae (Macro/micro algae)	AlgKnit (keel Labs)	Kelp-based yarns and polymers	Pilot projects (Apparel/ footwear)	Renewable, biodegradable, minimal land/ freshwater
Algae (micro algae pigments)	Vollebak	Algae-based dyes	Experimental garments	Natural dye alternative reduces synthetic dyes pollution
Mycelium (fungal hyphae)	Bolt Threads (Mylo) MycoWorks	Mycelium leather Mylo	Adidas (Stan Smith Mylo) Stella McCartney Lululemon	Rapid growth in waste, low energy, biodegradable, replaces leather

However, the adoption of bio-based and bio-fabricated textiles remains restricted due to several obstacles. High production costs, lack of efficient fermentation factories and problems in scaling up remain as the major obstacles. The quality and performance perceptions of biomaterials are the main reasons for the slow consumer acceptance, as these perceptions influence the consumers' choice of buying or not buying the product. On the other hand, more cooperative ventures between biotech companies and big international brands are showing that the market is slowly opening to the idea of commercialising bio-based products. The Fashion for Good coalition and other similar initiatives are revealing how far the collaboration between different sectors is taking 'the biomaterials' entry to the supply chain [30].

In summary, bio-based and bio-fabricated materials are among the most revolutionary areas of sustainable fashion. Using bacteria, yeast, algae and mycelium, the sector can move away from an extractive towards a regenerative material production model. Examples from Adidas, Stella McCartney, Lululemon and AlgiKnit showcase the commercial potential of biomaterials, while ongoing research ensures further refinement in performance and price. With increasing investment in synthetic biology, bioengineering and sustainable design, bio-fabrication is on the verge of becoming a pillar of the textile economy of the future that is ecologically embedded, technologically driven and culturally attuned to the values of circularity and mindful consumption. A comparative overview of bio-based and bio-fabricated textile innovation is shown in Table 5.1.

5.4 Waste valorisation: transforming industrial and agricultural by-products into textile fibres

Waste valorisation is one of the most promising ways to develop a circular and sustainable-textile economy. Waste valorisation is the process of transforming low-value or waste materials from industries other than textiles, such as agriculture, food processing and seafood, into high-value textile fibres and fabrics. This technique not only reduces land use and incineration of garbage, but it also addresses resource shortages by providing substitute raw materials with low virgin resource inputs. Waste valorisation, on the other hand, represents the very heart of a closed-loop system, whereby the products of one sector are rechannelled as precious inputs to the next one, thus minimising environmental externalities and creating stronger economic connections [31]. This process has been widely accepted in the fashion and textile industry, where researchers and companies are eager to put an end to the environmental harm caused by conventional fibres while at the same time making a profit on waste streams that are of low value today.

5.4.1 Agricultural waste to textiles

The agricultural sector yields huge quantities of ligno-cellulosic waste, and a major part of it is either incinerated, dumped, or decomposed, which further contributes to the emission of greenhouse gases. The method of waste valorisation changes the character of these residues into a sustainable fibre raw material [32]. For example, pineapple leaves that are usually thrown away after harvesting fruit can be transformed into a robust, lightweight fibre named Piñatex. Piñatex was developed by Ananas Anam and is now commercially marketed as a leather substitute in footwear, fashion and accessories, with brand collaborations such as Hugo Boss, H&M and Nike [33]. In the same way, banana stems, yet another farm residue, have been used to generate banana fibres, a hard, stiff and biodegradable substance applied in textiles as well as composites. India, being the world's leading producer of bananas, has spearheaded efforts that recover banana fibre for local and export textile industries, opening further revenue sources for farmers [34, 35].

Other critical ones include grape skins and stem residual from the winemaking process, valorised into Vegea, a vegetable leather substitute that is now included in fashion ranges by H&M's Conscious Exclusive label [36]. Similarly, orange peels, a waste product of the juice industry, have been upcycled into Orange Fibre, a silky cellulose fabric developed in Italy and used by Salvatore Ferragamo for high-end clothing brands. These technologies not only minimise agricultural waste but also substitute eco-intensive fibres like cotton or leather, thus being useful in a sustainability context [37].

5.4.2 Food industry waste utilisation

Food processing, which is among the most significant producers of organic waste, presents other possibilities for textile use. For instance, waste milk (casein proteins from rotten or surplus milk) has been reinvented as fibres since the 1930s, but recent developments have restimulated it. QMilk in Germany is one of the companies that has reinvented casein-based fibres that are biodegradable, antibacterial and applicable to clothing, bedding and medical textiles [38]. Similarly, coffee grounds, a high-volume waste stream of international beverage consumption, have been converted into fibres via a series of processes that infuse used coffee particles into polyester matrices, such as observed in S.Café fabric produced in Taiwan. These fabrics have natural odour blocking, UV protection and quick-drying functionality, demonstrating how waste valorisation can produce not just sustainable but also functional materials [39].

5.4.3 Seafood waste as a fibre resource

Marine and aquaculture production has high levels of shell waste, mainly from shrimp, crabs and lobsters. These are chitin-rich wastes with the potential to be deacetylated to form chitosan, a material with wide-ranging applications, including antibacterial, biodegradable and film-forming effects [40]. Chitosan has been used in combination with fibre and textile coatings to improve their ability to resist bacteria, dyeing and moisture management. In addition, the integration of chitosan with both synthetic and natural fibres has been the subject of research by companies looking for an ecological fabric that is fit for medical, sports and hygienic uses. The use of collagen-rich fish scales has been considered as a potential source for biopolymer fibres and composites. These instances of waste valorisation show that the process is not limited to plant waste but also includes marine resources; hence, it expands the range of sustainable-textile sources.

5.5 Closed-loop and circular-economy perspectives

Waste valorisation is not just about replacing traditional fibres, but also it has additional advantages. The process fits perfectly into larger circular-economy models by reducing waste, extending resource life cycles and decreasing raw material exploitation. The Food-to-Fibre method maps out the sustainability of the system by the use of transdisciplinary innovation in Orange Fibre, Piñatex and Vegea, who have linked food and fashion, promoting sustainability [41]. Through the conversion of waste into valuable fibres, landfill pressure is relieved, methane loss due to decomposition is reduced and environmental risks connected to the burning of crop residues are avoided. Table 5.2 provides a comparative overview of textiles made from waste.

5.6 Advanced recycling and regenerative system

A major issue faced by the textile and fashion industry in their sustainability endeavour is the closure of the textile loop. The current production process is mostly linear, meaning resources are extracted, turned into clothes, consumed and then either thrown away in landfills or incinerated. The Ellen MacArthur Foundation estimates that less than 1 per cent of all textiles are recycled into new textiles and the others are downcycled into goods of lower value, like insulation, rags, or fillers. Transitioning from linear to circular economy will require the implementation of more sophisticated recycling and regenerative systems. These techniques aim not only to recycle fibres from post-consumer textiles but also to regenerate them into the feedstock for new clothing, thus, reducing the need for virgin raw materials and environmental impact [42].

5.6.1 Chemical recycling of textiles

Chemical recycling, which has the capability of dissolving fibres at the molecular level and producing virgin quality materials from textile waste, is regarded as one of the most exciting technologies. On the other hand, the mechanical process, which involves shredding and respinning the fibres, may lose their quality; however, the chemical recycling process preserves or even enhances the material properties. This difference becomes very relevant, particularly in the case of blended fabrics such as cotton-polyester blends, which are among the main contributors to global textile waste and are very difficult to recycle by the standard methods [43, 44].

5.6.2 Polyester recycling

Polyester, the world's most widely used fibre, can go through a process of chemical breakdown that eventually results in its monomers, pure terephthalic acid (PTA) and monoethylene glycol (MEG), after which it can be recycled into polyester (rPET) that exhibits equal characteristics as the virgin PET. The advantage of this method is that the entire process can be performed again without the quality being compromised, thus leading to practically limitless recycling. Worn Again Technologies has also been involved in this technology by creating methods to separate, depolymerise and recover polyester from textile mixes like cotton-polyester waste. Its process

Table 5.2 Comparative overview of waste-derived textiles.

Waste source	*Product/Material*	*Company*	*Brand collection*	*Sustainability benefit*
Pineapple leaves	Piñatex (leather alternative)	Ananas Anam	Hugo Boss, H&M, Nike	Valorised agricultural waste: replace animal/synthetic leather, lightweight, biodegradable
Banana stems	Banana fibre textile	Various local/SME initiatives (India, Philippines, East Africa); (Alternative Fibre Solutions, Fibsols) Pakistan – a start-up working on the development of banana-stem–derived fibres	Regional crafts, export markets and emerging sustainable-textile supply chains in Pakistan	Strong and durable natural fibre; creates alternative income streams for farmers; reduces agricultural residue waste and burning; supports circular-economy models in Pakistan's textile sector
Grape stem	Vegea (plant-based leather)	Vegea	H&M Conscious Exclusive	Upcycles wine industry waste, displaces animal leather and petro-based synthetics
Orange peels	Orange fibre (cellulose-based silky fabrics)	Orange fibre (Itlay)	Salvatore Ferragamo, H&M Conscious Exclusive	Reduce citrus peel waste, produce luxury biodegradable textiles, support a circular Food-Fibre economy
Milk waste (casein)	Casein (Qmilk)	QMilk (Germany)	Limited apparel/ medical textile	Used spoiled/excess milk, biodegradable, antibacterial fibres
Coffee ground	S.Cafe performance fabric	Singtex (Taiwan)	Outdoor and sportswear brands, Timberland, The North Face	Embed coffee ground in polyester, odour control, UV resistance, moisture wicking and reduces coffee waste
Seafood shells	Chitosan fibres and functional finishes	Various Biotexh firms	Function Textiles (sportswear, medical, hygiene sector)	Antibacterial, biodegradable, enhances dye uptake and moisture management
Fish scales	Collagen/ biopolymer-based fibres	Emerging startups, research labs	Experimental prototypes	Convert aquaculture waste into biopolymers, with potential for medical and apparel use

has since been integrated into partnerships with Renewcell's Circulose platform, showing how chemical recycling can be linked to larger regenerative textile systems. Global fashion brands such as Adidas and H&M have expressed strong interest in chemically recycled polyester, seeing it as a key to replacing virgin fossil-based polyester at scale [45].

5.6.3 Cotton and cellulose recycling

Chemical recycling technology breaks down cellulose-rich waste such as cotton into new cellulosic pulps, which are reformed into high-quality fibres. Renewcell, a Swedish company, has created Circulose, a cotton textile pulp derived from end-of-life cotton textiles (such as old jeans, T-shirts and manufacturing scraps). Circulose can be utilised as a direct replacement for virgin wood pulp or virgin cotton to make viscose, lyocell and other regenerated fibres, presenting a scalable route to circularity [46]. Concurrently, Infinited Fiber (Finland) has created Infinna™, a regenerated cellulosic fibre as soft as cotton, produced from dissolving and respinning textile waste using a proprietary carbamate process [47]. Contrary to downcycled fibres, Infinna and Circulose preserve characteristics usually superior to their virgin counterparts, such as increased durability, moisture absorption and dyeability. These technologies also solve the problem of cotton-polyester blends once and for all by breaking the two apart and value-adding both into novel feedstocks.

5.7 Regenerative systems and closed-loop models

Aside from the recycling process itself, newer recycling technologies operate within greater regenerative systems that imitate natural cycles where waste is a resource. Such a system approach focuses not only on fibre recovery but also on the redesign of textile value chains to reduce leakage and increase circular flows. Regenerative textile systems incorporate three key principles:

1. **Resource recovery**: recovering fibre, dye and chemical inputs for repeated reuse.
2. **Circular design**: product development with recyclability, e.g., mono-material clothing or modular construction that simplifies disassembly.
3. **Industrial symbiosis**: connecting textile recycling with other industries, e.g., packaging, bioenergy, or the construction sector, to optimise material use between industries.

For example, Renewcell's industrial-scale recycling facility in Sundsvall, Sweden, has demonstrated how post-consumer garments can be processed at tonnage scales into Circulose pulp, which is then supplied to global brands such as H&M, Levi's and Zara. These collaborations signal a growing industrial capacity to operate closed-loop systems. Similarly, Infinited Fiber's partnerships with Adidas, Inditex and Patagonia indicate strong commercial interest in scaling regenerated textiles as mainstream alternatives to virgin cotton and synthetics. Comparative overview of advanced recycling technologies in textiles [48].

5.7 Insight from an industry expert and research

The move towards bio-based, bio-fabricated, and circular apparel systems is characterised not only by technical advancement but also by system-related issues that require attention from both the industry and the academy. Industry captains and scholars amplify that the process of exiting the laboratory as a concept to become a common practice entails more than the actual development of new fibres. The conceptual framework for sustainable-textile innovation is shown in Figure 5.1.

Table 5.3 Comparative overview of advanced recycling technologies in textiles.

Recycling type	*Company*	*Technology*	*Output material*	*Brand collaboration*	*Sustainability benefit*
Chemical (Polyester depolymerisation)	Worn Again Technologies	Depolymerisation into PTA and MEG monomers	Virgin quality (PET, rPET)	Collaboration with H&M, Sulzer	Closed-loop PET recycling
Chemical (cellulose dissolution)	Renewcell (Circulose)	Mechanical and chemical dissolution of cellulose pulp	Circulose (regenerated cellulose pulp)	H&M, Levi's, Bestseller	Replace virgin cotton and viscose
Chemical (carbamate process)	Infinited Fiber (Infinna)	Convert cellulose into Carbamate-based fibres	Infinna fibre (Viscose-like textile)	Adidas, Inditex, H&M	Produce high-quality soft, durable fibre
Mechanical recycling	I:CO SOXE Group	Sorting. Shredding, re-spinning	Lower-grade yarns, insulation wipes	Multiple global brands	Reduce landfill waste, cost-effective, but with limited fibre quality
Chemical (polyester recycling)	Carbios (France)	Enzymatic depolymerisation of polyester	Virgin PET usable for fibres and bottles	Puma, Patagonia, L'Oreal	Enzymatic process with low energy use
Chemical (textile regeneration)	Evrnu (NuCyclTM)	Molecular regeneration	NuCycl fibre (Cellulose viscose-like)	Adidas, Levi's, Stella McCartney	High-strength circular fibre reduces reliance on virgin cotton
Blended textile recycling	BlockTexx (Australia)	Separation and recovery (PET and cellulose)	Recycled PET pellets and cellulose powders	Ongoing commercial pilots	Diverts blend fabrics from landfill, produces feedstock for textiles and packaging

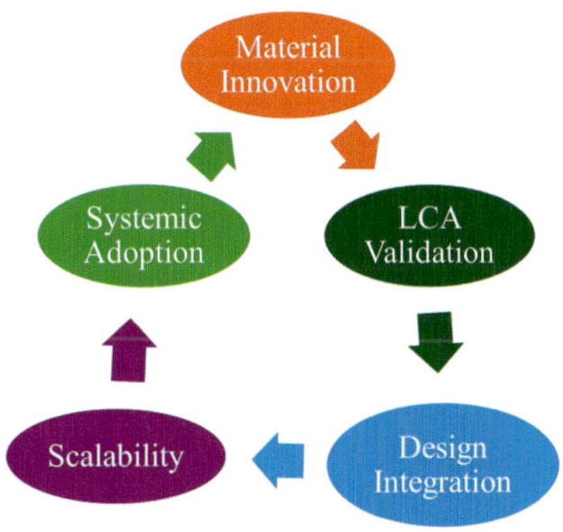

Figure 5.1 Conceptual framework for sustainable-textile innovation.

Dr Amanda Parkes, Pangaia's chief innovation officer, highlights the importance of systems thinking in material innovation: 'The actual innovation isn't the material itself, but the system it's a part of. We need to design for the whole life cycle, where it comes from, how it's made and where it goes. A biodegradable fibre can only be sustainable if disposed of in the proper environment.' This line of thinking highlights the too-often-forgotten fact that material sustainability must be considered on its own terms within the larger context of production, consumer consumption and end-of-life disposal. For instance, biodegradable materials like polylactic acid (PLA) or algae-based fibres could be low-impact production, but if they find their way into municipal waste streams without the industrial composting infrastructure, their benefits are lost.

Scholarly work provides support to this argument by emphasising the dangers of greenwashing and greenhushing terms, which refer to the distortion or purposeful downplaying of environmental qualities in fashion and textile production [49]. A lack of robust life-cycle assessment (LCA) studies may hide latent trade-offs. Although algae cultivation requires little freshwater and arable land, energy-intensive procedures such as drying, extraction and polymerisation can offset some of the perceived environmental benefits. A comparable and transparent LCA versus traditional fibres such as polyester and cotton, is consequently required to support environmental claims and inform industry adoption.

A representative from the industry voiced his opinion during the Future Fabrics Expo, saying, 'Still, a lot of plastic like "eco" innovations are very much in the pilot stage.' The problem is to do eco-friendly and ethical labour practices at every production step when moving from a limited-edition handbag made of a niche product to one that is about to be used in global clothing. This situation of the fashion supply chain is such that most textile advancements take place either in research labs or start-up incubators, but do not reach the market due to the inability to scale. Bringing the production of lab-grown mycelium leather from kilograms to the thousands of metric tonnes that are necessary for clothing aimed at the mass market requires not only technology but also large cash flow, strong chains of supply and seamless integration with the current production processes.

The scientific world has also pointed out the design aspect that is crucial for the circular economy. A circular-design and innovation professor, Dr Kate Goldsworthy, asserts that it is time to practice 'design for disassembly'. From the very first sketch, we must think of future use for clothes and at the same time we must look for 'mono-material' or easily accessible materials. The way her fantasy goes, even the greenest materials will not do the environment any good if they are used in garments whose treatments, blends, or accessories put recycling off. For example, a polyester that is recyclable can stand the test of time if elastane is added to it or if coated with polyurethane lamination instead of a simple blushing method. Thanks to Goldsworthy's contribution, the designing aspect is now the foremost factor that impacts the recycling of materials. The conditions suggest that the development of sustainable textiles in the future cannot be determined by the material alone. Rather, it should be characterised by a comprehensive approach that integrates design foresight with industrial scalability, technological performance and life-cycle integrity. Consequently, one needs to cross the usual frontiers and fuse biotechnology, materials science, industrial engineering, design and policymaking together to create a holistic system. Otherwise, even the best bio-based or recycled fibres may only be used for specialised applications rather than being able to change the present textile sector. This is particularly true when no such multi-stakeholder approach is in place.

References

[1] A. Kozlowski, M. Bardecki and C. Searcy, 'Environmental Impacts in the Fashion Industry: A Life-Cycle and Stakeholder Framework', *Journal of Corporate Citizenship*, 45 (2012): 17–36.

[2] V. Gupta and S. Gupta, 'Cotton Fiber and Its Sustainable Alternatives', in *Novel Sustainable Raw Material Alternatives for the Textiles and Fashion Industry*, Cham: Springer, (2023), pp. 139–60.

[3] H. A. Srinivasan, 'The Sustainable Practices of Small "Sustainable" Clothing Companies', University of California, Berkeley, CA, USA, (2025), https://escholarship.org/uc/item/4fr861kx.

[4] A. Šaravanja, T. Pušić and T. J. M. Dekanić, 'Microplastics in Wastewater by Washing Polyester Fabrics', *Materials* (Basel), 15 (7) (2022): 2683.

[5] J. Hyvärinen, 'Utilization of Cotton and Polyester Blend Textile Waste in Soil Applications', Bachelor's Thesis, Tampere University of Applied Sciences (TAMK), Tampere, Finland, (2022).

[6] X. Huang, Y. Tan, J. Huang, G. Zhu, R. Yin, X. Tao, and X. Tian, 'Industrialization of Open-and Closed-Loop Waste Textile Recycling Towards Sustainability: A Review', *Journal of Cleaner Production*, 436 (2024): 140676.

[7] Z. Chen, H. Sun, W. Kong, L. Chen and W. J. G. C. Zuo, 'Closed-Loop Utilization of Polyester in the Textile Industry', *Green Chemistry*, 25 (11) (2023): 4429–37.

[8] S. Oyedeji, N. Patel, R. Krishnamurthy and P. O. Fatoba, 'Agricultural Wastes to Value-Added Products: Economic and Environmental Perspectives for Waste Conversion', *Agricultural Waste to Value-Added Products*, Singapore: Springer, (2024).

[9] S. Plakantonaki, K. Kiskira, N. Zacharopoulos, V. Belessi, E. Sfyroera, G. Priniotakis and C. Athanasekou, 'Investigating the Routes to Produce Cellulose Fibers from Agro-Waste: An Upcycling Process', *ChemEngineering*, 8 (6) (2024): 112.

[10] S. V. Mohan, K. Amulya and J. A. J. B. T. Modestra, 'Urban Biocycles: Closing Metabolic Loops for Resilient and Regenerative Ecosystem: A Perspective', *Bioresource Technology*, 306 (2020): 123098.

[11] V. Rognoli, B. Petreca, B. Pollini, C. J. S. S. Saito, 'Materials Biography as a Tool for Designers' Exploration of Bio-Based and Bio-Fabricated Materials for the Sustainable Fashion Industry', *Sustainability: Science, Practice and Policy*, 18 (1) (2022): 749–72.

[12] E. Selamet, 'Biotechnology, Bio-fabrication, Bio Fashion: From Living Organisms to Garments', in *Proc. 24th IFFTI Conf*, (2022), vol. 5.

[13] Lorena Trebbi. *Evolving Matter: The Future of Materials and Design in the Biofabrication Era*, (2024), Milano: Franco Angeli.

[14] T. G. Volova, S. V. Prudnikova, A. G. Sukovatyi, E. I. J. A. M. Shishatskaya, 'Production and Properties of Bacterial Cellulose by the Strain *Komagataeibacter xylinus* B-12068', *BioTechnology*, 102 (17) (2018): 7417–28.

[15] H. Hansell, 'BIOmatters: A New Age of Bio-Synthetic Technology in the Textile and Apparel Industry', master's thesis, University of Brighton, (2018).

[16] S. S. Muthu and M. Ramchandani, *Vegan Alternatives for Leather*, Cham: Springer, (2024).

[17] X. Liu, X. Zhang, X. Wang, O. Yue and H. Jiang, 'Engineered, Environmentally Friendly Leather-Like Bio-Based Materials', *Trends in Biotechnology*, 43 (5) (2025): 1104–15. https://doi.org/10.1016/j.tibtech.2024.11.006.

[18] H. Poddar, R. Breitling and E. J. E. B. Takano, 'Towards Engineering and Production of Artificial Spider Silk Using Tools of Synthetic Biology', *Synthetic and Systems Biotechnology*, 4 (1) (2020): 1–6.

[19] A. van Rooyen, Sustainable Innovation and Entrepreneurship, 'Sustainable Fashion Tech Innovation: A Case Study of Bolt Threads' Redefinition of Fashion Systems, Design Activism, and Social Entrepreneurship', *International Journal of Design for Social Change, Sustainable Innovation and Entrepreneurship*, 5 (2) (2024): 1–13.

[20] F. J. S. Qua, 'A Qualitative Study on Sustainable Materials for Design through a Comparative Review of Leather and Its Modern Alternatives', Master of Science in Engineering and Management (S.M. thesis), Massachusetts Institute of Technology, (2019).

[21] M. Janarthanan and M. Senthil Kumar, 'Extraction of Alginate from Brown Seaweeds and Evolution of Bioactive Alginate Film Coated Textile Fabrics for Wound Healing Application', *International Journal of Biological Macromolecules*, 49 (3) (2019): 328–51.

[22] Claudia E. Henninger, Taylor Brydges, Aurelie Le Normand, Shuchan Luo, Luis Quijano, Jane Wood and Songyi Yan, 'How Do Companies Communicate Their "Sustainable" Material Innovations on Company Websites?', *International Journal of Sustainable Fashion & Textiles*, 2 (2) (2023): 163–88.

[23] P. Franzo and C.Moradei, 'From Sea to Fashion: Seaweeds as Material for a Sustainable Transition', *Industrial Design*, 21 (2023): 190–7.

[24] A. Arulprakasam and V. Balasubramanian, 'Algae Materials for Textile Industries', in K. Arunkumar et al. (eds), *Algae Materials: Applications Benefitting Health*, Amsterdam: Academic Press (Elsevier), (2023), pp. 231–46.

[25] J. Barnes, 'Coloration and Dyeing Trends', *AATCC Review*, 24 (2) (2024).

[26] S. Mallon, 'Materialities of "Mushroom-Leather": A Critical Appraisal of the Material Dimensions and Sustainability of Fungus-Based Alternatives to Leather for Use in Fashion and Textiles', *Ethnologia Fennica*, 51 (1) (2024): 81–110.

[27] E. Williams, K. Cenian, L. Golsteijn, B. Morris and Scullin, 'Life Cycle Assessment of MycoWorks' Reishi™: The First Low-Carbon and Biodegradable Alternative Leather', *Environmental Sciences Europe*, 34 (1) (2022): 120.

[28] Walter Sampa Chisumbe, Peter Mukalula, Clinton Aigbavboa, Erastus Mwanaumo, Thandiwe Mbewe, Kelvin Dekesa, Josiline Mabuto, 'Redefinition of Fashion Systems, Design Activism and Social Entrepreneurship', *International Journal of Design for Social Change, Sustainable Innovation and Entrepreneurship*, 14 (2024): 34.

[29] B. Bhandari, 'Sustainable Practices in Textile Design', in Sadhna, R. Kumar and S. Greeshma (eds), *The Art and Craft of Modern Textile Design: Woven Whimsy*, Springer, (2025), pp. 31–47.

[30] A. Patti and D. J. P. Acierno, 'Towards the Sustainability of the Plastic Industry through Biopolymers: Properties and Potential Applications to the Textiles World', 14 (4) (2022): 692.

[31] N. Reddy and Y. Yang, 'Biofibers from Agricultural Byproducts for Industrial Applications', *Trends Biotechnol*, 23 (1) (2005): 22–7.

[32] M. N. Khan, T. A. Sial, A. Ali and F. Wahid, 'Impact of Agricultural Wastes on Environment and Possible Management Strategies', in *Frontier Studies in Soil Science*, Cham: Springer, (2024), pp. 79–108.

[33] Q. T. Islam, R. Sanjana and F. Zabir, *SDG Synergy: Ananas Anam's Pineapple Textile Trail*, London: Sage, (2025).

[34] B. Mushtaq, Y. Nawab, S. Ahmad and F. Ahmad, 'An Eco-Friendly Enzymatic Treatment to Prepare Spinnable Banana Fibers as an Alternative to Cotton for Textile Applications', *International Journal of Biological Macromolecules*, 278 (2024): 134630.

[35] H. Othman, M. M. Ragab, Hassabo, Coloration and P. Science, 'Utilization of Regenerated Cellulose Fiber (Banana Fiber) in Various Textile Applications and Reinforced Polymer Composites', *Journal of Textiles, Coloration and Polymer Science*, 21 (3) (2024): 51–71.

[36] R. B. Malabadi, K. P. Kolkar, R. K. Chalannavar, H. J. W. J. o. A. E. T. Baijnath, 'Plant-Based Leather Production: An update', *World Journal of Advanced Engineering Technology and Sciences*, 14 (1) (2025): 031–059.

[37] M. E. Shaikh, 'Articles Processing of Orange Fibers in Textile Industries', *World Journal of Advanced Engineering Technology and Sciences*, (2021).

[38] Ahmed G. Hassabo, Nadeen Khaleed, Sohaila Shaker, Neaama A. Abd El-Salam, Nourhan A. Mohamed, Nehad Z. Gouda, and Eman Abd El-Aziz, 'Significance of Casein Fiber in Textile Technology', 21 (1) (2024): 63–73.

[39] A. K. Pattanayak, 'Coffee Fibres from Coffee Waste', in R. Nayak (ed), *Sustainable Fibres for Fashion and Textile Manufacturing*, Amsterdam: Woodhead, (2023), pp. 287–307.

[40] Mohammadali Kiehbadroudinezhad, Homa Hosseinzadeh-Bandbafha, Sunita Varjani, Yajing Wang, Wanxi Peng, Junting Pan, Mortaza Aghbashlo and Meisam Tabatabaei, 'Marine Shell-Based Biorefinery: A Sustainable Solution for Aquaculture Waste Valorization', *Renewable Energy*, 206 (2023): 623–34.

[41] M. Pešić, I. Nemeša, N. Buhonka and V. Bozokifryt, 'Fruit Based Sustainable Textile Materials', *9th International Joint Conference on Environmental and Light Industry Technologies*, (2023), pp. 91–9.

[42] P. Wojciechowska, 'Fibres and Textiles in the Circular Economy', in M. I. H. Mondal (ed.), *Fundamentals of Natural Fibres and Textiles*, Amsterdam: Woodhead, (2021), pp. 691–717.

[43] M. Ribul, A. Lanot, C. T. Pisapia, P. Purnell, S. J. McQueen-Mason and S. J. J. o. C. P. Baurley, 'Mechanical, Chemical, Biological: Moving towards Closed-Loop Bio-based Recycling in a Circular Economy of Sustainable Textiles', *Journal of Cleaner Production*, 326 (2021): 129325.

[44] S. Rani and Z. Jamal, 'Recycling of Textiles Waste for Environmental Protection', 4 (1) (2018): 164–8.

[45] I. Nemeša, M. Pešić and V. J. T. I. T. I. Bozoki, 'Chemical Recycling of Textile Waste', *Tekstilna industrija*, 73 (1) (2025): 56–61.

[46] S. Karlsson, 'Mechanical Properties of Viscose Filament: Made from Recycled Cotton Textile and Softwood Dissolving Pulp', M.Sc. thesis, Karlstad University, Karlstad, Sweden, (2022).

[47] E.-K. Uusi-Tarkka, 'Processing and Properties of Wood Fibre Textile All-Cellulose Composites', Ph.D. dissertation, University of Eastern Finland, Joensuu, Finland, (2024).

[48] M. Dahlbom and A. Martvall, 'Maturity of Fibre-to-Fibre Recycling in Europe: Assessment of Recycling Companies in Europe', IVL Swedish Environmental Research Institute, (2025).

[49] H. W. Van Bommel, 'A Conceptual Framework for Analyzing Sustainability Strategies in Industrial Supply Networks from an Innovation Perspective', *Journal of Cleaner Production*, 19 (8) (2011): 895–904.

6 Responsible textile production techniques

Raja Muhammad Waseem Ullah Khan and Abdel-Fattah M. Seyam

6.1 Introduction

The textile sector is important for both people and the global economy, even if it is linked to big social and environmental problems. The old ways of making textiles use a lot of resources and release untreated waste that contains harmful chemicals, heavy metals and toxic colours. They also use a lot of water, especially when growing cotton and wet processing. These kinds of pollutants impact people's health and biodiversity by damaging ecosystems and waterways (rivers, seas and oceans). The textile sector also causes severe air pollution, including greenhouse gases (GHGs) and volatile organic compounds, because the methods used to make them utilise a lot of energy [1]. The accumulation of solid waste, particularly of non-biodegradable synthetic fibres, lasts for many decades in the ecosystem, creating an additional burden on the ecological system by filling landfills and remaining there. The social impact of the textile industry is also quite alarming, as well as the environmental impact. In most of the developing countries, the industry is typified by labour exploitation as workers are often subjected to inhuman working conditions, poor wages and the renunciation of basic human rights. Health risks due to exposure to hazardous chemicals and fibre dust are also serious and problems such as child labour and gender inequality remain universal in certain parts of the supply chain. These interdependent environmental and social problems bring out the urgent necessity of switching to more ethical and ecological methods of textile production. The responsibility now has to be adopted, especially via corporate social responsibility (CSR) [2]. CSR is needed in order to promote moral labouring practice, lessen environmental harm, promote community participation, as well as improve the reputation of the business. Finally, ethical textile production is not merely what should be done, but it is also a strategic move towards the sustainability of the textile business in the long term.

6.2 Principles of responsible textile production

A set of basic principles directs the development of responsible textiles to transform the textile industry into a socially and ecologically responsible industry. The basis of these principles is the life-cycle assessment (LCA), which calculates the overall impact of the product (from development to disposal) on the environment and society at every step [3]. The manufacturers of these products can reduce the overall impact of the product on the environment by assessing the complete life cycle at each step of its life cycle. Based on these principles, the circular economy focuses on the reduction of waste through the overall development of reusable and recyclable items from the cradle to the cradle design concepts. Along with these aspects, transparency and traceability are other essential items for ethical production. Businesses may more effectively diagnose and manage ethical, social and environmental risks by increasing the visibility of their

DOI: 10.4324/9781003638148-8

supply chain. A key component of ethical textile production is ensuring fair compensation, protecting working conditions and observing workers' rights across the supply chain. Furthermore, it is widely acknowledged that sustaining human dignity and advancing fairness in the workplace are essential components of any sustainable company practices [4].

6.3 Sustainable raw material sourcing

Currently, the textile industry practices resource-intensive production techniques and synthetic fibres, which severely degrade the environment. More than 54 per cent of the world's human-made fibre output in 2022 was polyester (PET), which is the most popular synthetic fibre [5]. PET, a non-renewable substance that is non-biodegradable, is a major source of pollution and the source of microplastics in ecosystems. To solve these issues, the industry is progressively using biodegradable and bio-based substitutes. These materials decrease the long-term effects since they decompose under environmental circumstances. Among these natural biopolymers with strong mechanical strength and biodegradability are polylactic acid (PLA) [6], which is made from renewable resources like maize starch and sugarcane and polyhydroxyalkanoates (PHAs). However, its usage in high-performance textiles is constrained because of its brittleness and low heat resistance. To improve its performance and extend its range of uses, PLA has recently been combined with other biodegradable polymers or natural fibres. PHA polymer, which is made by microbiologically fermenting carbohydrates and fats, is completely biodegradable and may be turned into fibres and films. These materials are a preferable option for sustainable textile applications since they break down more rapidly than conventional plastics, especially in soil [7].

Biopolymers and natural fibres such as hemp, bamboo, flax, jute, banana, creeper fibres [8] and organic cotton provide a renewable and biodegradable alternative to synthetics. In addition to being environmentally valuable, these fibres utilise less energy and water during production and processing. Natural fibres are now more viable in terms of quality and performance due to developments in fibre processing methods. Because of their low cost, lightweight and low density, natural fibres are becoming more and more popular among industry and designers as environmentally friendly alternatives, particularly plant-based fibres (categorised as straw, grass, fruit, leaf, seed and bast fibres), each of which has special properties and advantages [9]. However, their broad use is hindered by higher production costs, poor mechanical performance and a lack of infrastructure for recovering biodegradable constituents. The trend towards ethical raw material procurement is not only required but also unavoidable due to the rising consumer demand for sustainability and environmental concerns [10].

6.3.1 Organic, naturally coloured and regenerative cotton

Compared to conventional cotton, organic cotton is more environmentally friendly and sustainable since it is cultivated without the use of pesticides, which have a range of disadvantages, including environmental pollution, water, soil and air contamination, and humans and wildlife. Organic cotton with natural colouring, which comes in brown and green hues, has been used for thousands of years and is becoming more famous due to its advantages for the environment [11]. Although its yields and fibre strength are often lesser than those of white cotton, naturally coloured cotton utilises less water, chemicals and energy during processing because it doesn't need to be dyed. Breeding efforts are ongoing to improve the quality and colour range of cotton fibres, which are appropriate for a variety of textile goods, including clothing and home furnishings.

Regenerative cotton is the most suitable option for sustainability to enhance soil health and restore ecosystem functions by adopting a holistic farming method. Certain methods can raise the water retention, erosion, chemical discharge and carbon storage for land, e.g., crop rotation, composting, decreased digging and the control grazing [12]. In addition to producing cleaner cotton, farmers who use regenerative techniques improve soil biodiversity, decrease their dependency on artificial inputs and improve working conditions, all of which contribute to long-term sustainability. Regenerative cotton is becoming a viable option for a climate-smart and ethical textile sector due to increasing customer demand and brand initiatives for ethical sourcing.

6.3.2 Hemp, linen and other low-impact natural fibres

The use of clothes has increased significantly in recent decades, leading to serious environmental degradation and societal complications. Synthetic fibres, which currently account for more than two-thirds of textile materials and exist in more than half of all textiles, are primarily responsible for this increase. With its high CO_2 emissions, excessive water usage, chemical pollution, microplastic contamination and massive waste output, the textile sector has a significant environmental impact. Pre- and post-consumer textile waste has increased because of these effects being intensified by fashion practices. The major business partners and legislators have started actively responding to these matters, e.g., the European Union (EU) has given a plan to transform the textile industry to a circular economy that will be a carbon-neutral industry [13]. This approach places a strong emphasis on durable, recyclable textiles that lack hazardous materials, have a high percentage of recycled fibres and are made in an equitable social and environmental setting. Due to the increasing concerns about health and the environment, natural and recycled materials are becoming increasingly popular nowadays. This trend is also induced in the composite materials [14, 15] where flax, jute, hemp and kenaf fibres are used at a large scale as a reinforcement [16]. As discussed earlier, these fibres are the most suitable option due to their low cost, lightweight, easy availability and biodegradability. Among all these natural fibres, hemp and flax are the most eye-catching and studied fibres. Some of the major attributes of flex fibre, like low cost, lightweight, better mechanical properties and environmental friendliness, make it the most suitable and desirable option as compared to the human-made fibres [17].

6.3.3 Recycled fibres (post-industrial and post-consumer waste)

Studies show that up to 75 per cent of the textiles that are produced end up in landfill or are burned out. Only 23 per cent of waste clothes are collected for recycling, and of these only 1 per cent of the fibre is used to make new clothes, which shows that the fibre-to-fibre recycling is very limited. Since 1980, there has been about an 80 per cent increase in the production of textile materials, but the recycling ratio is still the same, which increases the waste of textile materials. It is an assessment that up to 2050, there will be 28 million tonnes of textile waste from the American region only, which mostly goes to landfill [18]. In some studies, the LCA revealed that recycled materials can decrease the overall impact on the environment. For instance, denim produced from 100 per cent recycled cotton shows that water consumption can be reduced, as well as the overall environmental impact.

Most studies focus on the environmental benefits instead of the practical properties, e.g., abrasion resistance, tear strength, shrinkage, or fibre release during the laundering process [19, 20]. It is still a big question that if the denim produced from recycled cotton can exhibit the same properties as compared to the denim produced from 100 per cent virgin cotton, especially regarding durability and longevity. It is required to understand how the recycled material composition and repeated washing will affect the service life of the recycled denim produced from the waste material [21]. This knowledge is very important for the manufacturers and consumers, who want sustainable products that have long service [22, 23].

6.3.4 Development of fibres from natural sources for textile and 3D printing

To get the desired properties like silk and wool, scientists achieved alternatives to costly animal fibres by producing the regenerated protein fibres, e.g., zein, soy and casein, in the early twentieth century. Soy protein has emerged as a promising, tuneable biopolymer for the development of fibre with renewed focus on sustainability and waste valorisation, which is an abundant by-product of the soybean oil extraction process [24]. These fibres are also used for 3D printing, along with textile applications. 3D printing gives a transformative potential for the fashion industry by enhancing sustainable development through waste minimisation, ethical production and customisation. It has a huge potential to create sustainable jobs, enable a circular economy and inspire innovation across the global fashion supply chain by aligning with the United Nations Sustainable Development Goals [25].

3D printing is a fast-growing technology that enables the development of complex high-performance materials along with significant design flexibility. Recent developments, specifically the incorporation of fibre reinforcement into plastic resins, enhanced the mechanical properties of 3D printed composites, promoting further research and development into the structure, performance and emphasising the future potential [26, 27]. In a study, a researcher explored the transformative potential of 3D printing in the development of fibre-reinforced composites by assessing the fibre polymer distribution, fibre volume fraction and material properties used in commercial printers. It also calculated the abilities and limitations of the current 3D printing software and hardware, offering insights for further improvement in the printing quality, structural integrity and guiding the future advancements next next-generation composite materials development [28].

6.4 Eco-friendly processing techniques

Eco-friendly processing techniques are basically a variety of approaches that focus on reducing negative impacts on the environment at each stage of the production process. To attain a more sustainable and resource-efficient method of development and processing, these strategies place a high priority on cutting the use of hazardous chemicals and energy used especially in weaving, knitting and finishing processes. The growing decrease in environmental pollution has enhanced the probability of eco-friendly alternatives to plastics and other synthetic materials [29]. From 2010 to 2024, there is a literature of about thirty-two studies that focused on sustainable practices for improving efficiency and reducing the environmental impact [30]. Eco-friendly packaging and waste reduction are also the most important for reducing plastic pollution. Advanced technologies like artificial intelligence (AI) play a significant role in reducing operations and lowering energy utilisation. Sustainable practices are appreciated by regulatory frameworks, for example, the Paris Agreement [30]. Reducing the environmental impact needs to be achieved by all textile processes, but sustainable dyeing and finishing techniques are mainly important. Innovative techniques that reserve or improve fabric performance while dropping water use, chemical load and energy use have been the subject of recent studies. For PLA textiles, a new, environmentally friendly dyeing technique has been put forth that uses a non-aqueous solution that does not require hydrolysis.

This method omitted the hydrolytic breakdown of the conventional PLA dyeing technique and explicitly reduced water consumption. These techniques provide long-lasting colours while maintaining the PLA's mechanical properties and biodegradability [7]. Also, different plants were used for the extraction of natural dyes, which were applied with the help of different weaving groups in Oinlasi Village (South Central Timor, Indonesia) and Taekas Village (North Central Timor, Indonesia) [31]. In this research, twenty plants were selected that provide the colours from different sections, including leaves, stems and roots. Some other plants, like turmeric and hibiscus, also provide natural dyes. These dyes are extracted from different processes, like boiling, non-boiling and fermentation, based on the requirements of the dye extraction. These dyes also have some bioactive compounds, for example, flavonoids, alkaloids and tannins, which can be used in food, cosmetics and pharmaceutical applications. Although natural dyes have different issues such as colour fading and fixation, the researchers are more focused on the improvement of the durability and diversity of natural dyes to make them a suitable choice for environmentally conscious textile products [31].

Conventionally, disperse dyes are used for polyester dying at enhanced temperatures (around 130 °C), including the different carriers, which enhance the dye uptake, but they also produce environmental and health issues due to their toxicity, unpleasant odours and wastewater pollution. Researchers have explored the different dyeing methods, which are carrier-free, to overcome these issues. At the moment, nano-TiO_2 has been used to pretreat the polyester fabrics by increasing the dye uptake and colour fastness without the need for these toxic carriers. Also, the microencapsulated disperse dyes showed a very effective dyeing without the requirement of these dispersants, the other auxiliaries and the levelling agents. The final dyed fabrics exhibited comparable fastness and shade depth as compared to the conventionally dyed fabrics, which eliminated the requirement of reduction clearing. Along with these concepts, the natural alternatives have also been investigated [32].

6.4.1 Water conservation

Water sustainability is a critical issue nowadays globally, because, overall, 60 per cent of the world's population has very low access to clean water. The processing of textile products requires nearly 100 to 150 litres of clean water for one kilogram of fibre. In this regard, the supercritical carbon dioxide ($scCO_2$) provides the best alternative solution for waterless dying as compared to the conventional finishing and dying techniques, which use a high amount of water. These techniques not only save water but also provide fewer chemicals and less hazardous processes, which are both environmentally friendly and the most sustainable process [33]. There is also another option, which is the enzymatic dyeing, which is a more sustainable and less water-consuming technique. E.g., Laccase enzymes have been found in different applications across a range of textile processes like bio-bleaching, scouring, neps removal, dye synthesis and effluent treatment.

In the laccase-based discharge printing, there's no need for harmful reducing agents, for example, sulphoxylic acids and tin salts, as well as bio-bleaching, with the laccases, which reduces the effluent load and as well as maintains fibre integrity. Additionally, the laccase-assisted washing process also decreased the requirement of synthetic detergents, which further contributes to the eco-friendly textile processing [34]. The textile industry's renewed interest in natural dyes provides a basis for its sustainability benefits. Natural dyes are less toxic and biodegradable as compared to synthetic dyes, but they have certain challenges, like colour fastness and limited colour range. The current ongoing research on natural dyes aims to fulfil these limitations by the utilisation of bio-based mordants, optimising dyeing parameters and discovering new natural dye sources, as well as expanding the feasibility and functionality of the natural dyes in the upcoming future textile applications [35].

6.5 Waste management and recycling

The environmental footprint of the transportation sector, specifically on road transportation, is significant because of huge energy consumption and GHG emissions. These impacts can be reduced by the adoption of sustainable practices and decreasing the transportation and utilisation of virgin material as compared to waste materials. Along with this, the EU aggressively advocates the utilisation of recycled waste materials, including construction and demolition waste, as well as textile waste, to develop the most sustainable infrastructure, following the circular economy principles. Adopting circular economy principles in pavement construction and rehabilitation offers multiple environmental and economic benefits, such as reduced consumption of natural resources by replacing virgin materials with recycled ones; minimised landfill contributions by repurposing waste materials that would otherwise be discarded; lowered energy and water usage through more efficient material processing; decreased GHG emissions, helping to mitigate climate change [1].

The textile industry remains the most polluting industry globally due to the generation of a huge amount of waste during both the manufacturing and processing of the clothes and the fabric during the dying process, etc. Poor waste management also contributes the environmental degradation due to the huge amount of junk. The European Directive 2018/851/EU addresses this problem by applying strict rules that encourage proper waste management in the textile industry. The extended producer responsibility principle directs businesses to handle their own waste, as well as they are also responsible for waste management; hence, they are not allowed to landfill or burn the leftover textiles [36]. Hence, the production units enforced by this strategy invest in the recycling practice to adopt sustainable production technologies and create circular business models that reduce waste production. The addition of textile waste into the circular economy is not only possible but is also mandatory to attain the most general sustainability objectives and resource efficiency. This environmental protection is the result of government actions and the adaptation of sustainable practices by the industry [37].

6.5.1 Zero-waste design strategies

The zero-waste supply chain concepts prioritise environmental sustainability and resource efficiency, along with estimating the financial consequences. The objectives of the zero-waste strategy are to minimise waste generation, improve the recycling output, encourage the reuse of material throughout the production steps, from manufacturing, transportation, to distribution and its end of life [38]. It also supports increasing consumer awareness about recycling and enforcing the legal and regulatory framework, as well as also emphasising providing financial incentives for sustainable practices. The zero-waste strategy is based on the adoption of circular economy principles, integration of smart technologies to improve material and waste tracking and management for all businesses [39]. Hence, in developing countries, it is more difficult to cross the barriers on the way to adoption of these circular practices, as well as the inadequate recycling infrastructure and limited technology access, also create a hindrance [40].

6.5.2 Case study on decarbonisation and circular carbon economy

Drawing on 139 expert interviews and 124 site visits across the USA, Norway and the United Arab Emirates (UAE), a study investigated national strategies for industrial decarbonisation through the circular carbon economy (CCE) and defossilisation [41]. While defossilisation focuses on eliminating fossil-fuel use, CCE provides a more flexible and realistic pathway, specifically for oil and gas-dependent economies. Key insights from the case studies include:

- The USA employs a technology portfolio approach, combining renewable energy, low-carbon fuels and carbon capture.
- The UAE leverages its existing fossil-fuel infrastructure while investing in cleaner technologies.
- Norway focuses on renewable energy, carbon capture and storage and green hydrogen, capitalising on its oil and gas expertise.

This study recommended and emphasised leadership in technology innovations and regulation, that there must be a public-private collaboration to scale up the CCE technologies, and also, it should be ensured that the social equity in the transition to net-zero industries is not compromised. These studies also highlighted the need for international cooperation to enhance research on integrated CCE with the de-fossilisation and to develop cohesive strategies and road maps [41].

6.5.3 Recycling: chemical, mechanical and thermo-mechanical

The recycling of textile waste is divided into two categories, the first one is the mechanical recycling and the other one is chemical recycling. There are also some allied recycling techniques with these primary recycling techniques, including the thermo-mechanical, thermochemical and hybrid techniques, depending on the waste type and the desired outcome. Mechanical recycling works on the shearing principle, which means that the textile waste is broken down into small, short fibres, which results in downcycling. The recycled fibres are generally of reduced quality; specifically, they have short length and low strength as compared to the virgin fibres [42]. Mechanical recycling has some advantages in maintaining the colour and material composition of the original waste. The output of mechanical recycling is generally used in the production of

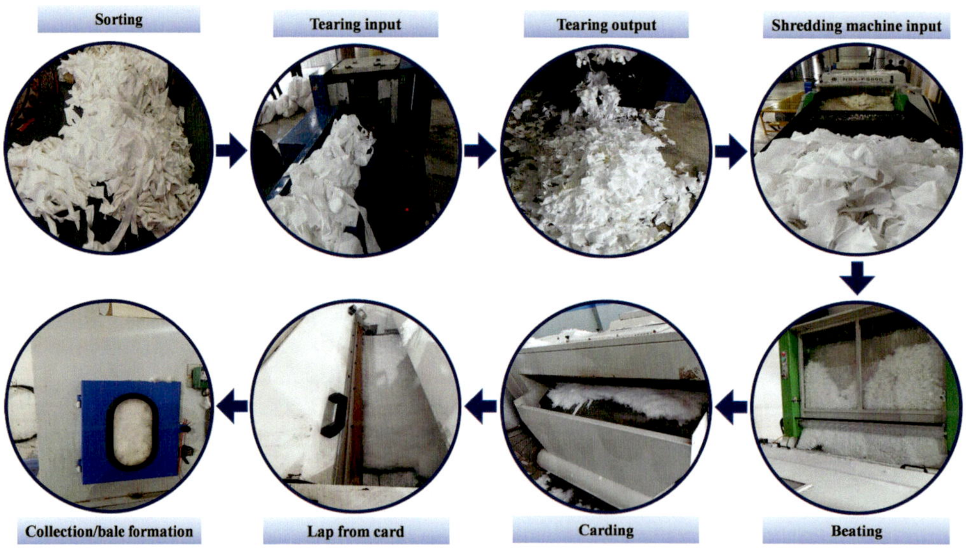

Figure 6.1 Process flow chart of mechanical recycling of textile waste.

non-woven sheets, which are further utilised as thermal insulation material, acoustic material and vibration insulation in the automotive, furniture and other construction industries. These non-woven sheets are also incorporated in composite materials to decrease the dependence on the vision materials [43].

On the other side, chemical recycling has some advantages and offers some more advanced solutions by breaking down waste fibres like cotton and viscose and synthetic fibres like polyester through depolymerisation. This also enables utilisation of this waste material for upcycling, specifically for textiles of the same quality or even sometimes for the superior quality textiles, as compared to the original materials. Chemical recycling plays a significant role in the fibre-to-fibre systems for establishing a true circular economy. Chemical recycling can change the overall composition of the raw material, which is basically the main drawback of this technique as compared to mechanical recycling [44]. It can also be hazardous for the environment, as well as it is also costly compared to mechanical recycling [45]. Nowadays the mechanical recycling is mostly utilised for the recycling of textile waste, and it is applicable for both post-consumer and post-industrial wastes, regardless of the origin of the fibre. However, the quality of the recycled output is highly dependent on the fibre type and the structural characteristics of the textile material, such as whether it is knitted or woven, twisted yarn and weave type [46]. The mechanical recycling process typically involves several key steps: shredding the textile waste, separating hard waste, blending the shredded material, preparing it for processing, tearing the fibres using specialised machinery and finally pressing the recycled fibres (Figure 6.1). Many companies provide the technological backbone for efficient mechanical recycling systems, supporting the global shift towards more sustainable textile waste management [47].

Thermo-mechanical textile recycling combines heat and mechanical processes to recycle thermoplastic materials such as polyester and polyamide into textile fibres.

The most notable innovations in this category are UNIFI, REPREVE and REPREVE Take-back, commercially recycled fibres made from post-consumer PET plastic bottles, as well as post-consumer and post-industrial textile waste. REPREVE, first introduced in 2007, is the first major branded recycled fibre, recognised for its use by leading global brands in apparel, automotive, home and other textiles categories. Since its inception, REPREVE has diverted over 42 billion post-consumer bottles from the waste stream and 950 million T-shirt equivalents (refers to the weight of material equal to that of a single polyester T-shirt) of textile and yarn waste transformed into sustainable fibres. The REPREVE process from bottle to fibre includes:

- **Collection**: Post-consumer PET plastic bottles are collected through municipal recycling programmes and partners.
- **Sorting and cleaning**: Bottles are sorted and cleaned, and labels are removed.
- **Flake**: Clean bottles are shredded into plastic flake, which is further cleaned and prepared for melting.
- **Melting and extrusion**: Clean flake is melted and converted into high-quality recycled polyester resin. Proprietary performance enhancers (such as stretch, moisture-wicking, or UV protection) and traceability additives are incorporated. UNIFI's FiberPrint technology is added to ensure brands and consumers can verify the recycled content.
- **Spinning into fibre**: The melted resin is then extruded and spun into REPREVE yarn, then cut or textured depending on the end use (e.g., staple fibre, filament yarn).
- **Textile manufacturing**: RERPEVE fibre is then knitted or woven into fabric, ready to be used in fabrics for apparel, footwear, automotive, home furnishings and more.

Figure 6.2 summarises the flow chart of REPREVE pictorially.

Figure 6.2 REPREVE: from post-consumer recycled plastic bottles transformed into sustainable recycled fibre.

Figure 6.3 REPREVE Takeback: from post-industrial and post-consumer textile waste transformed into sustainable recycled fibre.

UNIFI started the Textile Takeback programme in 2011 with major expansions in 2022 and 2024. The process allows post-consumer and post-industrial fabric waste to be recycled into REPREVE Takeback fibre and ThermaLoop insulation products. The REPREVE Takeback process from textile to fibre includes:

- **Collection of textile waste**: UNIFI partners with brands, mills and manufacturers to collect post-industrial and post-consumer fabric waste, including offcuts, misprints and end-of-life garments.
- **Sorting and preparation**: Collected textiles are sorted with a proprietary system for high-volume grading, purification and strengthening, to carefully prepare and manage materials throughout the recycling process. Contaminants such as labels and zippers are removed.
- **Melting and extrusion**: Textiles are melted down into resin and may be blended with post-consumer plastic bottle flake. UNIFI's FiberPrint technology is embedded at this stage to ensure full traceability and product integrity.
- **Fibre production**: Melted polyester is extruded, cooled and spun into fibre, then cut or textured depending on the end use (e.g., staple fibre, filament yarn, insulation).
- **Textile manufacturing**: New REPREVE Takeback fibres are transformed into fabrics or utilised in products, completing the textile-to-textile loop.

Figure 6.3 summarises the flow chart of REPREVE Takeback.

Select UNIFI products are backed by global LCA data to verify the product's impact on GHG emissions, energy consumption, fossil-fuel consumption, freshwater consumption and other impact categories that show the environmental savings versus virgin polyester production. The REPREVE and REPREVE Takeback LCA data are third-party peer-reviewed to ensure quality, transparency and scientific rigour. Leading brands globally utilise REPREVE's products to support ongoing sustainability goals, reduce their carbon footprint and source recycled materials, such as REPREVE's Champions of Sustainability Award Winners, which include Nike, Target, Walmart, Under Armour, Levi's, New Balance, the North Face, Costco, Kate Spade, Lovesac and more.

6.6 Digital tools for responsible production

The EU action plan for the circular economy includes different strategies for decreasing the environmental pressure and resource consumption by reducing waste generation, fostering sustainable production and consumption models, influencing consumer behaviour and encouraging the recycling and utilisation of secondary raw materials. This change highlights the significance of the upgradation of recycling setups and technologies to reduce the overall environmental effects further. Additional investment in the recycling of textile waste is crucial to reduce the disposal costs and enhance the effectiveness of waste management. Due to the problems created by the textile manufacturing and disposal industry, research on sustainable alternatives has increased intensively [49]. The methods for textile waste management are being changed day by day due to enhanced public awareness and legislative actions because the conventional methods, for example, landfilling and incineration, are not supported due to environmentally and economically unsustainable outputs. With almost 75 per cent of textiles ending up in landfills or being burned, fast fashion plays a key role in this issue.

Due to the ineffective collection methods, this problem is becoming severe day by day. As a potential solution, recycling and reuse have a huge positive impact on the environment, especially when they are adopted instead of incineration and landfill disposal. The reusing of waste textiles eliminates the need for virgin manufacturing, although its performance is affected by different variables, for example, transportation. To understand sustainable textile waste management completely, it is necessary to do some intensive research on collecting, sorting and the application of renewable energy at the end-of-life treatments for textiles. Better traceability, automatic sorting, data-driven decision-making and better communication throughout the textile value chain are all made possible by digital tools and smart technologies, which may help facilitate this shift [50].

6.6.1 Life-cycle assessment software

The main technique for evaluating a process's environmental performance is LCA, which also includes circularity indicators and computational techniques that may help assess circular design strategies [51]. LCA is a comprehensive methodology used to evaluate the environmental impacts of a product or system throughout its entire life cycle, as outlined by the International Organization for Standardization (ISO 14040) [52, 53] shown in Figure 6.4. LCA plays a critical role in determining the environmental performance of buildings in Europe, following the guidelines which are provided in the EN 15978 standard by the European Committee for Standardization in 2011. Although LCA covers many environmental indicators, it specifically emphasises reducing GHG emissions, aligning with the climate change mitigation goals, specifically for the construction industry. A vital metric used in the LCAs is the global warming potential, which evaluates the overall GHG emissions of a product or process. The emissions that are generated during the production, delivery, installation and disposal of the building materials at the end of their life cycle are known as embodied carbon (EC) [51]. EC is responsible for about 11 per cent of the worldwide emissions, according to the World Green Building Council 2019. There is another terminology, which is called upfront carbon, that refers to the emissions directly linked with the initial phases of the building, transportation and manufacturing of the product

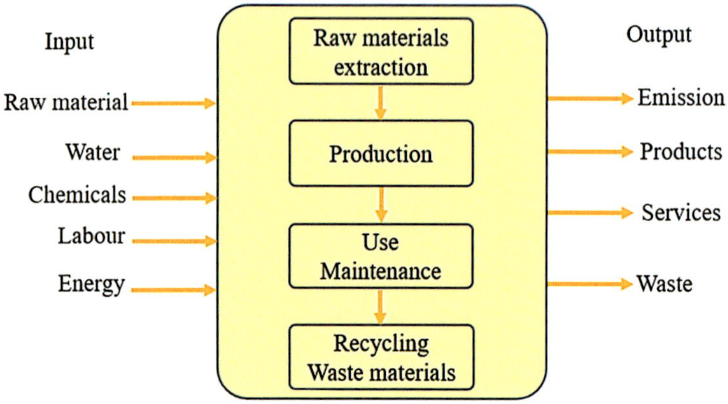

Figure 6.4 Inputs and outputs of LCA.

as a subset of EC. Due to the energy utilisation during the building's operation phase, overall 28 per cent of the worldwide CO_2 emissions are generated, called Operation Carbon. Whole-life carbon provides a detailed picture of the product's carbon footprint during its overall lifespan by incorporating both the embodied and operational carbon emissions. Nowadays, the digital LCA tools are a basic need for supporting data-driven decision-making, enhancing material selection performance and encouraging ethical production methods, both in the textile and construction industries [13].

6.6.2 *Blockchain for traceability*

The businesses are now more concerned about the digital transformation to remain a part of the business stream and enhance efficiency due to the Fourth Industrial Revolution (Industry 4.0) [54]. The four main drivers of Industry 4.0 (that is, a revolutionary stage for technological advancement) are:

1. the surge in data volume, increased computing power and connectivity;
2. the development of analytics and business intelligence;
3. improved human-machine interaction;
4. the emergence of robotics and 3D printing technologies.

The installation and adaptation of advanced machinery systems and workflows by digitalisation throughout the production chain and allied activities is a basic principle of Industry 4.0. However, the industry is facing huge pressure to enhance its competitiveness and innovation because of the high pace of technical improvements and the increased worldwide competition. Many textile sectors, like weaving, knitting, dyeing, finishing and upstream industries, for example, spinning and fibre production, remain susceptible because of the global economy and competition.

Poorly skilled labour is a main hindrance to upgrading the textile sector. There are also some other difficulties, for example, the limited technical skills, the lack of technological competence and a poor organisational learning culture. Hence, the digital transformation is necessary to enhance production efficiency and competitiveness. It is required by Industry 4.0 that the digital technologies, which enhance customer experiences, streamline processes and facilitate the development of novel businesses, must be adopted. These technologies include AI, smart systems and cloud computing, which are gradually gaining their position as essential tools in achieving the digital transformation. However, the organisational attitudes and the perceived ease of implementation play a vital role in the adoption and effective utilisation of these technologies. If the companies do not have a clear commitment to produce innovative solutions aligned with evolving market demand, they will not be able to adopt the technological advancements and systematic execution of digital strategies. Digital transformation also needs human capital with the write digital competencies. In the developing digital landscape, blockchain technology has considerable potential to increase the traceability and transparency within the textile supply chain. Blockchain system gives a real-time tracking of the materials and products by making it safe, verifiable data interchange among the different stakeholders, when aligned with other digital technologies. This supports compliance with sustainability regulations, enhances consumer trust and strengthens the overall resilience of the textile value chain [54, 55].

6.6.3 ***Digital product passports***

Supply chain management (SCM) plays a critical role in the textile manufacturing industry for the timely raw material acquisition, efficient inventory control and the continuous finished product distribution, which affect the overall production efficiency. Due to the severe competition in the textile business, effective SCM is a basic need for maintaining product quality, increasing customer satisfaction and providing an immediate response to customers, along with streamlining operational procedures. SCM in the textile industry is still defined by the manual process, basic forecasting models, which are nowadays incapable of dealing with current issues, including quick changes in customer preferences, unstable market behaviour, raw material shortage and disruptions in logistics.

As the textile industry is going through the digital transition, the digital product passports (DPPs) have appeared as a promising innovation to enhance the supply chain sustainability, traceability and transparency. A DPP is a unique digital record, containing detailed information about the life cycle of a product from the sourcing of raw material to distribution, use and its end of life. From this DPP, the manufacturer, suppliers and other regulators, including the consumers, can get reliable information related to the product's origin, its composition, its overall environmental impact and the observance of sustainability standards. By integrating AI and machine learning (ML) into the SCM system, the utility of DPP is further increased. Furthermore, the textile industry can also implement a circular economy strategy by using AI-driven DPP systems. Aligning with the sustainability objectives and environmental protocols, the DPPs enable the recycling, reuse and repair of waste textile items by digitally documenting the product specifications and material compositions. For instance, the extended regulatory frameworks in the EU, like the Eco-design for sustainable products regulations, are expected to make the DPPs mandatory for different product categories, like textiles. Hence, it can be assumed that, in the direction of more intelligent and transparent supply networks and the integration of DPPs into the textile SCM, supported by AI and ML technologies, is a key step. In the age of Industry 4.0, DPPs may support the textile sector in achieving both financial and ecological goals by providing end-to-end visibility and traceability [56].

6.6.4 ***Supply chain optimisation in textile manufacturing***

The textile manufacturing sector has different challenges in the supply chain environment, which is highly complex from the fibre development to the final garment or product mentioned in Figure 6.5, because it is marked by ever-changing consumer expectations, variations in prices

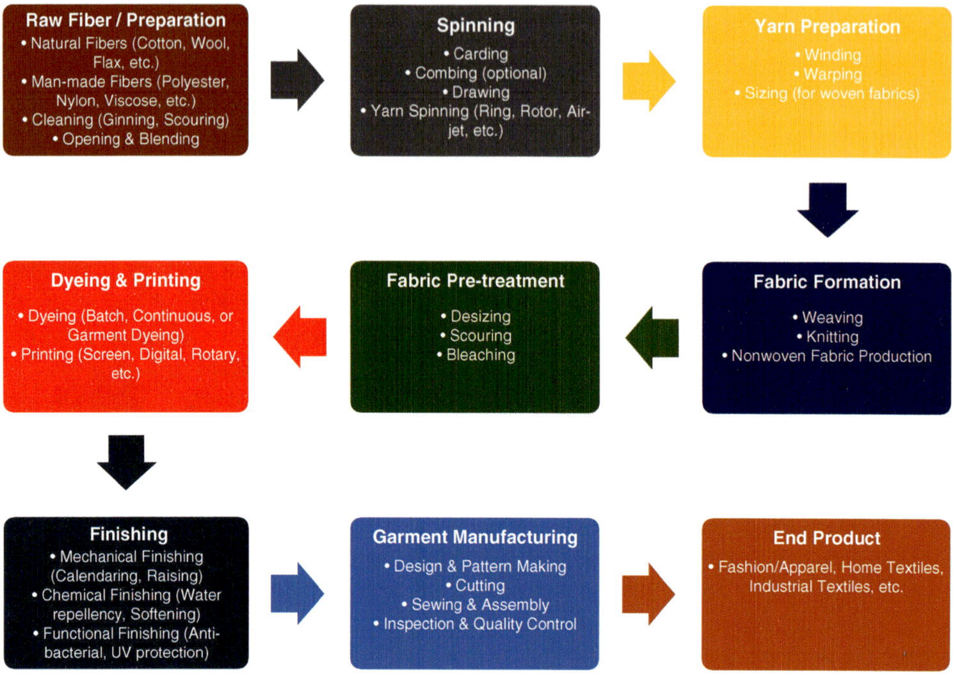

Figure 6.5 Flowchart of processing from fibre to product.

of raw material, along with different logistic challenges. Due to the dependence on historical data and a manual decision-making process, conventional supply chain models normally fail to adapt to these unpredictable circumstances. Such inefficiencies result in some negative outputs like overproduction, excessive inventory, delays in deliveries and disruption of production, ultimately resulting in increased operational costs and decreased overall competitiveness. In this scenario, AI has evolved as an emerging tool for the upgrading of operations of the textile industry supply chain. AI technologies also support real-time decision-making along with advanced features, for example, inventory optimisation, automated demand forecasting and resource management.

These AI technologies use statistical methods, deep learning methods, analytics and ML algorithms to process a huge, diverse data set. These technologies can predict possible disruptions through the supply chain, forecast demand changes, predict hidden trends and offer action insights to minimise business risk. AI systems, like real-time demand signal analysis, can dynamically make production plans that result in minimising the lead times and enhancing the service levels. Additionally, AI can improve the inventory control to maintain minimum stock levels, reducing both the stockouts and overstocking by analysing the seasonal trends, sales patterns and historical data usage [57].

Along with other benefits, AI-based supply-chain alignment and route optimisation techniques can improve accurate delivery and minimise the lead times related to transportation and logistics, resulting in improved economic performance and financial stability. Additionally, when it is joined with blockchain technology and the Internet of Everything, AI provides a higher visibility throughout the supply chain, ensuring traceability, responsiveness and transparency. To overcome the global uncertainties, these integrations are necessary for the development of resilient and sustainable textile supply chains. In summary, textile producers can attain strategic improvements that enhance responsiveness, minimise waste, streamline their operations and support the data-driven sustainability goals by applying AI in the whole supply chain for optimisation [58].

For a sustainable production process, are lot of inventions have been made in the field of textiles, for example, based on sustainability, ITEMA developed an instrument that eradicated the requirement of auxiliary selvage yarns and cutters and reduced the fringes on the fabric's left-hand side. USTER developed an on-loom inspection system to indicate defects or faults in the fabric at the loom stage and suggest the root cause, which avoids production of B-grade fabric, resulting drop in the fabric waste [59]. The global networks regarding the supply chains are becoming highly interconnected and volatile, which requires the adaptation of such digital advancements as a basis for maintaining competitiveness in the textile business [58].

6.7 Future directions and innovations

Two main factors play an important role in the future of sustainable textiles, for example, raw material production techniques and the structure processing methods. The most urgent requirement is the development of biopolymer research centres, specifically identifying the alternative raw materials, for example, agricultural and industrial wastes. Improving the biodegradability, strength and flexibility requires the basic chemical modification and addition of reinforcing agents or monofilaments. These basic improvements are necessary to enable the biodegradable textile materials to be usable for a diverse range of applications, for example, functional and technical textile applications. To evaluate the environmental performance of these biodegradable textiles, it is necessary to develop international standards and certification that will ensure customers' trust, by improving sustainable supply-chain transparency and helping businesses to make sustainable products, claiming more credibility and being understandable for consumers. The demand for verifiable sustainability improvements continues to grow, driven by increasing consumer awareness [7].

Integration of biodegradable textiles into the circular economy system is another gap. For the future, the closed-loop solutions must be the centre of research, development and commercialisation, whereby degradable materials can be recycled, industrially composted and regenerated into the value chain at the end of their life cycle. This may include the development of advanced recycling technologies that effectively separate and recover the biodegradable fibres from different textile waste streams. To address the current challenges, for example, mixed material waste and low recycling rates, the adaptation of these technologies will be essential. It's also very important to avoid making textile products with different fibre blends, for example, cotton/polyester, because it can increase the recycling cost at the separation stage. Policymakers, manufacturers, retailers and academic collaborators must focus on the creation of a supportive regulatory framework that offers incentives for sustainable innovation to make these advancements achievable [60]. This cooperative approach can help transform the textile industry into a more circular system by closing the gap between research discoveries and commercial application [61].

6.8 Conclusion

The increasing need for environmental conservation and sustainable resource management has led to a significant emphasis on responsible textile production. This chapter extensively explored various strategies and technologies focusing on minimising the environmental footprint of textile production. It also evaluated the potential of post-industrial waste and post-consumer waste as a viable raw material for the development of sustainable textile products with special attention to their reusability, recyclability and life-cycle impacts. Mechanical, chemical and thermo-mechanical recycling methods have been elaborated in detail, highlighting their roles in the transformation of textile waste into valuable products. The integration of LCA as a decision-making tool throughout the chapter demonstrated its vital role in the assessment of environmental impacts and guiding sustainable design. Sustainable design, material innovation, emerging technologies and circular economy principles collectively play a role in the more responsible textiles production system, which aligns with the global sustainability goals. For the future of the textile industry, mainly based on adopting holistic science-based and eco-efficient technologies. By leveraging industrial symbiosis, adopting waste as a resource and fostering innovation, the textiles sector can move towards a sustainable circular economy. This chapter provides a foundation for industry professionals and policymakers desiring to adopt and promote responsible textile production at all levels of the supply chain.

References

[1] A. Hasanbeigi and L. Price, 'A Technical Review of Emerging Technologies for Energy and Water Efficiency and Pollution Reduction in the Textile Industry', *J Clean Prod*, 95 (May 2015): 30–44, doi: 10.1016/j.jclepro.2015.02.079.

[2] M. Heale, 'Corporate Social Responsibility in the Fashion Industry', (December 2013), www.eco-business.com/id/opinion/corporate-social-responsibility-fashion-industry/.

[3] M. Prinçaud, C. Aymonier, A. Loppinet-Serani, N. Perry and G. Sonnemann, 'Environmental Feasibility of the Recycling of Carbon Fibers from CFRPs by Solvolysis Using Supercritical Water', *ACS Sustain Chem Eng*, 2 (6) (2014): 1498–502, doi: 10.1021/sc500174m.

[4] C. Zimdars, A. Haas and S. Pfister, 'Enhancing Comprehensive Measurement of Social Impacts in S-LCA by Including Environmental and Economic Aspects', *Int J Life Cycle Assess*, 23 (1) (2018): 133–46, doi: 10.1007/s11367-017-1305-z.

[5] S. Ahmed, S. Kanchi and G. Kumar, *Handbook of Biopolymers*, vol. XVI, Boca Raton, FL: CRC Press, 2018. doi: 10.1201/9780429024757.

[6] Y. Song, J. Liu, S. Chen, Y. Zheng, S. Ruan and Y. Bin, 'Mechanical Properties of Poly (Lactic Acid)/Hemp Fiber Composites Prepared with a Novel Method', *J Polym Environ*, 21 (4) (December 2013): 1117–27, doi: 10.1007/s10924-013-0569-z.

[7] M. Rahman, 'Innovations and Challenges in Biodegradable Textile Materials: A Review of PLA, PHA and Natural Fibers in Sustainable Fashion', *Int J Text Sci*, 15 (1) (January 2025): 1–4, doi: 10.5923/j.textile.20251401.01.

[8] Khubab Shaker, Raja Muhammad Waseem Ullah Khan, Madeha Jabbar, Muhammad Umair, Asra Tariq, Muhammad Kashif, Yasir Nawab, 'Extraction and Characterization of Novel Fibers from Vernonia Elaeagnifolia as a Potential Textile Fiber', *Ind Crops Prod*, 152 (September 2020): 1–9, doi: 10.1016/j.indcrop.2020.112518.

[9] J. Sawant, R. Guru, D. Grewal, S. C. Talekar and S. P. Kulkarni, 'Sustainability in Textiles: A Critical Review of Eco – Friendly Practices and Materials', *ShodhKosh J Vis Perform Arts*, 5 (2) (2024): 85–97, doi: 10.29121/shodhkosh.v5.i2.2024.891.

[10] M. chares Subash and P. Muthiah, 'Eco-friendly Degumming of Natural Fibers for Textile Applications: A Comprehensive Review', *Clean Eng Technol*, 5 (October 2021): 1–11, doi: 10.1016/j.clet.2021.100304.

[11] G. R. Gamble, 'The Effect of Bale Ageing on Cotton Fiber Chemistry, Processing Performance, and Yarn Quality', *J Cotton Sci*, 11 (2007): 98–103.

[12] A. P. Periyasamy, J. Wiener and J. Militky, 'Life-Cycle Assessment of Denim', in S. Senthilkannan Muthu (ed), *Sustainability in Denim*, Sawston: Woodhead, (2017), pp. 83–110. doi: 10.1016/B978-0-08–102043-2.00004–6.

[13] P. Tekin, H. Alıcı and T. Demirdelen, 'A Life Cycle Analysis of a Polyester–Wool Blended Fabric and Associated Carbon Emissions in the Textile Industry', *Energies*, 17 (2) (2024): 1–22, doi: 10.3390/en17020312.

[14] R. M. W. U. Khan, A.-F. M. Seyam, M. Hussain and Y. Nawab, 'Valorizing Denim and Polyethylene Sheet Waste to Fiber-Reinforced Composites', *Polym Compos*, 46 (13) (2025): 1–21, doi: 10.1002/pc.29730.

[15] Khubab Shaker, Muhammad Umair, Salma Shahid, Madeha Jabbar, Raja Muhammad Waseem Ullah Khan, Muhammad Zeeshan, Yasir Nawab, 'Cellulosic Fillers Extracted from Argyreia Speciose Waste: A Potential Reinforcement for Composites to Enhance Properties', *J Nat Fibers*, 19 (11) (2022): 4210–22, doi: 10.1080/15440478.2020.1856271.

[16] A. Ari, M. Karahan, M. Kopar, M. Ahrari, R. M. Waseem Ullah Khan and M. Hussain, 'Comparative Analysis of Natural Fibres Characteristics as Composite Reinforcement', *Ind Textila*, 74 (4) (August 2023): 403–11, doi: 10.35530/IT.074.04.2022110.

[17] A. S. Sigaard and K. Laitala, 'Natural and Sustainable? Consumers' Textile Fiber Preferences', *Fibers*, 11 (2) (January 2023): 1–23, doi: 10.3390/fib11020012.

[18] R. B. Baloyi, O. J. Gbadeyan, B. Sithole and V. Chunilall, 'Recent Advances in Recycling Technologies for Waste Textile Fabrics: A Review', *Text Res J*, 94 (3–4) (February 2024): 508–29, doi: 10.1177/00405175231210239.

[19] S. Z. Haider, Y. Nawab, B. Qadir and M. Hussain, 'Impact of Post-industrial and Post-consumer Waste on the Characteristics of Denim Fabric Made from Recycled Yarns', *J Ind Text*, 55 (December 2025): 1–27, doi: 10.1177/15280837251322951.

[20] A. Telli and O. Babaarslan, 'Commercialized Denim Fabric Production with Post-industrial and Post-consumer Wastes', *Tekst ve Konfeksiyon*, 26 (2) (2016): 213–20.

[21] Ravindra B. Malabadi, Kiran P. Kolkar, Raju K. Chalannavar and Himansu Baijnath, 'Plant-Based Leather Production: An Update', *World J Adv Eng Technol Sci*, 14 (1) (January 2025): 31–59, doi: 10.30574/wjaets.2025.14.1.0648.

[22] C. K. Chow, 'Innovations for Sustainable Textile Production', *Text Asia*, 44 (9) (2013): 8–9.

[23] Y. Arafat and A. J. Uddin, 'Recycled Fibers from Pre- and Post-consumer Textile Waste as Blend Constituents in Manufacturing 100% Cotton Yarns in Ring Spinning: A Sustainable and Eco-Friendly Approach', *Heliyon*, 8 (11) (November 2022): 1–19, doi: 10.1016/j.heliyon.2022.e11275.

[24] M. Tahir, A. Li, M. Moore, E. Ford, T. Theyson and A.-F. M. Seyam, 'Development of Eco-friendly Soy Protein Fiber: A Comprehensive Critical Review and Prospects', *Fibers*, 12 (4) (March 2024): 31, doi: 10.3390/fib12040031.

[25] A. Armstrong, S. M. F. Kabir, K. Mathur and A.-F. M. Seyam, '3D Printing, a Road to Sustainable Fashion', in Subramanian Senthilkannan Muthu (ed.), *Novel Sustainable Alternatives for the Textiles and Fashion Industry*, Cham: Springer Nature, (2023), pp. 1–27. doi: 10.1007/978-3-031-35451-9_1.

[26] S. M. F. Kabir, K. Mathur and A.-F. M. Seyam, 'A Critical Review on 3D Printed Continuous Fiber-Reinforced Composites: History, Mechanism, Materials and Properties', *Compos Struct*, 232 (January 2020): 1–24, doi: 10.1016/j.compstruct.2019.111476.

[27] M. Tahir and A.-F. Seyam, 'Greening Fused Deposition Modeling: A Critical Review of Plant Fiber-Reinforced PLA-Based 3D-Printed Biocomposites', *Fibers*, 13 (5) (May 2025): 64, doi: 10.3390/fib13050064.

[28] S. M. F. Kabir, K. Mathur and A.-F. M. Seyam, 'The Road to Improved Fiber-Reinforced 3D Printing Technology', *Technologies*, 8 (4) (September 2020): 51, doi: 10.3390/technologies8040051.

[29] V.-T. Nguyen and N.-K. Nguyen, 'Method for Weaving Basket Using Eco-friendly Materials in Industrial Production', *MethodsX*, 13 (October 2024): 1–9, doi: 10.1016/j.mex.2024.103025.

[30] N. Sembiring, F. A. Khalif, C. Herlim, E. B. Tarigan and M. Puspitasari, 'Redefining Food Processing: Smart Innovations for Efficiency and Eco-friendly Production', *BIO Web Conf*, 169 (March 2025): 1–6, doi: 10.1051/bioconf/202516901007.

[31] L. Mukkun, H. J. D. Lalel, A. V. Simamora, Y. L. Kleden and M. Bano, 'The Application of Local Plants as Environmentally Friendly Dyes on Timorese "Ikat" Weaving', *IOP Conf Ser Earth Environ Sci*, 1417 (1) (December 2024): 1–14, doi: 10.1088/1755-1315/1417/1/012026.

[32] S. M. T. Shahin, M. Montazer and M. Mahmoudirad, 'Low-Hazardous Multi-functional Textile through One-Step Nano-assisted Processing', *Clean Technol Environ Policy*, 27 (1) (January 2025): 171–89, doi: 10.1007/s10098-024-02844-9.

[33] H. A. Eren, O. Avinc and S. Eren, 'Supercritical Carbon Dioxide for Textile Applications and Recent Developments', *IOP Conf Ser Mater Sci Eng*, 254 (8) (October 2017): 1–4, doi: 10.1088/1757–899X/254/8/082011.

[34] A. N. Garje, 'Green Revolution in Textile Processing by Using Laccases', (July 2014), www.fibre2fashion.com/industry-article/5480/green-revolution-in-textile-processing-by-using-laccases.

[35] S. Shokrzadeh, M. Razbin, F. S. Ghaheh, M. Tehrani and M. Sadrjahani, 'Optimizing the Eco-friendly Dyeing of Wool and Nylon Fabrics with Prangos Ferulacea (L.) Lindl Using Artificial Intelligence', *Sci Rep*, 15 (1) (April 2025): 1–19, doi: 10.1038/s41598-025-97551-w.

[36] J. P. Juanga-Labayen, I. V. Labayen and Q. Yuan, 'A Review on Textile Recycling Practices and Challenges', *Textiles*, 2 (1) (March 2022): 174–88, doi: 10.3390/textiles2010010.

[37] C. Alonso-Troyano, D. Llopis-Castelló and B. Olaso-Cerveró, 'Incorporating Recycled Textile Fibers into Stone Mastic Asphalt', *Buildings*, 15 (8) (April 2025): 1–17, doi: 10.3390/buildings15081310.

[38] D. Çınar, 'Zero Waste Supply Chain: Conceptual Framework, Application Areas, and Future Perspectives', *Uluslararası Batı Karadeniz Sos ve Beşeri Bilim Derg*, 8 (2) (December 2024): 262–79, doi: 10.46452/baksoder.1572222.

[39] A. D. Aleksic, D. Gordic, V. Vukasinovic, M. Josijevic and J. Nikolic, 'Achieving Sustainable Carbon Neutrality of Food Processing: Toward a Net-Zero Margarine Production Facility', *Environ Prog Sustain Energy*, 44 (2) (March 2025): 1–15, doi: 10.1002/ep.14570.

[40] P. R. Barnett, N. S. Hmeidat, B. Zheng and D. Penumadu, 'Toward a Circular Economy: Zero-Waste Manufacturing of Carbon Fiber-Reinforced Thermoplastic Composites', *npj Mater Sustain*, 2 (1) (March 2024): 1–12, doi: 10.1038/s44296-024-00006-y.

[41] S. Griffiths, B. Sovacool, M. Iskandarova and H. J. Walnum, 'Bridging the Gap between Defossilization and Decarbonization to Achieve Net-Zero Industry', *Environ Res Lett*, 20 (2) (February 2025): 1–16, doi: 10.1088/1748–9326/adaed6.

[42] I. Nemeša, M. Pešić and V. Bozoki, 'Mechanical Recycling of Textile Waste', *Tekst Ind*, 72 (4) (2024): 24–8, doi: 10.5937/tekstind2404024N.

[43] A. Mamun, F. Kuntz, C. Golle and L. Sabantina, 'Biotechnological Solutions for Recycling Synthetic Fibers', *The 4th International Electronic Conference on Applied Sciences*, 56 (1) (November 2023): 1–7. doi: 10.3390/ASEC2023–16301.

[44] B. Piribauer and A. Bartl, 'Textile Recycling Processes, State of the Art and Current Developments: A Mini Review', *Waste Manag Res J a Sustain Circ Econ*, 37 (2) (February 2019): 112–19, doi: 10.1177/0734242X18819277.

[45] K. H. D. Tang, 'State of the Art in Textile Waste Management: A Review', *Textiles*, 3 (4) (December 2023): 454–67, doi: 10.3390/textiles3040027.

[46] E. G. Sproul, S. A. Khalifa and B. L. Ennis, 'Environmental and Economic Assessment of Wind Turbine Blade Recycling Approaches', *ACS Sustain Resour Manag*, 2 (1) (January 2025): 39–49, doi: 10.1021/acssusresmgt.4c00256.

[47] M. P. Todor, M. Rackov, I. Kiss and V. G. Cioata, 'Composite Solutions with Recycled Textile Wastes', *J Phys Conf Ser*, 2212 (1) (February 2022): 1–9, doi: 10.1088/1742–6596/2212/1/012030

[48] Repreve, 'Recycling Post-consumer and Post-industrial Textile Waste', (n.d.), https://repreve.com/textile-takeback.

[49] T. Šimonová, M. Malindzakova and D. Malindzak, 'Integrating Management Strategy and Porter's Five Forces Model for the Sustainable Recycling of Textile Waste', *Sustainability*, 16 (21) (October 2024): 1–16, doi: 10.3390/su16219283.

[50] M. Kathirvel and M. Chandrasekaran, 'Predictive Maintenance and Energy Optimization with AI-Driven IoT Framework in Textile Manufacturing Industry', *Int J Comput Exp Sci Eng*, 11 (2) (April 2025): 2408–22, doi: 10.22399/ijcesen.1584.

[51] A. Dervishaj and K. Gudmundsson, 'From LCA to Circular Design: A Comparative Study of Digital Tools for the Built Environment', *Resour Conserv Recycl*, 200 (January 2024): 1–19, doi: 10.1016/j.resconrec.2023.107291.

[52] ISO 14040, 'Environmental Assessment: Life Cycle assessment – Principles and Framework', *Int Stand Organ*, pp. 1–20, (1997).

[53] R. M. W. U. Khan and K. Shaker, 'Life Cycle Assessment of Textile Products', in S. R. Batool, S. Ahmad, Y. Nawab and M. Hussain (eds), *Circularity in Textiles*, Cham: Springer Nature, (2023), pp. 159–76. doi: 10.1007/978-3-031-49479-6_6.

[54] R. Wahyuningtyas, 'The Effect of Digital Competence and Knowledge Sharing on Digital Transformation in the Greater Bandung TPT (Textile and Textile Products) Industry', *J Soc Sci*, 4 (2) (2025): 100–15.

[55] K. Bisht, M. Gurusamy and S. Ghosh, 'Digital Printing in Textiles: New Frontier of Customization and Sustainability', *Asian Text J*, 34 (2) (2025): 1–13.

[56] M. Busari, 'AI and Machine Learning Algorithms for Predictive Supply Chain Management in Textile Manufacturing', (March 2025), pp. 1–14.

[57] A. Noor, M. A. Saeed, T. Ullah, Z. Uddin and R. M. W. Ullah Khan, 'A Review of Artificial Intelligence Applications in Apparel Industry', *J Text Inst*, 113 (3) (2022): 505–14, doi: 10.1080/00405000.2021.1880088.

[58] M. Clement, 'AI-Driven Predictive Analytics for Supply Chain Optimization in Textile Manufacturing', (April 2025), pp. 1–14.

[59] A.-F. M. Seyam, 'Innovation in Weaving Lights the Road to Industry 4.0', *Textile World*, (September/October 2023).

[60] T. Olasehinde, 'Strategies for Reducing Waste in Textile Retail', (November 2024).

[61] I. Wojnowska-Baryła, K. Bernat, M. Zaborowska and D. Kulikowska, 'The Growing Problem of Textile Waste Generation: The Current State of Textile Waste Management', *Energies*, 17 (7) (March 2024): 1–12, doi: 10.3390/en17071528.

7 Tools and methods for sustainable design in fashion and textiles

Awais Khatri, Danyal Rashid Khan, Muhammad Dawood Husain, and Sami Ur Rehman

7.1 Introduction

Sustainable design is critical to the fashion and textile business due to the intense pressure of environmental degradation, scarcity and ethical issues with resources usage, as these problems threaten the long-term sustainability of the business. The chapter is a practical guide to designers, engineers, educators and professionals who want to incorporate sustainability into their practice by applying the proven concepts to the real-world challenges.

The explanation starts with the conceptual bases of sustainable design instruments (Section 7.2), and it explains how design thinking has been modified since focusing on products to a system-oriented model that integrates aspects of environmental, social and economic factors. This gives the concept on how tools allow the structured decision-making by systems thinking, circular-economy principles, which slow resource loops and prefer preventative strategies over reactive or regenerative approaches.

Enhancing this, adherent to Section 7.3, the environmental-assessment and decision-support technologies such as life-cycle assessment (LCA), footprint tools and traceability systems are reviewed. These tools measure the effects of raw materials to end of life and work beyond data complexity challenges in small businesses. They allow making well-informed decisions in the field of dyeing and finishing, where evaluation systems enhance the minimisation of pollution and better process performance in line with the recent findings on sustainable fashion applications.

Section 7.4 deals with materials in sustainable design, including with a concern on renewability, recyclability, low toxicity and ethical sourcing, which are backed by databases such as the Higg Index. Systems of material replacement are described and explained with examples such as bio-based alternatives to traditional auxiliaries, which strengthens the idea of a life cycle of field to textile to future so that environmental footprints can be minimised and the role of material selection in the process of creating and maintaining a circular value chain can be explained.

Section 7.5 covers design approaches that enhance sustainability such as product durability through construction of durable and repairable products, recyclability through mono-materials and modularity and upcycling in closed-loop systems. The case of modular clothes that can be used in many ways, or other similar examples, is an illustration of how design-led innovation can reduce the wasteful nature of fast fashion and promote a more conscious and slower approach to consumption.

Section 7.6 deals with practice digital and educational tools such as computer-aided design (CAD)-based simulation and artificial intelligence (AI)-driven optimisation of physical waste reduction. It also brings into focus training programmes that inculcate sustainability in curriculums that underscores human-based strategies to empower workers, learners and communities.

DOI: 10.4324/9781003638148-9

To connect theory to practice, Section 7.7 includes case studies and applications to clarify South Asian adaptation of small to medium-sized enterprises (SMEs) and changing of materials as a global brand. These are instances of contrasting the results of the processes with tool aided with the traditional ones as this provides some insight on how the process can be applied practically.

Lastly, Section 7.8 covers the new trends, direction and challenges of the future such as AI-enabled traceability, cost of digital transformation and the ongoing skills gaps. It promotes a sustained and dedicated action of holistic sustainability in the textile and garment industries. This chapter highlights the fact that tools are enablers but not the ultimate objectives and that there must be a real move of ensuring that compliance is replaced by commitment to change the industrial practice. Fashion and textile sustainability is not a fad but an instrumental need to handle pollution, waste and resource degradation.

The next sections discuss applicable methodologies and decision-support tools such as LCA, sustainable material selection and measures towards improving the durability and recyclability of the products. The tools are helpful in evaluation, monitoring and adaptive decision-making, such as LCAs and eco-index are used to evaluate the performance of materials; traceability applications show a fabric journey, farm to shelf and digital-twin or 3D-modelling technologies lessen sampling waste and save resources [1].

Nevertheless, they are not automatons. Their performance is based on the ability and motivation of those who use them. The philosophy of people before tools highlights how people who are motivated and competent are essential, as without knowledge or moral devotion, the state-of-the-art AI will not be useful. When the fashion and textile industry is viewed as over dependence on technology without a corresponding change of culture, the result will be underutilised systems instead of sustainable change. Integration of tools and the human insight and mutual motivation is the key element in realising holistic success [2].

The current chapter thus discusses a great variety of methods, such as appraisal systems (LCA, systems of material selection, design styles with the focus on durability and digital or pedagogical techniques). All these are preventative, holistic and human-centred techniques of integrating sustainability into design and production. The case studies on dyeing optimisation and upcycling present real-world examples of textiles, both showing the opportunities and the challenges involved, both can provide lessons in terms of applied research, as well as experience in the industry [3, 4].

7.2 Fundamentals of sustainable design tools

Modern design strategies have been offering a way out of some of the challenges that the textile and fashion industries are experiencing, especially in terms of environment and social issues. To overcome the aesthetic future, the conceptual groundwork on which the development of design thinking took its course should be understood and applied to become a more sustainable subject. These strategies allow one to comprehend the effects of design decisions on the environment, social and economic consequences.

In the past, designers were mainly concerned with the visual and functional attributes of the product such as the form, colour or even the texture of the garment and omitted the effects of production, usage and disposal. With the increasing importance of the environmental concerns, especially in the fashion and textile industries, the necessity of the development of the products ecologically was born. The early conceptualisations of the 1970s and 1980s of green design revolved around the application of greener materials. This developed, by the 1990s, to eco-design, a product life cycle, considering the full product life cycle, both during the extraction of raw materials and at the end life. Design thinking today is an element that incorporates social, economic and ecological systems which represent a balance among people, profit and the planet. In the case of textiles, this can be explained as the selection of fibres which do not pollute the ecosystems and do not misuse labour [5].

Sustainable textile design is meant to lessen environmental degradation and enhance social welfare and economic viability. The sustainable design tools help the product developers and designers to develop environmentally responsible products. The underlining principles are based on environment science, economics and design theory, which are founded on the interrelation between the three pillars of sustainability, including people (social equity), planet (environmental protection) and profit (economic feasibility). The principles relevant to textile in the context of textiles include water pollution caused by dyeing and finishing facilities, carbon emissions caused by manufacturing facilities and low standards of labour in the supply chain. This is changing by the shift in the traditional linear-based model known as take–make–dispose into a circular-design model, where there is reuse, recycling, or upcycling of materials [6].

Sustainable design also encourages life-cycle thinking (LCT) whereby designers should look at the path a product takes once it is extracted as a raw material to the time when it is discarded or recycled. As an example, opting to use organic cotton will decrease the usage of pesticides and will also make recycling easier after the end of life of the product. In the same way, the circular-economy model focuses on ensuring the sustainability and usefulness of materials in the long term. It applies upcycling, waste reprocessing into new uses and zero-waste pattern cutting, which reduces the amount of waste in the creation of garments [6]. The following five areas that sustainability covers are interconnected and are also shown in Figure 7.1.

1. Environmental (low-impact or eco-friendly materials used)
2. Social welfare (providing equal wages and working conditions)
3. Economic (embracing cost-effective approaches)
4. Aesthetic (upholding design attractiveness)
5. Cultural (sensitive to the local traditions and diversity)

All these dimensions combined make sure that sustainable designs are not only environmentally friendly but also cost-effective, inclusive and socially sensitive to diversity and culture [7]. The theory driving sustainable design tools is eco-design that aims at reducing environmental impact on the life cycle of the product. The eco-design was initially a matter of incremental changes but has also since become systemic meaning that whole supply chains should be reconfigured.

Further to this, Design for Sustainability makes design an innovator of design at various scales (see Figure 7.2). The Cradle-to-Cradle (C2C) approach considers waste as a useful resource, and it creates products in such a manner that the materials can easily be recycled into natural or industrial cycles safely. In textiles, this amounts to biodegradable fabrics or even totally recyclable synthetics. An illustration of this is a designer who recycles a jacket using polyester yarn, which can further be reused to produce more fibres and in turn minimise the landfill waste and conserve raw materials.

Such practices are increasingly being favoured by European sustainability policies that focus on a longer, more sustainable and a circular textile future. Systemic thinking assists designers in seeing the interrelationships between the value chain, such as in the case of cotton farming, where it is discovered that the farming requires too much water, designers may contemplate using a different type of fibre, including hemp, bamboo or recycled fibres to alleviate the ecological burden [8]. These theoretical approaches emphasise system-level strategies of minimising global environmental effects, including fibre growing, to consumer consumption.

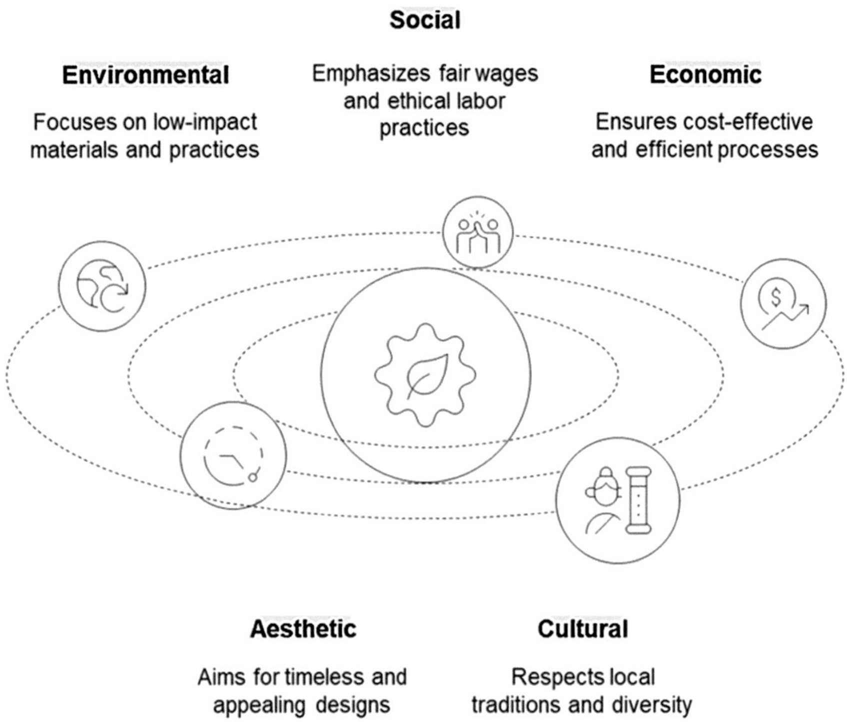

Figure 7.1 The five key dimensions of sustainable design: environmental, social, economic, aesthetic and cultural.

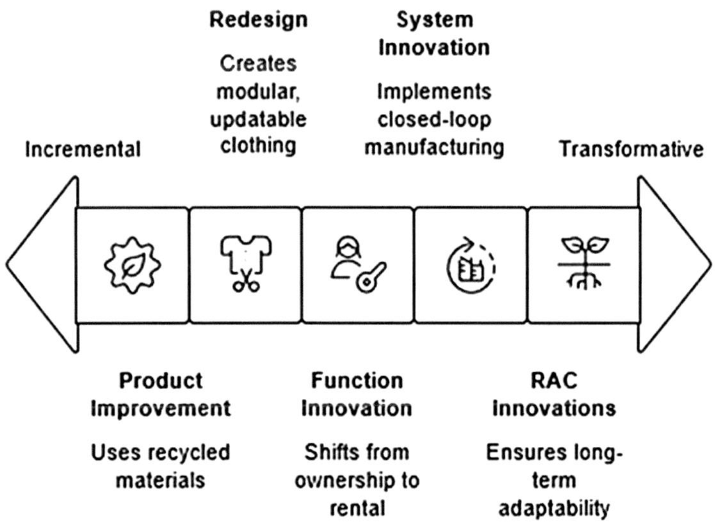

Figure 7.2 Level of innovation from incremental to transformative change for sustainability design.

The practical implementation of these principles is made possible through sustainable design tools which can be classified into three general categories namely the universal, participatory and assessment tools [7].

Generalised tools such as TEN Sustainable Design Strategies framework and Sustainable Design Cards, provide strategic advice in all levels of the product development. They encourage greener choice of materials, reduction of waste and energy and are therefore more specifically applicable to SMEs. Participatory tools engage the users into the designing process. Indicatively, Sustainable Fashion Bridges Ideation Toolkit enables co-design, where emotional attachment between the consumers and the products is promoted and this increases the lifespan of the product.

Environmental performance is measured using the assessment tools such as LCA software or the Higg Index. The LCA approach evaluates the effects throughout the entire life cycle of a product (cradle to grave) and makes useful comparisons between materials and processes. The Higg Index is used to assess fabrics based on various sustainability standards such as water consumption, greenhouse gases (GHGs), energy consumption, chemical effects, eutrophication, wastes and efficiency. There are also other tools including the Nike Maker App that allow designers to see the sustainability performance in real time.

Pattern optimisation, virtual prototyping and minimising wastes are growing in new digital solutions with the use of AI. To the educator and researcher, these tools are significant to teach and simulate LCT in the classroom. Nevertheless, these problems still exist, especially the cost factor, which is difficult to afford by small businesses. The solution to these limitations can include the creation of open-source systems, easier user interfaces and sharing knowledge networks [9].

7.3 Environmental assessment and decision-making tools

It is one thing to learn what sustainable design entails and quite a different thing to quantify what sustainable design has achieved. This part also presents items that will contribute to sustainable design through gauging the environmental performance of a product and making the process of selecting a material eco-friendlier. It is aimed at introducing real-world techniques that facilitate the detection of the possible issues at an early stage to take preventative, not curative measures.

7.3.1 Life-cycle thinking

LCT is the method that promotes making a global view on the existence of a product, including extracting it as raw material and its ultimate disposal. In the textile industry, this would mean tracking cottons, yarns, cloths dyeing, clothing, transportation, domestic consumption, recycling and disposal at the end (see Figure 7.3). LCT facilitates holistic decision-making to minimise environmental degradation throughout the whole product development chain without necessarily having to have large datasets.

In comparison, LCA is a formal and quantitative instrument that integrates with LCT, as it assesses quantifiable effects, e.g., carbon emissions, water usage and energy usage, in a detailed manner and by use of specified standards. Whereas LCT offers a qualitative and adaptable structure to be used in preliminary conceptualisation or classroom consultation, LCA deals with a data-driven analysis with the help of specialised software to produce validated outcomes to be reported, certified and benchmarked. Due to the example, a designer might use LCT to conceptually re-examine the life cycle of a garment but apply LCA in generating quantitative data to prove that there are less environmental effects.

In summary, LCT contributes to conceptual awareness, whereas LCA supports it by practical data, which is inseparable in sustainable fashion where the creative motive should be supported

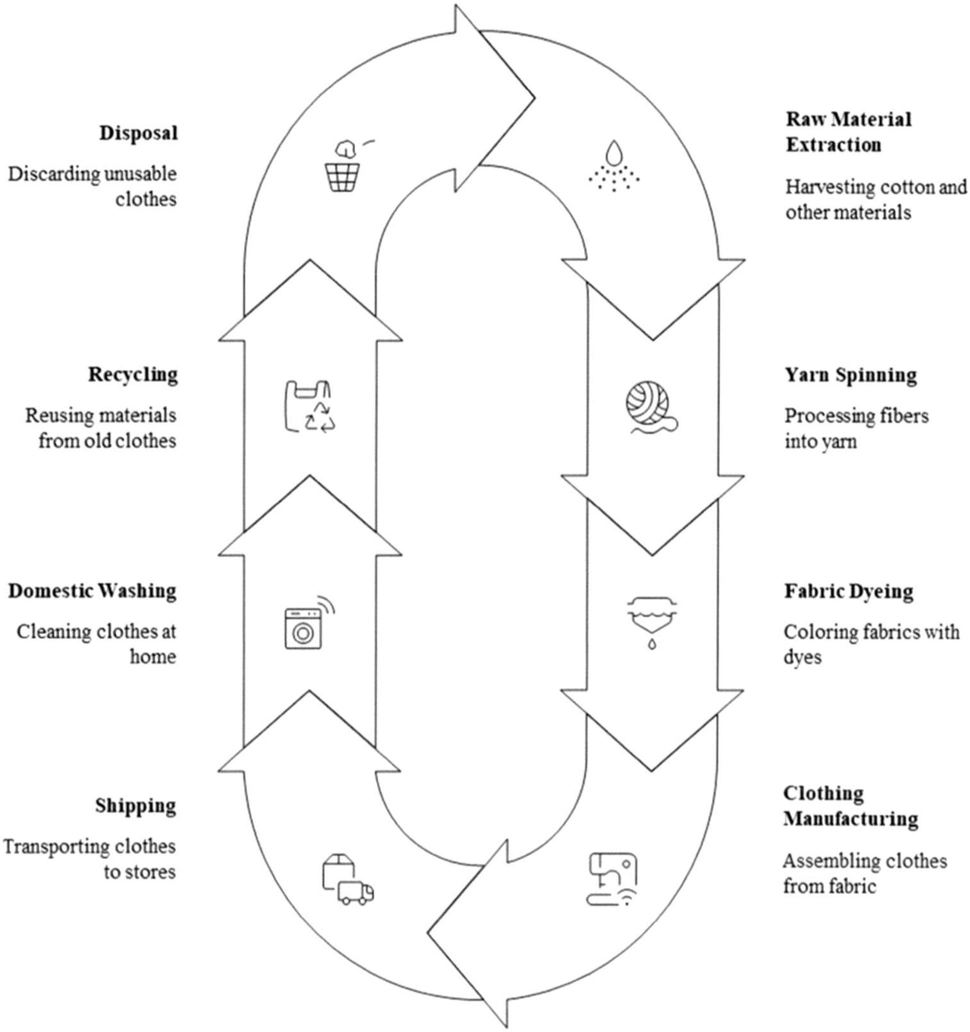

Figure 7.3 Life-cycle approach in textiles.

by empirical data. The initial exposure of students to LCT in the educational setting aids in the development of intuitive knowledge prior to moving onto the technical rigour of LCA [10, 11, 12].

7.3.2 Environmental-assessment tools

Numerous assessment frameworks are employed to assess and risk environmental impacts in the textile and fashion industries to inform decision-making and aid sustainability goals. The LCA is one of the primary tools that examines the entire path of a product: raw materials to the end of its life and measures such parameters as energy consumption, emissions and generated waste. The Product Environmental Footprint is a complimentary standardised methodology in use in Europe to evaluate categories such as climate change, resource depletion and ecosystem impact. It can be used together with LCA to make comparative assessments between similar products [13].

The Higg Index is an integrated system that has modules such as the Materials Sustainability Index (MSI), which measures the effect of materials; the Facility Environmental Module, which represents the site level performance; and the Brand and Retail Module, which measures the performance of corporate sustainability. This benchmarking and self-assessment system assist in brands following their progress on the environmental grounds and declaring the outcomes openly [14].

Verification and market communication are also very important in eco-labelling systems. The Global Organic Textile Standard certifies that organic fibre is used and the manufacturing is ethical; Oeko-Tex checks the safety of the product in terms of harmful chemicals and Bluesign certifies that the production process is clean and does not emit much pollutants. These certifications offer fast appraisals procedures and boost consumer confidence via third parties' certification [15].

Specific assessment tools are applied to measure different environmental metrics. The carbon footprint measures the chain of GHG emission whereas the water footprint measures the consumption and pollution, especially the dyeing and finishing (which is a water-intensive process). Combined, these indicators aid specific reduction strategies [16]. Environmental Product Declarations provide confirmed, LCA-based data on the environmental properties of a product, which fosters transparency and facilitates sustainable procurement choices [17].

In the case of SMEs, a simplified digital calculator, i.e., a spreadsheet tool or web-based application, can be used to conduct simple sustainability assessment processes without expensive software. The material and process efficiency is most analysed using sector-specific models, including the Higg Index of textiles. As an example, a designer can utilise a water-footprint tool to compare the sustainability performance of cotton or polyester mix [13].

7.3.3 Strengths, limitations and accessibility of the tools

In choosing the tools of sustainable design, one should take into consideration their benefits as well as limitations. These tools offer necessary information that can facilitate sound decision-making. They are also strong in their capability of determining hotspots of the environment, which are critical specially for high-temperature dyeing and test alternative alternatives that can save cost, energy and chemicals and comply with the regulations. They also facilitate transparency, which makes brands prove environmental claims by quantifiable information.

Nevertheless, several challenges have remained. The best barriers are the availability and reliability of supply-chain data especially in the globalised production networks. Extensive LCA studies need technical knowledge, a lot of time and detailed information gathering, which may limit its feasibility. Another obstacle is cost: more advanced software and service is not always affordable to a small company. Big companies such as Nike and H&M have the means to do full LCAs and involve suppliers, yet the SMEs usually must use simplified or partial evaluation because of the complexities and lack of budget. Positively, open-source software, shared databases and commercial data-sharing sites are making access more accessible and expenses lower, allowing sustainability measurement at various levels of operation to become feasible [18, 19].

7.3.4 A case study

An analysis of textile dyeing and finishing processes based on LCA and carbon footprint revealed that dyeing was a significant contributor to environmental cost particularly because of chemical-contaminated effluent and excessive amount of water [18]. In comparative analysis of organic cotton yarn, researchers found out that reactive dyeing was a process that consumed much energy and produced large quantities of wastewater unlike low-impact dyes which minimised water pollution by about 30 per cent and the use of energy by 15 per cent. A comparative LCA of vat and reactive dyeing of cotton fabrics showed that vat dyes were less ecotoxic, but they had greater potential to cause acidity through the production of more gas emissions [20, 21].

Carbon-footprint tools also helped discover areas of the process of finishing that could be enhanced, including the better utilisation of steam or the use of natural dyes. This would be translated into concrete gains to the manufacturers in terms of benefits to the environment, saving on labour and labelling and cost of operation. The challenge of collecting data with suppliers is no longer an issue; however, the advantages of using these tools, including greater transparency and competitiveness, prove the worth of using them to convert the sustainability goals into their practical policies.

7.3.5 Traceability and transparency

Traceability systems allow the tracking of a product continuously and ascertain sustainability assertions and reinforce accountability. Digital product passports are electronic labels on clothes that hold information about material, sources and effects of production that is available as QR codes or through specific apps. The blockchain-based technology can supplement this and offer secure and tamper-proof records of each stage involved, such as sourcing of raw materials to retail processes, limiting the risk of counterfeiting and greenwashing.

Tracing systems such as TextileGenesis trace fibres using blockchain and other companies such as Kering rely on it to ensure ethical sourcing. Another innovation is FibreTrace: luminous pigment markers are introduced into uncooked fibres and can be authenticated in real time as they are passed through the supply chain. These systems are known to check the sustainability attributes such as organic certification or low-impact production and are also made to match with both natural and artificial fibres. Notably, the handheld scanners have enhanced the ability of SMEs to conduct traceability audit without expensive installations [22].

By 2025, the EU regulations will demand digital product passports (DPPs) in the textile industry, which require end-to-end transparency to facilitate the recycling and circularity efforts. Nevertheless, it is important to mention that technology is not sufficient to ensure sustainability. Unless such traceability systems are driven by a sense of true ethical motivation, e.g., by an interest in fair labour and environmental responsibility, the systems will be a shallow change agent, instead of an effective marketing tool. According to one critic, a blockchain can show the path of cotton, but when the farm mistreats workers, it will have no sense of being transparent [23].

To sum up, the tools of environmental assessment and traceability can be regarded as a crucial element of sustainable design as they allow decision-making that can be transparent and preventative, and it is data-driven. They can be used in the form of LCAs, digital traceability and so on; this application can minimise environmental impact while cultivating a sense of accountability and innovation. In the end, to have sustainability in the textile industry means a balance between design and responsible action that is supported by both powerful tools and human-commitment.

7.4 Materials for sustainable design

Sustainable design in textiles and apparel is based on the choice of proper materials. It goes well beyond beauty or comfort; it involves evaluation of the impact of materials on the environmental, social and economic consequences of the life cycle of the product. The section presents the main principles and tools to assess and choose materials with references to the practical examples. Through environmental analysis, social issues and economic factors, designers can make sound decisions in their material selection by establishing value to all the stakeholders in terms of production and consumption.

7.4.1 Criteria for sustainable material selection

The key points to pay attention to when evaluating materials on sustainability are those that will reduce environmental and social impacts. Renewability is the use of materials that come out of fast replenishing resources such as bamboo, as opposed to the use of synthetics that are petroleum-based and finite. The recyclability is used to show the ability of a material to be re-processed and reused without losing its integrity for example in the case of polyester, product of recycled bottles, is a greener material as compared to virgin plastic, which is usually discarded in landfills. The subject of toxicity deals with the application of toxic substances; non-toxic substitutes do not use dyes or finishes that either pollute water or become harmful to the health of the workforce (see Figure 7.4).

Biodegradability is also very critical. The natural fibres such as wool or hemp break down without causing harm to the soil and hence, the waste products would not accumulate with time, as compared to synthetic polymers which could take centuries to decompose. Energy and re-source intensity is another important parameter that is critical, as it measures water, energy and land used to produce a fibre. An example is cotton that is rated low because of the high-water irrigation needs and thus the adoption of low water-consuming crops such as lyocell (Figure 7.5).

The social and ethical sourcing also determine sustainability. This consists of equitable wages, work safety and absence of child labour in the farms and industries [21, 22, 23]. Fairtrade Cotton and other certification schemes make sure that the communities directly gain out of production, incorporating a social justice and environment loss reduction strategy. By ignoring these aspects, there will be a risk of greenwashing in which materials are promoted as green without reveal-ing the human exploitation. Balanced sustainability demands authenticity: getting sustainability

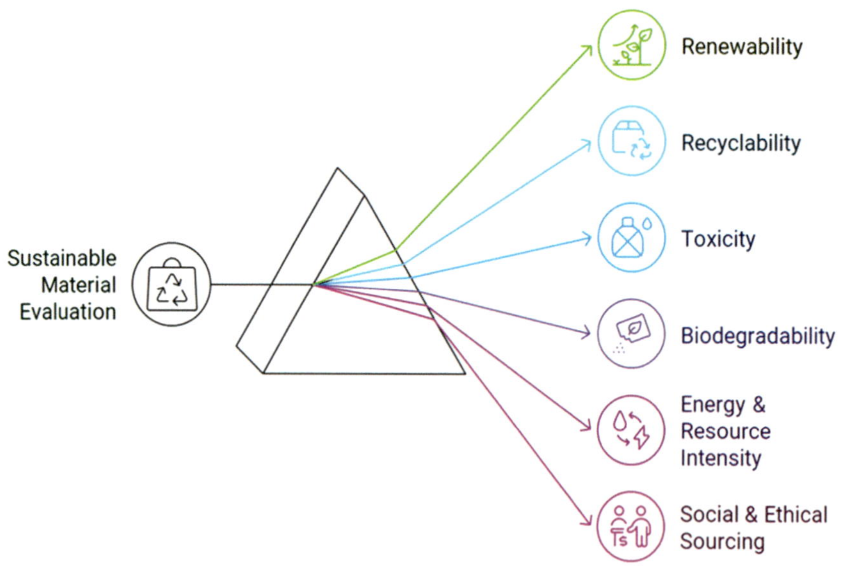

Figure 7.4 Dimensions of sustainable material evaluation.

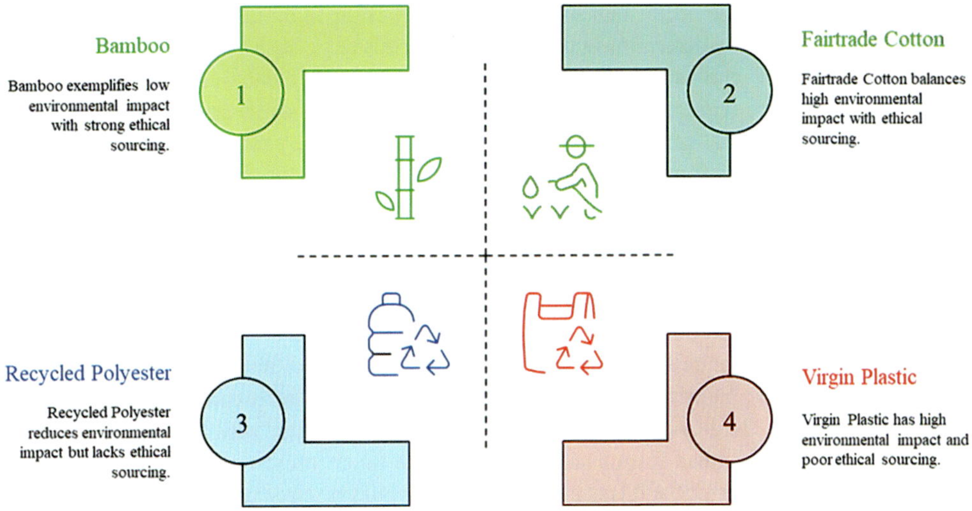

Figure 7.5 Evaluating material sustainability.

using renewable fibres and checking supply-chain integrity. It has been empirically discovered that ethical sourcing promotes brand credibility and long-term resilience [24].

7.4.2 Databases and indices for sustainable material selection

Different databases and indices help professionals to compare materials in terms of sustainability criteria, simplifying the process of decision-making. Higg MSI is a collection of fabrics assessed in various areas of impact, potential global warming, water and energy consumption and chemical intensity, production to delivery [25].

The Textile Exchange Preferred Fibre and Materials Benchmark (PFMB) [26] is a measure of the share of materials used by a company that are certified organic or recycled that tracks global trends like increasing the use of recycled polyester and assists companies in tracking sustainability programmes [27]. Any product developer and compliance team can find these platforms useful as they offer standard comparisons and access to baseline environmental information and compatibility with software such as LCA software. Their organised structures promote innovation, such as Higg scoring has brought an impetus to synthetic-fibre performance.

However, such systems may be complicated to SMEs. Most Higg MSI requires technical training to decode its data architecture and metrics. Therefore, participation might be difficult in smaller companies, which do not have sustainability departments, whereas big corporations can use significant resources to do reporting and analysis. This imbalance has the potential to cause competitive differences in the supply chain.

7.4.3 Frameworks for sustainable material substitution

The substitution of traditional materials with materials that are environmentally favourable needs a systemic approach. Decision-support systems (DSS) employ the multi-criteria analysis to rank and pick the most appropriate options. Another approach, the Analytic Hierarchy Process (AHP) [28] is a weighted algorithm that uses 40 per cent sustainability, 30 per cent performance and 30 per cent cost to weigh and rate alternatives systematically.

Engineers in the textile sector can use SimaPro with Higg to model material replacement. This is usually initiated by determining gaps in the materials (e.g., high toxicity of some auxiliaries) and then alternative materials are filtered using sustainability databases. Durability is guaranteed by performance criteria; affordability is guaranteed by life-cycle costing and ecological improvement is verified by environmental indicators based on LCA scores. An example would be where in case a dye is found to have excess energy intensity, the framework goes through benchmarking, pilot testing and stakeholder review and then adopted.

Although organic cotton may require higher cost than traditional fibre in the short run, its water efficiency and reputation could result in long-term benefits. A study on the Mauritian textile SMEs [29] observed that DSS tools enhanced materials prioritisation and minimisation of waste on prototyping and trial production.

More sophisticated models are incorporating neural networks and optimisation methods based on evolutionary algorithms to optimise the accuracy of the prediction of sustainability situations [30]. A field-to-textile-to-future approach is a holistic way to look at the entire continuum: field would mean responsible sourcing with no deforestation; textile would mean manufacturing that is efficient and less polluted; future would be circular-design initiatives such as recyclability and end-of-life recovery. Mapping of supply chains to detect environmental hotspots, including chemical runoff and land-use stress, is also noted by the United Nations to support the relevance of considering the whole in the material substitution process.

7.5 Design methods for sustainability

In environmentally friendly textile and fashion design, aesthetics is not the only approach applied but the development of a circular system that reduces waste production and decreases the adverse effects of the environment. In this method, priority is based on extending the resources consumption, disposal is postponed, and the negative impacts of overproduction and overconsumption are dealt with. These tenets can offer an operational advice to students, professionals and researchers in re-examining the whole product development cycle concept to end of life. Investigating the opportunities of the long-term perspective, recyclability, upcycling and actively opposing the culture of fast fashion, designers can transform the ideas of the circular design into the feasible and practical approaches that respond to the issue of sustainability in the industry.

7.5.1 Designing for longevity

Longevity in design can be understood as producing clothes that can stand the test of time both physically and aesthetically thus avoiding an untimely discarding. The first step to physical durability is to use high quality fibre, such as long-staple cotton or fine wool and then rigorous spinning and finishing methods help it to endure wear and be reused through repeated washing. Repairability is also of equal importance. It can be done by using visible mending, modular elements (e.g., removable collars), or by provision of spare materials (matching thread or patches). The aesthetic longevity means to create classic shapes and colour schemes, which would not be crowded with the fashion fads, but would remain relevant and useful.

The easy-care finishes such as enzyme-based wrinkle-resistant ones cut down on dry-cleaning expenses and improve care of the garments with the least environmental damage. The slow-fashion movement further confirms these values with made-to-order manufacturing, small batches and craftsmanship quality. As the case of a denim brand can be used, to make the jeans more durable the brand will put the reinforced knee areas and replaceable zippers and years in place of months.

Research has indicated that the extension of garments life cycle significantly minimises the environmental impacts, especially water and energy consumption in the manufacturing process. Studies also mention the impact of the choice of colourfast materials and customisable fittings in long product life and customer satisfaction [31].

7.5.2 Designing for recyclability

The design should be made with its recyclability so that the materials can be reused and processed easily at the expiry of their life cycle. The design of mono-material, such as designing all-cotton clothes, makes sorting and reprocessing easier by removing fibre blends that make decomposition or reprocessing challenging. The modular design, as well as contributes to recyclability, enables parts (sleeves or panels) to be removed and installed using zippers, snaps, or ties. Equally, disassembly-friendly construction does not use adhesives or thermos-fusion and instead uses stitches or mechanical fasteners which allow materials to be easily separated.

The C2C philosophy reflects such a policy and considers waste as a resource that undergoes either biological (biodegradable) or technical (recyclable) loops. As an example, a jacket made of recycled bottles of PET can be re-extruded into new yarn several times with minimal degradation. Modular in smart textiles allow detachable electronic parts, which minimises e-waste, through reuse and repair. These principles turn waste into resource potential, which complies with the rules of circularity and supports the idea of design as a regenerative and not extractive, practice [32, 33].

7.5.3 Upcycling and closed-loop design

Upcycling has been described as the reuse or reuse of discarded or used textiles to manufacture a more valuable product whereas closed-loop systems guarantee that materials do not go to waste. Reuse and redesign should be designed to enable materials to be reused, such as fabric panels should be pre-cut to be reorganised into new clothes. Typical uses are transforming shirts into quilts or industrial symbiosis; whereby waste product is raw material of another process. In this regard, modular garments are especially useful: a dress made with removable skirts can be redesigned to change the silhouette over time, or brands that use magnetic fasteners can be redesigned to convert an old dress into a bag (and the other way around). Such strategies, experiments demonstrate, can decrease the demand of virgin material by 30 per cent and industrial partnerships, including the usage of cotton scraps to insulate or pad, can create zero-waste networks that will benefit several parties [34, 35].

7.5.4 The fast-fashion trap

The fast-fashion approach speeds up the consumption and disposal process and overwhelms markets with cheap clothing disposed of quickly. To reverse this trend, it is necessary to develop strategies that will reduce the speed of fashion cycles and encourage more valuable consumption. The first solution is to focus on classics and versatile design, such as multi-way wraps or adjustable ones, which are not limited by the season. Narrative-driven marketing can also be used by brands, and it can be used to create emotional attachment and persuade customers to appreciate and use their garments longer.

Designers ought to focus on quality and longevity using the quality of materials, high thread-count fabrics that age beautifully and rental or resale services that prolong products life. Waste and overproduction are mitigated further through on-demand manufacturing and garment care education to consumers. Studies prove that slow fashion models decrease the use of resources and human exploitation and positively influence sustainable activities. Multifunctional collections that are called capsule wardrobes have proven to reduce consumption by up to 50 per cent, supporting the mindful purchasing behaviour [36, 37].

Sustainability can also be achieved with the help of emerging technologies, including 3D knitting and zero-waste pattern cutting. The innovations have the potential to cut down the sampling waste and assist in small-scale production, which can result in a 40 per cent decrease in material waste and inventory expenses. The adoption of these practices on a scale is associated with major emission cuts and efficiency in supply chain [38].

All these design approaches enable the fashion and textile industries to develop a more sustainable future. The major lessons to be learned underscore the need to focus on durability to reduce waste, design to be recycled to bridge material cycles, adopting upcycling as a means of creative innovation and slowing our consumption habits in order to redesigning the value of the fashion system.

7.6 Digital and educational tools for sustainable practice

The fashion and textile sectors are being digitised and provide a significant number of tools that can be used to render sustainability ideas into practice [39]. CAD and 3D technologies make virtual prototyping and digital twins less material-wasteful and increase the speed of design iterations. Blockchain systems are increasing supply-chain transparency and traceability, whereas Industry 4.0 tenets are redefining the production process by automating, connecting and making it efficient [40]. AI facilitates innovation along the value chain, including predictive analytics and intelligent material development and optimisation of resources and consumer-facing solutions, including sustainability-oriented apps and interactive solutions, can enable individuals to make informed purchasing choices. In the meantime, advancement of CAD/computer-aided manufacturing (CAM) systems and digital printing of textiles has been moving to simplify the production process lessening mistakes and environmental burden. Collectively, the innovations are transforming the world textile ecosystem, establishing a smarter, more transparent and more sustainable industry [41, 42].

7.6.1 Virtual prototyping and digital twins

The textile and fashion sector is experiencing a significant change through the incorporation of digital twins and 3D technologies which not only make the industry more sustainable but also more engaging to the customers [43]. These systems connect all the process of value chain, which includes design and conceptualisation to production and waste management, consequently leading to creativity, productivity and environmental friendliness.

During the design phase, virtual prototyping allows designers to experiment with different styles, colours and materials within a simulated environment meaning they do not have to use physical samples, and it also saves on the development time. Such a process enhances efficiency and ensures sustainability through reduction of wastes. Virtual and augmented reality (VR/AR) also expand the possibilities of design because of realistic visualisation of garments, before the manufacturing process, customers and designers can see how the garments fit, feel and look.

The 3D body scanning tools will also guarantee individual fit and enhance consumer satisfaction through fewer returns. In addition to the manufacturing sector, digitalisation can also benefit marketing and retail with VR-powered showrooms and AR fitting tools allowing consumers

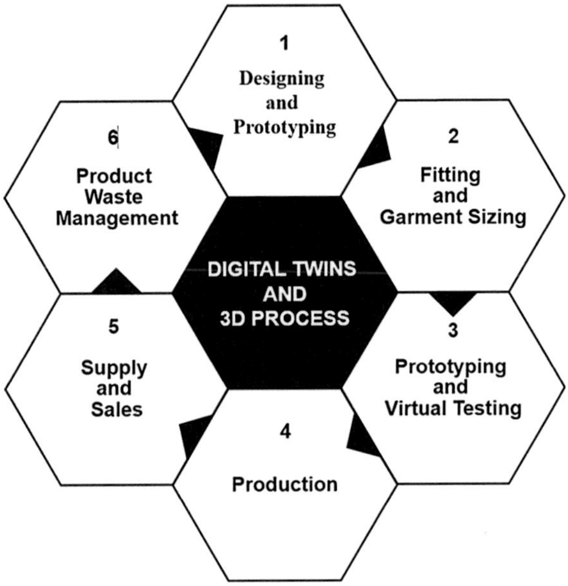

Figure 7.6 Digital twins and 3D virtual prototyping process.

to interact in a better way and close the gap between digital and physical shopping experiences [44, 45]. Key processes digital twins and 3D virtual prototyping are shown in Figure 7.6, which include:

- **Design and prototyping**: 3D software is used to create garments designed and variations in colour, form and texture can be tested without physical sampling.
- **Fitting and garment sizing**: 3D scanning of the body to give a perfect fit and sizing.
- **Virtual testing**: Simulation of drape, movement and colour response to assess performance pre-production.
- **Production**: With 3D printing, customised clothes with reduced waste can be produced with the use of the material required.
- **Supply and sales**: AR/VR virtual fitting rooms can improve customer interaction and decrease the number of returns.
- **Waste management of products**: Digital twins enable recycling data acquisition, with the aim of circular-fashion goals.

7.6.2 *Blockchain-based traceability tools*

Blockchain technology is changing supply-chain transparency whereby it will allow safe, decentralised and verifiable exchange of data [46]. With the adoption of traceability being a key need in the textile industry [47], companies are re-architecting their blockchain-based supply chains and strategy to incorporate new digital potentials and promote accountability throughout all production processes [48]. Digitisation enables organisations to monitor materials and product characteristics between raw fibre to finished garment to enable an end-to-end visibility through the whole manufacturing and distribution network (Figure 7.7).

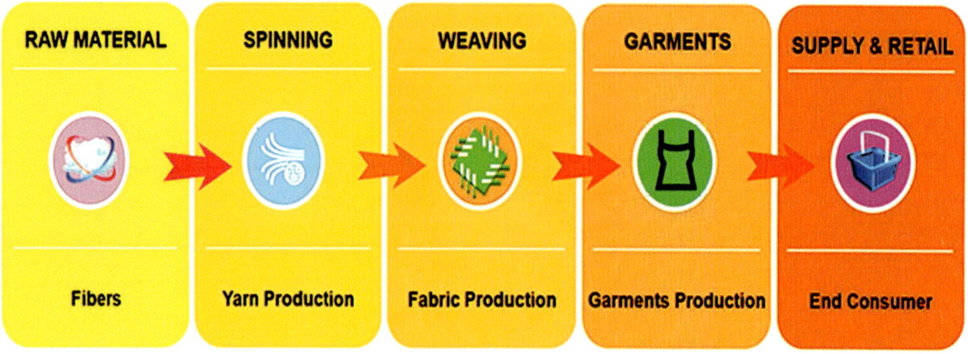

Figure 7.7 Tracing supply at different stages of textile manufacturing.

The supply chain may provide industry with a global look at their products throughout the manufacturing process and have a total overview of the supply chain. Such innovations in supply chain allow organisational decisions to be made in response to supply-chain changes in a short period. These innovations help bring the management of production processes and the work of supply chains under real-time control so that companies could react promptly to interruptions and maximise productivity.

The theory of supply-chain management presupposes four fundamental principles traceability, transparency, accountability and predictability as fundamental elements of an efficient and sustainable system.

1. Traceability is used to guarantee complete visibility of material sources and flows.
2. Transparency enhances sharing of data among the stakeholders, achieving consumer confidence.
3. Responsibility will guarantee ethical behaviour and environmental responsibility in the chain.
4. Predictability helps in making proactive decisions giving firms a chance to foresee the risks like shortage of raw materials or fluctuation in demand.

When supported with digitalisation, these principles would form an efficient, ethical and future resilient supply chain [49]. Embracing blockchain with AI, Internet of Things (IoT), Big Data and cloud computing (the main attributes of Industry 4.0) enhances traceability, authenticity and consumer confidence and results in a smarter and more sustainable textile ecosystem [50].

7.6.3 Industry 4.0 in the textile industry

Industry 4.0 principles are gaining popularity in the textile and apparel industries in the quest to realise a digital transformation. This transformation displaces the traditional and labour-intensive models of manufacturing with integrated, data-driven and smart systems. In this regard, the Industry 4.0 model is not merely a technological upgrade but also a change of attitudes, the willingness to apply smarter, more connected and sustainable processes. These principles have revolutionised the old form of manufacturing, as over the past years, the limited remote monitoring and manual maintenance has developed to automated manufacturing networks with the aid of digital trade fairs, virtual showrooms and direct-to-consumer business models [51]. This digital transformation is based on the IoT and the big data architecture that interconnects the software, sensors and machines throughout the value chain.

Radio Frequency Identification (RFID), cyber-physical systems (CPS), CAD/CAM platforms, robotic sewing units and CLO3D pattern tools have become the key to the transformation. Other uses are defect detection, colour matching and printing automation, which are also improved using AI systems that constantly improve production and quality [52]. The combination of these technologies is what makes up the smart textile industry, which can deliver more productivity, traceability and sustainability.

7.6.4 Artificial intelligence in textile development

AI is transforming the production of textile and apparel by incorporating intelligence into all the value-chain steps. Its application improves creativity, operational efficiency as well as sustainability and minimises the wastage of resources as well as production mistakes. Applications in AI can be seen not only in the design and generation of patterns but also in colour matching, defect detection, assure quality and customer services [53, 54].

Software used to design and produce textiles is increasingly using AI algorithms to aid in 3D modelling, pattern development and automated cutting to enable designers to visualise the results of their design prior to making. Applications such as CLO3D provide real-time simulating features and assist in the reduction of the number of prototypes and eliminating waste [55]. This combination of AI and textile creates a smarter production ecosystem, a system that learns data, constantly evolves and changes to fit the ever-changing market and sustainability demands.

7.6.5 AI and big data for predictive sustainable fashion

In the context of the ethical and sustainable fashion, proper prediction of demand is crucial to ensure that production is matched with consumption and that there is minimal waste. Predictive analytics is a machine learning and data-science-powered predictive technology that helps fashion businesses to forecast customer behaviour, operate inventories effectively and minimise overproduction [56]. Predictive analysis is the synthesis of statistical applications, algorithms and data-based modelling to forecast future trends, demands and operational results depending on the past and current information [57]. Normally it entails the following three key aspects.

1. **Data collection**: Collection of data regarding customer preferences, previous sales, trends in the market, performance of the products, lead times, supplier reactions and sustainable usage of materials.
2. **Statistical modelling**: The use of analytical tools such as time-series analysis, regression analysis and Auto Regressive Integrated Moving Average (ARIMA) models to determine patterns and relationships between variables.
3. **Machine learning**: Machine learning involves the use of algorithms to enhance the accuracy of prediction over time such as the support vector machine, the decision tree, or the neural network.

With the use of these strategies, textile and fashion companies make evidence-based decisions that enhance sustainability, can predict future market changes and maximise resource utilisation.

7.6.6 Consumer engagement tools for sustainable fashion

The consumer engagement is an essential element in the development of the circular-economy practices and the promotion of sustainable business models [59]. Citizens have a leading role in this system as they are involved in and contributing to sustainable consumption and reuse processes. Many digital applications have been created to promote interaction, such as gamification applications, wardrobe management systems and carbon calculators. Gamification also brings in game interfaces which create awareness of sustainability by use of challenges, leader boards and awards to motivate consumers to embrace eco-friendly practices [60].

Wearable systems digitise and classify clothes and enable users to plan and optimise their wardrobes by visualising clothes [61]. The applications minimise the waste of fashion by providing a structured perspective of the wardrobe and promoting mindful consumption [62]. Equally, adding a carbon-footprint calculator to the retail or brand app can assist consumers to know the environmental price of their purchases. The calculator considers extracting of raw materials, manufacturing and logistics to determine GHG emissions [63]. Through technology and transparency, brands promote responsible decisions and participation.

7.6.7 CAD/CAM based digitisation of fashion and textiles

CAD and CAM have transformed the way fashion and textile are made [64]. Such systems can be used to design the 2D and 3D models very quickly thus making it easy to do the design iterations and be able to control workflow. The CAD structures allow several product changes, virtual simulation and high-level visualisation prior to the actual physical production [65]. CAD/CAM helps to save energy and material wastage, making it more sustainable, whereas increasing the level of customisation by minimising the duration of sampling. It is now technology that is used in conjunction with VR, AR and Mixed Reality platforms to provide immersive environments in real-time visualisation and design refinement [66]. CAD/CAM in textile is applied in weaving, knitting and development of garment pattern. Several software such as Starfish enables knitted fabric parameters to be optimised and ProCAD is a common programme used in creating complex knit patterns. These tools combine creativity and efficiency to make designs sustainable, accurate and market ready.

7.6.8 Digital textile printing

Digital textile printing is one of the most important developments in sustainable textile manufacturing, which is mainly done by inkjet technology. The ongoing development of printheads, ink formulas and process regulation has made print quality and efficiency in production significantly better [67]. Inkjet systems today have high resolution printheads and micro-droplet accuracy and versatility with textile materials, such as natural (cotton, silk, jute) and synthetic materials (polyester, nylon). One of them is the piezo-driven injection system that has become prominent because it employs a high-voltage waveform on a piezoelectric membrane to force ink droplets into a drop accurately [68]. New trends in ecologically friendly ink chemistry including water-based or biodegradable made formulations have further reduced environmental impact with no loss of colour intensity or longevity. These new technologies affirm the use of digital printing as a smarter, cleaner and more adjustable technology that can benefit the sustainability of the textile industry [69].

7.7 Case studies and applications

The textile and clothing industry is a multi-trillion-dollar business with close to 2 per cent of the global GDP [64]. Although it is of economic significance, it is among the resource-consuming and polluting industries in the world [70]. Research has shown that a large proportion of waste generated in the world is that of textiles comprising about 20 per cent of all waste materials [71]. Further, the present world economy is less than 10 per cent circular, which means that most of the material is wasted after use. The reduction of the textile waste requires the application of technological advances, reduction losses in production and improvement of material reuse and recycling at the advantage of producers and consumers alike [72].

7.7.1 ***Zara: Revolutionising fast fashion***

The flagship brand of the Inditex Group, Zara, has transformed fast fashion through its ability to respond quickly to market trends and can respond fast with its supply-chain management. Having more than 2,000 stores in more than ninety countries, the operational success of Zara is attributed to its capability to reduce the time lead times of the design to store but keep the trend relevancy. This speed is a challenge though: how to strike a balance between responsiveness and sustainability [73].

To deal with this challenge, Zara has adopted digital technologies in its supply-chain network. The development of data systems, cloud computing and networked information platforms will allow the company to track materials and inventory in real time, hence enhancing traceability and operational efficiency [74]. One of the innovations is the adoption of RFID which is powered by AI-based analytics and cloud technology [75]. With this system, the stock in shelfs is automatically notified when it is 25 per cent full and thus, replenishment will occur as a way of reducing waste.

Moreover, the slow-design philosophy that Zara adopts, where undyed cloths are used in designing prototypes has reduced the prices of the raw material and cut down the procurement period. It means that, in addition to optimising logistics and inventory, RFID and AI will contribute to transparency and thus sustainability is an objective element of speed and efficiency.

7.7.2 ***Simulation-based production optimisation***

The textile industry has always had a challenge to do with efficient scheduling and use of resources especially when it comes to handling the fabric-production schemes. To solve this, the manufacturers are gradually embracing the use of computational simulation tools that help to optimise the workflow without interfering with what is currently taking place. Of these, one of them, Discrete Event Simulation (DES), has been particularly beneficial. Widely applied in the production industry, DES models real-life production lines, finds out inefficiencies and experimentally tests solutions [76]. DES apps in garment factories are used to create a sense of balance in production lines so that machines and personnel are utilised efficiently without wasting money on downtime.

Simulation tools can also be applied to risk-free testing since they help decision-makers to simulate production situations and then initiate physical changes [77]. Other studies document that when used in combination with RFID-based data collection, DES can properly monitor process times and help in shifting activities through the production line with the aim of maximising throughput [78]. All in all, the simulation-based optimisation enables manufacturers to meet deadlines, minimise wastes and enhance decision-making.

7.7.3 ***Digital transformation of luxury fashion***

The area of digital transformation, which involves the use of digital technologies in all business processes, has highly changed the marketing of luxury fashion. Given the growth in social media, the emergence of e-commerce and data-driven personalisation, luxury brands struggle with the dual task of preserving exclusivity and reaching out to a more technologically oriented audience.

AI is a technology that has become part of this evolution. Virtual assistants and recommendation engines powered by AI provide hyper-personalised customer experiences resembling the tailor-made service that used to be provided in luxury boutiques. Louis Vuitton and Hermes (LVMH) are top houses that use AI technology to analyse purchase history and use browsing behaviour to provide personalised product recommendations. These technologies will increase user interactivity, boost brand loyalty and doubled up sustainability by producing with more focused areas and less waste [79].

In that way, the case of the digital transformation of the luxury fashion can be viewed as the illustrative example of how a data-driven approach may help retain craftsmanship, but at the same time, be in line with the sustainability and contemporary consumer needs.

7.7.4 Adidas Speedfactory: A model of digital sustainability

The digital transformation is now the focal point of the sustainability and operational approach of Adidas. The retail digital platform of the company is well aligned to its corporate vision which is to create better consumer relations via digital platforms. With consumers moving across all levels of social media and e-commerce platforms, Adidas is using digital channels to interact with them throughout the buying experience, reducing the necessity of having physical contact with retailing [80]. The Speedfactory project is a step towards a combination of sophisticated digital manufacturing processes and consumer and production information. This innovation facilitates mass production that is customisable and real-time supply and demand matching. Other strategies used by Adidas to bridge the gap between sustainability and culture and innovation include storytelling and celebrity partnerships. Adidas has shown the next generation of sustainable consumers how digital systems, automation and data integration can make operations more efficient, less harmful to the environment and redefine brand interactions through Speedfactory [81].

7.8 Emerging trends, future directions and challenges

The textile and fashion sectors are in a critical junction in the redefinition of the sustainability of the future, functionality and innovation. Speedy changes in technology and changing consumer demands are transforming the production and consumption trends [82, 83]. The contemporary textile is being made not solely for comfort and beauty but are also responsive, functional and eco-friendly [84]. In the past, clothing was only used with utilitarian purposes, as a barrier against the outside factors such as cold and rain. With time, the needs changed to visual sophistication and multifunctionality [85]. There has been an advent of smart and interactive textiles in recent years as the succeeding generation of fabrics with sensors, adaptive materials and intelligent systems that incorporate performance and sustainability [86]. These new generation textiles are disrupting the world production with improved material development, digitalising and circular-design concepts [87].

The textile and apparel industry is however plagued with challenges. The differences in quality of raw materials, labour-intensive tasks and reliance on big engineering make the process less efficient. Issues of excessive wastes, production flaws and overproduction through faulty forecasting and excessive energy and resource utilisation remain a challenge to sustainability. Such problems are usually grouped as the seven wastes of lean, overproduction, waiting, transport, over-processing, excess inventory, motion and defects. Their resolution needs a powerful, versatile strategy combining lean manufacturing and quality-management systems based on data [88]. The subsections below point out the emerging trends that will likely make the world a more sustainable, more productive and the utilisation of resources in an optimised manner in the coming years.

7.8.1 *Textile manufacturing technological innovation*

Advancement in technology has transformed textile and apparel production, allowing it to be more flexible, efficient and creative. The key elements of this change include automation, digitalisation and advanced manufacturing processes. Smarter production lines can be created using technologies such as computer-integrated manufacturing, robotics, additive manufacturing and AI-enabled process monitoring, thus minimising human error and waste. The innovations also encourage customisation and agility; this gives companies an opportunity to respond to constantly evolving market needs without affecting environmental performance. Technologies based on new materials, including biopolymers, nanofibers and bio-engineered fibres, also increase the functionality of the products and reduce ecological footprints [89]. Both innovations put the industry in a position to attain competitiveness and sustainability in the international market.

7.8.2 *Sustainability initiatives in textile industries*

With the growing environmental concerns and heightened regulatory interest, sustainability is a strategic requirement in the world of textile and apparel manufacturers. The industry is embracing various efforts that focus on reduction of wastes, energy conservation and resource cycling. These include the following:

- optimisation of processes through cleaner-production techniques and real-time systems of monitoring
- green choice of materials, where dangerous chemicals are substituted with biodegradable or water-soluble ones
- waste-minimisation strategies, such as post-industrial waste recycling and effluents reuse
- low-energy consuming methods of production, such as low-temperature dyeing or foam finishing
- recycling and remanufacturing models in a circular economy.

These projects are also effective to not only reduce ecological effects but also improve brand reputation and adherence to new sustainability standard [90].

7.8.3 *Circular economy in the textile industry*

The circular economy is a transition to the old paradigm of take–make–dispose and focuses more on durability, resource-efficiency and material recovery. Circularity in the textile and fashion industries is concerned with the design of products to be made with a long life cycle, reused, repaired and recycled. Key enablers include the following:

- constructing to last, to insure maximum wear cycle
- resource use, by minimising the impact process with low impact, using renewable inputs
- installing end-of-life systems, which would allow recycling or fibre regeneration.

New business model, including product-as-a-service, renting, resale market and closed-loop partnerships, grows fast, especially in Europe and East Asia [91]. Circularity increases sustainability by considering waste as an asset, as well as generating new sources of revenue.

7.8.4 *Shifting consumer preferences and market demand*

The textile value chain is experiencing transformational change due to the changes in consumer attitudes. Ethical sourcing, product transparency and sustainability credentials are becoming of growing concern to modern consumers as a factor that influences purchasing choices and brand loyalty. These are dynamic tastes defined by the global culture and internet connectivity that pose a challenge as well as opportunity to manufacturers. With the shift towards slow fashion, multifunctional clothes and locally made products, companies will have to change accordingly by enhancing traceability, minimising lead times and designing strategies. The answer to these shifts will be to make creative direction more aligned to the sustainability values, making sure that style and ethics coexist [92].

7.8.5 *Geopolitics and global textile trade*

International business in clothing is highly influenced by geography, trade policies and regional blocs. Sourcing, pricing and market accessibility can be transformed by political instability, tariff changes and changing international relations. Indicatively, trade arguments or supply-chain disturbances can motivate changes to regionalisation or reshoring of production. These geopolitical pressures have been found to affect production strategies, supply-chain resilience and the investment decisions of the textile industry [93]. The development of flexible and diversified trade networks will hence become crucial in long-term competitiveness and sustainability.

7.8.6 *Resilience and optimised supply chains*

The modern unstable international environment needs a strong supply chain as the basis of business continuity and sustainable development. The current supply-chain approaches focus on traceability, transparency and adaptability facilitated with the use of technologies, including blockchain, IoT sensors and AI-driven analytics. These tools increase both efficiency and sustainability by helping to optimise inventory, forecast disruptions and enhance coordination among the stakeholders. Collaborative systems enable manufacturers, suppliers and stores to exchange information safely and create trust and responsiveness in the network. Studies are ongoing to define best practices in developing agile, data-integrated and low-carbon supply chain that can endure in face of global uncertainties whilst contributing to environmental and social responsibility [94].

7.9 Conclusions

The chapter has shown how the vision of sustainable design in fashion and textiles can be put into practice with the help of practical tools and systematic approaches. Through the idea base, we discussed how these instruments can help in the decision-making process by enabling us to reach to well-organised reasoning, evidence-based decision-making and the elevated level of responsibility in our profession. LCT and LCA are key methods among them and therefore designers and engineers can learn about the effects of products on the environment, a trade-off and choose a solution that reduces the harm during the life of a product.

Another important pillar of the sustainable practice was material selection. Among the points raised in the discussion were evaluation criteria, renewability, recyclability, toxicity and ethical sourcing and the utilisation of global databases and indices to make environmentally conscious decisions. Some of the strategies analysed in the design approach method included durability improvement, recyclability and upcycling, as combined means to overcome the waste production and the unsustainable cycle of fast fashion.

Another aspect that was highlighted in the chapter was the increased presence of digital and educational tools. The examples of how technology can enable waste reduction, make processes more transparent and efficient are virtual prototypes, digital twins, traceability based on blockchain and AI-driven predictive systems. Similarly, online learning and training packages enable both designers and consumers to be active participants in the sustainability practices.

The case studies provided in the industry, such as those by Zara, Adidas and large luxury brands, demonstrated how organisations are applying these principles in their real-world contexts, changing supply chains through innovation and accountability. The discussion of the trending topics also demonstrated the continuous development of the sector, which is both advanced in terms of technology and follows a cyclical business and where consumer values shift more in favour of responsibility and authenticity.

Finally, these tools require cooperation and dedication to work. Green change needs to include the work of designers, technologists, manufacturers, educators and consumers. Through integrative action between creative vision and environmental responsibility the fashion and textile industries will be able to shift the paradigm of compliance to commitment with the creation of a future where sustainability is not a penalty but instead a cultural and professional value.

7.10 Future directions

With sustainable design in textile and fashion continuing to evolve more, there are several emerging lines that provide the potential to advance the tools and approaches to systemic change within the entire value chain. To begin with, the demand to develop data-driven sustainability where life-cycle-based insights, material impact assessment and digital traceability systems are integrated into integrated decision-support platforms is increasing. This integration would allow designers, manufacturers and policymakers to cease discrete evaluations and shift to real-time sustainability optimisation. The second future development is the scaling of the circular-design practices, specifically with fibre-to-fibre recycling technologies, modular product architectures and disassembly-based construction. Despite significant progress that has already been achieved, large-scale adoption would need cutting down on costs, building up of infrastructure and enhanced partnership between recyclers, brands and material innovators. Further increase of AI, machine learning and predictive analytics is also an opportunity. In addition to optimisation of operations, future systems can also be used in automated eco-design suggestions, early identification of environmental hotspots, personalised consumption modelling and scenario planning to minimise overproduction. Ethical considerations about ownership of data, transparency and inclusivity should also be taken into consideration as these tools change. The other direction is the human-centred sustainability, which is also a vital but has become an ignored aspect. The real-life performance of sustainability tools will be established through capacity building of workers, designers, production heads and SMEs. The next projects should focus on skills development, value-chain literacy and motivation based cultural change as well as technological upgrades. Lastly, the industry will enjoy greater policy convergence and global harmonisation, particularly as the emergence of longer producer responsibility, DPPs and global environmental reporting standards will have more to play. Greater regulatory channels, coupled with industry cooperation can hasten the uptake of sustainable design tools as well as provide uniformity and responsibility across markets. Together, these possibilities of the future identify the necessity of interdisciplinary collaboration, investment in enabling infrastructure and constant improvement of tools to develop a regenerative, transparent and resilient textile and fashion system. Sustainability will be an ever-changing process where it takes commitment and not a compliance process.

References

[1] M. Glogar, S. Petrak and M. Mahnić Naglić, 'Digital Technologies in the Sustainable Design and Development of Textiles and Clothing: A Literature Review', *Sustainability*, 17 (4) (February 2025): 1371, doi: 10.3390/su17041371.

[2] M. Dwyer, 'Fashion and AI: Technologies Geared towards More Sustainable Manufacturing Practices', *Premiere Vision*, www.premierevision.com/en/articles/529ed87a-75fa-ef11-90cb-00224888722c/fashion-and-ai-technologies-geared-towards-more-sustainable-manufacturing-practices.

[3] 'Module 5: Holistic Thinking and Quality Production in Fashion – Holistic Approach to the Sustainable Fashion Industry', *Tecofash*, (n.d.), https://tecofash.erasmus.site/courses/module-5-holistic-thinking-and-quality-production-in-fashion/lessons/holistic-approach-to-the-sustainable-fashion-industry/.

[4] M. J. Aladaileh and E. Lahuerta-Otero, 'A Holistic Framework for Assessing Risks in Sustainable Supply Chain Innovation in the Garment, Textile and Leather Industry', *Sustainable Technology and Entrepreneurship*, 4 (2) (May 2025): 100101, doi: 10.1016/j.stae.2025.100101.

[5] F. Ceschin and I. Gaziulusoy, 'Evolution of Design for Sustainability: From Product Design to Design for System Innovations and Transitions', *Design Studies*, 47 (November 2016): 118–63, doi: 10.1016/j.destud.2016.09.002.

[6] E. D'Itria and F. Vacca, 'Shaping Sustainable Solutions in Fashion through Design-Led Strategies, Approaches and Practices', *Discover Sustainability*, 5 (1) (December 2024): 409, doi: 10.1007/s43621-024-00624-5.

[7] A. Kozlowski, M. Bardecki and C. Searcy, 'Tools for Sustainable Fashion Design: An Analysis of Their Fitness for Purpose', *Sustainability*, 11 (13) (July 2019): 3581, doi: 10.3390/su11133581.

[8] T. Duhoux, K. Le Blévennec, S. Manshoven, M. Arnold, 'Textiles and the Environment: The Role of Design in Europe's Circular Economy', *European Environment Agency*, Report No. ETC/CE Report 2/2022 (2022).

[9] Markus Winkler, Dieter Stellmach, Guido Grau, Marcus Winkler, Meike Tilebein, 'Digital Solutions for SMEs' Circularity Transition: Examples from the textile industry', *Industry 4.0 Science*, 40(5) (2024): 26–33, doi: 10.30844/I4SE.24.5.26.

[10] B. Zamani, 'Life Cycle Assessments for Textile and Fashion Products', *Carbonfact*, (2023), www.carbonfact.com/blog/knowledge/lca-fashion-textile.

[11] E. Kaynak, I. S. Piri and O. Das, 'Revisiting the Basics of Life Cycle Assessment and Lifecycle Thinking', *Sustainability*, *17*(16) (August 2025): 7444, doi: 10.3390/su17167444.

[12] F. de Fátima Maciel, R. Gates, I. F. F. Tinôco, N. Pelletier, M. A. Ibarburu-Blanc, F. C. Sousa and N. S. Renato, 'Life Cycle Thinking and Its Importance in the Context of Sustainability Management: Review', *Research, Society and Development*, 13 (3) (March 2024): e2813345034, doi: 10.33448/rsd-v13i3.45034.

[13] N. Mutambo, A. Peirson-Smith, C. KeChi-Okafor, A. Irving-Munro· K. J. Sheridan, M. T. Prendergast-Miller, A. Namdeo, T. Stanton, M. D. Gallidabino and A. James, 'Mapping the Environmental Impact Assessment Landscape in the Fashion and Textile Industries: Critical Gaps and Challenges', *Sustainability*, 16 (19) (October 2024): 8377, doi: 10.3390/su16198377.

[14] K. J. Watson and S. G. Wiedemann, 'Review of Methodological Choices in LCA-Based Textile and Apparel Rating Tools', *Sustainability*, *11*(14) (July 2019): 3846, doi: 10.3390/su11143846.

[15] M. T. Rahaman and M. S. H. Khan, 'Green Merchandising of Textiles and Apparel in a Circular Economy', *Journal of Open Innovation*, 11 (1) (March 2025): 100457, doi: 10.1016/j.joitmc.2024.100457.

[16] S. G. Wiedemann, S. J. Clarke, Q. V. Nguyen, Z. X. Cheah and A. T. Simmons, 'Strategies to Reduce Environmental Impacts from Textiles', *Resources, Conservation and Recycling*, 198 (November 2023): 107119, doi: 10.1016/j.resconrec.2023.107119.

[17] A. M. Fet and C. Skaar, 'Eco-labelling, Product Category Rules and Certification Procedures Based on ISO 14025 Requirements', *The International Journal of Life Cycle Assessment*, 11(1) (January 2006): 49–54, doi: 10.1065/lca2006.01.237.

[18] K. B. Murugesh and M. Selvadass, 'Life Cycle Assessment for the Dyeing and Finishing Process of Organic Cotton Knitted Fabrics', *Journal of Textile and Apparel, Technology and Management*, 8 (2) (April 2013): 1–16, https://jtatm.textiles.ncsu.edu/index.php/JTATM/article/view/3396.

[19] Fashion Sustainability Directory, 'Sustainable Fashion Frameworks', (2026), https://fashion.sustainability-directory.com/term/sustainable-fashion-frameworks/.

[20] A. F. S. Costa, J. V. S. Aragão, A. D. Duarte, J. S. Macêdo, C. J. S. Galdino Jr., V. F. A. Milanez, G. L. Silva and L. A. Sarubbo, 'Analysis of the Environmental Life Cycle of Dyeing in Textiles', *Chemical Engineering Transactions*, 86 (2021): 727–32, doi: 10.3303/CET2186122.

[21] M. M. Islam, A. R. Aidid, J. N. Mohshin, H. Mondal, S. Ganguli and A. K. Chakraborty, 'A Critical Review on Textile Dye-Containing Wastewater', *Cleaner Chemical Engineering*, 11 (December 2025): 100165, doi: 10.1016/j.clce.2025.100165.

[22] FibreTrace, 'FibreTrace: Home', (n.d.), www.fibretrace.io/.

[23] A. Romano, 'Top Five Fashion Supply Chain Traceability Software in 2025', (2024), www.ettos.io/post/top-five-fashion-supply-chain-traceability-software-in-2024.

[24] H. Karaosman, P. Perry, A. Brun and G. Morales-Alonso, 'Behind the runway: Extending sustainability in luxury fashion supply chains, *Journal of Business Research*, 117 (September 2020): 652–663, doi: 10.1016/j.jbusres.2018.09.017.

[25] Apparel Resources News-Desk, 'Sustainable Apparel Coalition: Higg Index Product Tools', (2023), https://apparelresources.com/business-news/sustainability/sustainable-apparel-coalition-shares-review-higg-index-product-tools.

[26] Textile Exchange, 'Preferred Fiber and Materials', Market Report (2021), https://textileexchange.org/app/uploads/2021/08/Textile-Exchange_Preferred-Fiber-and-Materials-Market-Report_2021.pdf.

[27] Textile Exchange, 'Materials Benchmark', (2026), https://textileexchange.org/materials-benchmark.

[28] M. Siraj, 'Sustainable Performance Measurement in Manufacturing Process of Textile Industry Using AHP', *Quaid-e-Awam University Research Journal*, 21 (1) (June 2023): 46–53, doi: 10.52584/qrj.2101.06.

[29] C. Devkumar S., 'A Decision Support System for Environmentally Sustainable Strategies', *Heliyon*, 9 (1) (January 2023): e12939, doi: 10.1016/j.heliyon.2023.e12939.

[30] L. Y. Ljungberg, 'Materials Selection and Design for Sustainable Products', *Materials & Design*, 28 (2) (2007): 466–79, doi: 10.1016/j.matdes.2005.09.006.

[31] United Nations, 'Sustainability and Circularity in the Textile Value Chain: A Global Roadmap', *UN Environment Programme*, (March 2023), www.unep.org/resources/publication/sustainability-and-circularity-textile-value-chain-global-roadmap.

[32] M. G. Tadesse, M. T. Abate, J. F. Lübben and M. Rauch, 'Recycling and Sustainable Design for Smart Textiles: A Review', *Advanced Sustainable Systems*, 9(8) (August 2025): 2401072, doi: 10.1002/adsu.202401072.

[33] M. Braungart, W. McDonough and A. Bollinger, 'Cradle-to-cradle design: Creating healthy emissions – A strategy for eco-effective product and system design', *Journal of Cleaner Production*, 15(13–14) (September 2007): 1337–1348, 2007, doi: 10.1016/j.jclepro.2006.08.003.

[34] A. D. Pranta, M. T. Rahaman and M. R. Repon, 'Sustainable Waste Management of Textiles and Apparel in a Circular Economy: Recent Advances and Future Prospects of Life Cycle Assessment', *Materials Circular Economy*, 7(28) (December 2025): 1–28, doi: 10.1007/s42824-025-00167-8.

[35] M. Shamsuzzaman, M. Islam, M. A. Al. Mamun, R. Rayyaan, K. Sowrov, S. Islam and A. S. M. Sayem, 'Fashion and Textile Waste Management in the Circular Economy', *Cleaner Waste Systems*, 11 (June 2025): 100268, doi: 10.1016/j.clwas.2025.100268.

[36] P. Centobelli, S. Abbate, S. P. Nadeem and J. A. Garza-Reyes, 'Slowing the Fast Fashion Industry', 38 (December 2022): 100684, doi: 10.1016/j.cogsc.2022.100684.

[37] I. A. Millward-Pena, 'From Fast Fashion to Sustainable Slow Fashion', California State University, (2022), https://scholarworks.lib.csusb.edu/etd/1453.

[38] A. Pasricha and R. Greeninger, 'Exploration of 3D Printing for Zero-Waste Fashion', *Fashion and Textiles*, 5(30) (1 December 2018): 1–18, doi: 10.1186/s40691-018-0152-2.

[39] M. C. Bermeo-Giraldo, A. Valencia-Arias, E. M. Rojas, S. Cardona-Acevedo, P. A. Rodríguez-Correa, J. D. González-Ruiz, W Londoño-Celis and R. B. García, 'Digital Transformation in the Textile Sector', *Discover Sustainability*, 6 (1) (April 2025): 294, doi: 10.1007/s43621-025-01091-2.

[40] C. Roumeliotis, M. Dasygenis, V. Lazaridis and M. Dossis, 'Blockchain and Digital Twins in Industry 4.0: The Use Case of Supply Chain-A Review of Integration Techniques and Applications', *Designs*, 8 (6) (October 2024): 105, doi: 10.3390/designs8060105.

[41] W. Strielkowski, V. Grebennikova, A. Lisovskiy, G. Rakhimova and T. Vasileva, 'AI-Driven Adaptive Learning for sustainable educational transformation', *Sustainable Development*, 33 (2) (April 2025): 1921–1947, doi: 10.1002/sd.3221.

[42] M. Wang, R. Murphy and I. Christie, 'Sustainable Fashion Education', *Sustainability*, 17 (2) (January 2025): 631, doi: 10.3390/su17020631.

[43] B. Murugesan, K. B. Jayanthi and G. Karthikeyan, 'Integrating Digital Twins and 3D Technologies in Fashion: Advancing Sustainability and Consumer Engagement', in *Illustrating Digital Innovations Towards Intelligent Fashion*, (November 2024): 1–88, doi: 10.1007/978-3-031-71052-0_1.

[44] Y.-A. Lee and S. H. Koo, '3D Printing and Wearable Technology', *Fashion and Textiles*, 5 (1) (December 2018): 34, doi: 10.1186/s40691-018-0155-z.

[45] Theresa Riedelsheimer, Lisa Dorfhuber, Rainer Stark, 'User-Centered Digital Twin Development', *Procedia CIRP*, 90 (2020): 660–5, doi: 10.1016/j.procir.2020.01.123.

[46] E. K. S. Wong *et al.*, 'Supply Chain Traceability through Blockchain', *Green Intelligent Systems and Applications*, 4 (1) (February 2024): 11–28, doi: 10.53623/gisa.v4i1.355.

[47] Tarun Kumar Agrawal, Vijay Kumar, Rudrajeet Pal, Lichuan Wang and Yan Chen, 'Blockchain-Based Supply Chain Traceability', *Computers & Industrial Engineering*, 154 (April 2021): 107130, doi: 10.1016/j.cie.2021.107130.

[48] Liang Li, Fang Su, Wei Zhang and Ji-Ye Mao, 'Digital Transformation by SME Entrepreneurs', *Information Systems Journal*, 28 (6) (November 2018): 1129–57, doi: 10.1111/isj.12153.

[49] Rita Maria Difrancesco, Davide Luzzini and Andrea S. Patrucco, 'Purchasing Realized Absorptive Capacity as the Gateway to Sustainable Supply Chain Management', *International Journal of Operations & Production Management*, 42 (5) (2022): 603–36, doi: 10.1108/IJOPM-10–2021–0627.

[50] Nitin Liladhar Rane, Ömer Kaya and Jayesh Rane, 'Integrating IoT, Blockchain and AI', in *AI for Sustainable Industry 5.0*, Deep Science Publishing, (2024), doi: 10.70593/978–81–981271-8-1_6.

[51] P. Bertola and J. Teunissen, 'Fashion 4.0', *Research Journal of Textile and Apparel*, 22 (4) (November 2018): 352–69, doi: 10.1108/RJTA-03–2018–0023.

[52] Daniel Kerpen, Mario Löhrer, Marco Saggiomo, Maximilian Kemper, Jacqueline Lemm and Yves-Simon Gloy, 'Effects of Cyber-Physical Production Systems on Human Factors in a Weaving Mill', in *Proceedings of the 2016* IEEE *International Conference on Industrial Technology* (ICIT), (2016), pp. 2094–8, doi: 10.1109/ICIT.2016.7475092.

[53] Tesfay Welamo and Deng Sanpeng, 'Application of Artificial Intelligence in Textile Industry', *Journal of Emerging Technologies and Innovative Research*, 8 (10) (2001): 49–54, https://www.jetir.org/papers/JETIR2110307.pdf.

[54] Nuritdin Yuldoshev, Bobir Tursunov and Saidmuhtor Qozoqov, 'Use of Artificial Intelligence Methods in Operational Planning of Textile Production', *Journal of Process Management – New Technologies*, 6 (2) (2018): 41–51, doi: 10.5937/jouproman6–17221.

[55] Yan Hong, Xianyi Zeng, Xianyi Zeng, Pascal Brunixaux and Yan Chen, 'Evaluation of Fashion Design Using AI Tools', (2018), doi: 10.1007/978–81-13–0080–6_12.

[56] Oyebode Kayode, Ayuns Luz and Jude Herry, 'Predictive Analytics for Demand Forecasting in Ethical Fashion: Bridging Sustainability and Smart Technology', ResearchGate, Jun. (2025), doi: 10.13140/RG.2.2.39313.97720.

[57] M. Idowu, 'Machine Learning for Predictive Analytics in Fashion', www.researchgate.net/publication/392966655.

[58] L. K. Nwobodo, 'Importance of Data Analytics in Fashion', *Global Journal of Economic and Finance Research*, 1 (5) (October 2024), doi: 10.55677/GJEFR/04–2024-Vol01E5.

[59] Vivian S. C. Tunn, Ellis A. Van den Hende, Nancy M. P. Bocken and Jan P. L. Schoormans, 'Consumer Adoption of Access Based Product Service Systems: The Influence of Duration of Use and Type of Product', *Business Strategy and the Environment*, 30 (6) (2021): 2796–813, doi: 10.1002/bse.2894.

[60] Anton Gustafsson, Cecilia Katzeff and Magnus Bång, 'Evaluation of a Pervasive Game for Domestic Energy Engagement Among Teenagers', *Computers in Entertainment*, 7 (4) (2009): 1–19, doi: 10.1145/1658866.1658873.

[61] G. Jiang and L. Macintyre, 'Wardrobe Management Apps', *Sustainability*, 17 (9) (May 2025): 4159, doi: 10.3390/su17094159.

[62] Amy Limpitlaw, Michelle Spomer and Suzanne Estelle-Holmer, 'The Life-Changing Magic of Tidying Up', *ATLA Proceedings*, (January 2019), doi: 10.31046/proceedings.2018.116.

[63] David Andersson, 'A Novel Approach to Calculate Individuals' Carbon Footprints Using Financial Transaction Data: App Development and Design', *Journal of Cleaner Production*, 256 (2020): 120396, doi: 10.1016/j.jclepro.2020.120396.

[64] Shaik Vaseem Akram, Praveen Kumar Malik, Rajesh Singh, Anita Gehlot, Ashima Juyal, Kayhan Zrar Ghafoor and Sachin Shrestha, 'Implementation of Digitalized Technologies for Fashion Industry 4.0: Opportunities and Challenges', *Scientific Programming* (2022): 7523246, doi: 10.1155/2022/7523246.

[65] A. M. Țîțu and A. B. Pop, 'CAD/CAM/CAE Systems Implementation', *Proc Int Conf Business Excellence*, 18 (1) (June 2024): 3069–78, doi: 10.2478/picbe-2024–0253.

[66] P. Reipschläger and R. Dachselt, 'DesignAR', *Proc. ACM ISS*, (November 2019), pp. 29–41, doi: 10.1145/3343055.3359718.

[67] X. Qiao *et al.*, 'Inkjet Printing Pretreatment Effects', *Progress in Organic Coatings*, 165 (April 2022), doi: 10.1016/j.porgcoat.2022.106746.

[68] M. A. Shah *et al.*, 'Inkjet Printing Technologies', *IEEE Access*, 9 (21) (2021), doi: 10.1109/ACCESS.2021.3119219.

[69] D. S. T. Parveen Tomar, *Technological Innovation and Sustainability: Navigating the Future through Fashion.*

[70] Fu Jia, Shiying Yin, Li Chen, and Xiaojing Chen, 'The Circular Economy in the Textile and Apparel Industry: A Systematic Literature Review', *Journal of Cleaner Production*, 259 (2020): 120728, doi: 10.1016/j.jclepro.2020.120728.

[71] Binbin Xu, Qing Chen, Bailu Fu, Rong Zheng, and Jintu Fan, 'Current Situation and Construction of Recycling System in China for Post-Consumer Textile Waste', *Sustainability*, 14 (24) (2022): 16635, doi: 10.3390/su142416635.

[72] Weiqiang Fan, Yongzhen Wang, Rulin Liu, Jing Zou, Xiang Yu, Yaming Liu, Chao Zhi, Jiaguang Meng, 'Additive Manufacturing and Textile Waste Recycling', *Environmental Chemistry Letters*, 22 (4) (August 2024): 1–59, doi: 10.1007/s10311-024-01726-2.

[73] R. Adithya, Aditinanda Biswas, K. Anirudh Santhosh, S. Ankita, S. A. Ankita, and Pooja Nagpal, 'Revolutionizing Fast Fashion: How Zara-Inditex Uses Design Thinking to Stay Ahead', *International Research Journal of Education and Technology*, 7 (3) (2025).

[74] Ruojia Li, Wenxin Liu and Sunwen Zhou, 'Digital Transformation of Supply Chain Management in the Fast Fashion Industry: A Case Study of Zara', *SHS Web of Conferences*, 181 (2024): 04019, doi: 10.1051/shsconf/202418104019.

[75] R. Nayak, *Radio Frequency Identification (RFID) Technology and Application in Fashion and Textile Supply Chain*, Boca Raton, Fla.: CRC Press, (2019), doi: 10.1201/9781351238250.

[76] Luís M. S. Dias, António A. C. Vieira, Guilherme A. B. Pereira and José A. Oliveira, 'Discrete Simulation Software Ranking: A Top List of the Worldwide Most Popular and Used Tools', in *Proceedings of the 2016* Winter Simulation *Conference (WSC)*, (December 2016), pp. 1060–71, doi: 10.1109/WSC.2016.7822165.

[77] Chia-Nan Wang, Zhao-Hong Cheng, Phan Nguyen Ky Phuc, Phan Nguyen Ky Phuc, Nguyen Van Thanh and Nguyen Van Thanh, 'Scheduling Optimization Modelling', *Computer Systems Science and Engineering*, 38 (2) (2021), doi: 10.32604/csse.2021.016578.

[78] Woo-Kyun Jung, Hyungjung Kim, Young-Chul Park, Jae-Won Lee and Eun Suk Suh, 'Real-Time Data-Driven Discrete-Event Simulation for Garment Production Lines,' *Production Planning & Control*, 33 (5) (2022): 480–91, doi: 10.1080/09537287.2020.1830194.

[79] A. Dhingra, 'Digital Transformation in Luxury Brand Marketing', *International Journal of Social Science and Economic Research*, 9 (10) (2024): 1–6, doi: 10.46609/IJSSER.2024.v09i10.007.

[80] Teresa Devi Sepoetro, 'The Future of Services in Adidas Retail: Empowering the Creator through Co-Creative Service Experiences', master's thesis, Universidade da Beira Interior, Covilhã, Portugal, (2018), https://ubibliorum.ubi.pt/entities/publication/bc23e172-d4a3-4da9-ad61-c99a78ae84a2.

[81] Serlin Serang, Ardi Mahrifal Hi Taher, Miftahul Chair and Albani Mukti Adrias, 'Global Operations Strategy', *Journal La Bisecoman*, 6 (1) (January 2025): 1–13, doi: 10.37899/journallabisecoman.v6i1.1868.

[82] Amr Fouda, Mohammed F. Hamza, Th. I. Shaheen and Yuezhou Wei, 'Nanotechnology and Smart Textiles', *Frontiers in Bioengineering and Biotechnology*, 10 (August 2022), doi: 10.3389/fbioe.2022.1002887.

[83] A. S. Farooq and P. Zhang, 'Personal Thermal Management Textiles', *Composites Part A*, 142 (March 2021): 1–16, doi: 10.1016/j.compositesa.2020.106249.

[84] Nils-Krister Persson, Jose G. Martinez, Yong Zhong, Ali Maziz and Edwin W. H. Jager, 'Actuating Textiles', *Advanced Materials Technologies*, 3 (10) (October 2018): 1–12, doi: 10.1002/admt.201700397.

[85] Md. Mostafizur Rahman, Md. Abdus Shahid, Md. Shamsuzzaman Rasel, Mohammad Bellal Hoque, et al., 'Introduction to Textiles and Textile Fibers', in *Advanced Technology in Textiles: Fibre to Apparel*, Springer Nature, (2023), doi: 10.1007/978–981–99–2142–3_1.

[86] M. S. S. U. Patwary, 'Smart Textiles and Nanotechnology', *Journal of Textile Science & Engineering*, 5 (1) (2015): 1–7, doi: 10.4172/2165–8064.1000181.

[87] Md Tanvir Hossain, Md Abdus Shahid, Md Golam Mortuza Limon, Imam Hossain and Nadim Mahmud, 'Textiles for a Sustainable Future', *Journal of Open Innovation*, 10 (1) (March 2024): 1–17, doi: 10.1016/j.joitmc.2024.100230.

[88] Charlie Luca, 'The Role of Industry 4.0 Technologies in Textile Manufacturing Process Optimization', (March 2025).

[89] M. Glogar, S. Petrak and M. Mahnić Naglić, 'Digital Technologies in Sustainable Textile Design', *Sustainability*, 17 (4) (February 2025): 1–26, doi: 10.3390/su17041371.

[90] M. A. H. Shibly and M. M. U. Hoque, 'Sustainability Challenges of the Textile Industry', in *Textile Industry and the Environment*, IntechOpen, (2025), doi: 10.5772/intechopen.1009414.

[91] Neetu Singh, 'Circular Economy in Textile and Apparel Supply Chains', in *Innovation and Sustainability Through Circular Economy in Businesses*, IGI Global, (2025), pp. 1–40, doi: 10.4018/979-8-3693–4123-0.ch002.

[92] C. Paul, 'Consumer Behavior and Textile Merchandising', (2024), www.researchgate.net/publication/386142702.

[93] E. Zhuang, 'The Impact of Geopolitical Shifts on Global Trade Networks: A Comparative Analysis', *EDU Journal of International Affairs and Research*, 3 (2) (2024), https://edupublications.com/index.php/ejiar.

[94] N. Arora, B. C. Anant and K. Chaudhary, 'Optimizing Supply Chains in the Fashion & Textile Industry through AI', paper presented at the International Conference on Fashion Apparel and Textile (INCFAT23), (2023), Amity University, Noida.

8 Circular design and upcycling practices

Ali Raza Shafqat, Alberto Saccavini, and Yasir Nawab

8.1 Introduction

8.1.1 Rationale for circularity in fashion and textiles

The linear model of economy that relies on the take–make–waste model is ecologically disastrous, especially in the context of the fashion and textile industry. According to the World Bank, the fashion and textile industry is responsible for approximately 92 million tonnes of annual waste while consuming nearly 79 trillion litres of water every year [1, 2]. The fashion and textile sector contributes to 10 per cent of all the greenhouse gases (GHGs) emitted by industry; this percentage surpasses the combined emissions of the shipping and aviation industries altogether [3, 4]. Such huge carbon imprints not only cross the planetary limits but also introduce inequality among nations, especially between developing countries and developed countries. The rationale for the circular economy in the fashion and textile sector is demonstrated visually in Figure 8.1.

Low-income countries have to bear the consequences of environmental damage and social injustices [6]. Now the circular economy has emerged as a remodification of material systems that relies on integrating longevity, regeneration and reuse of materials with predefined and firmly integrated solutions into the product via preplanned execution right from the product designing phase [5].

The circular economy relies on three pillars for environmental regeneration, as shown in Figure 8.2 and described below:

1. Designing such systems that help rebuild natural systems rather than depleting these resources.
2. Enhancing resource efficiency by improving the material cycles and reducing waste while getting rid of dependence on virgin resources.
3. Introduction to circular business models that retain the value of materials from their end use to foster sustainable economic resilience along with profitability.

Figure 8.1 The rationale for the circular economy in the fashion and textile industry.

DOI: 10.4324/9781003638148-10

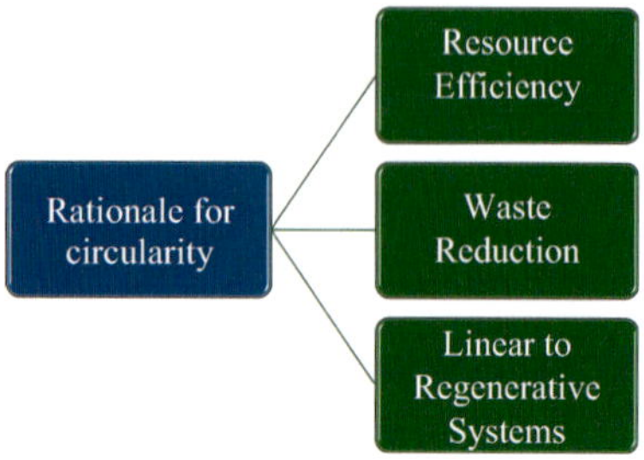

Figure 8.2 Three fundamental pillars of the circular economy.

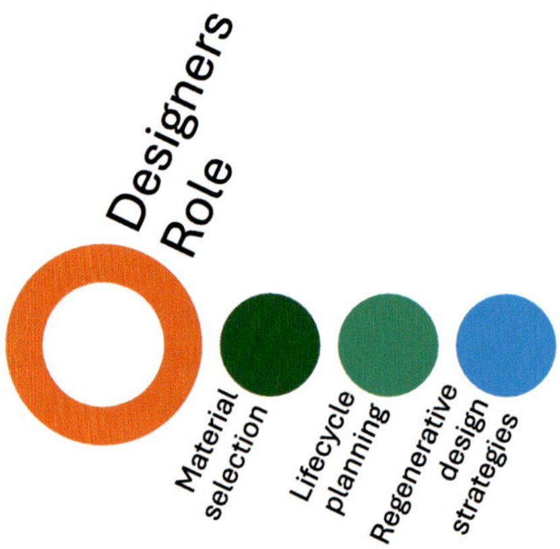

Figure 8.3 Circular economy aligns well with the following SDGs.

These key pillars help enhance the future of fashion and textiles in line with global imperatives of sustainability by getting support from the scientific consensus and tools like life-cycle assessment (LCA) [7].

8.1.2 Positioning circular design within sustainable development goals

Circular design aligns well with global sustainable development goals (SDGs) [8]. It helps meet multiple SDGs as demonstrated in Figure 8.3 and explained below:

- SDG 9 (Industry Innovation and Infrastructure): provide the steps to promote the development of infrastructure innovation and industry within planetary boundaries.
- SDG 12 (Responsible Consumption and Production): This involves efficient use of natural resources and implementation of sustainable material management systems.
- SDG 13 (Climate Action): By focusing on reducing carbon footprint and GHG emissions through recycling materials repeatedly.
- SDG 17 (Promoting Collaborations and Innovations): Promotes collaborations and innovations across multiple disciplines to build circular business models and circular economies.

The Make Fashion and Textile Sector Sustainable, Circular initiative of the Ellen MacArthur Foundation integrates global climate strategies with circular approaches and demonstrates how the material cycles at an industrial level can help improve planetary health [9].

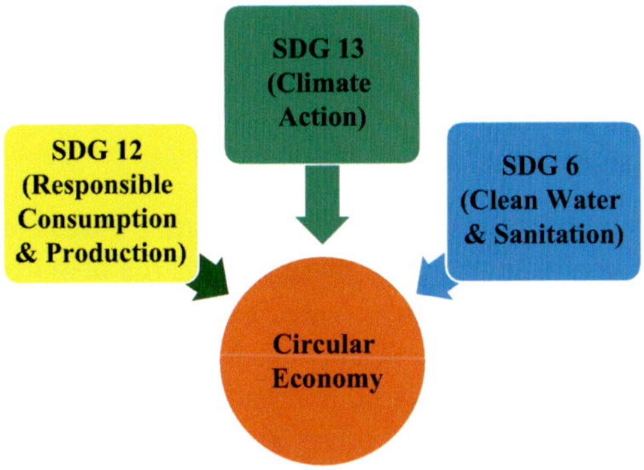

Figure 8.4 How designers can contribute to sustainability in the fashion and textile sector.

8.1.3 The role of designers in creating restorative and regenerative systems

Designers can play a vital role in the circular economy because their choices can influence the material recyclability, longevity and user affiliation [10]. They can contribute to sustainability in multiple ways, as demonstrated in Figure 8.4. The key responsibilities of designers encompass:

- **Embedding system thinking**: Incorporating the complete life-cycle impact of a product from the cultivation of fibre to the end of life of the material. Systems thinking can be implemented using tools like design briefs, digital twins and LCA [11]. The designers must understand the supply chain and influence the systematic flow of materials across it.
- **Rethinking materials**: The designers should select recyclable, regenerative, or bio-based materials that can help enhance closed-loop systems around the products, as they are at the forefront of adopting scientific advancements and innovations like using natural fibres and dyes or the use of recycled polyester, vegetable leather, etc. [12].
- **Enhancing function**: The designers must have the skills and abilities and options to design the garment with repairability, upgradability and modularity right from the very beginning. This can be accomplished by using adaptable sizing options, interchangeable components and other approaches that help extend the life cycle of the garment and enhance the user experience [13].
- **Collaborative creation with communities**: Taking into account the preferences of end users, repairers and local craftsmen, the collaborative design process can be implemented by the designers, which helps enhance the local circular systems and maintains cultural sustainability along with social inclusion [14].

By providing the proof of concept, designers can help the transformation of industry, they can advocate policies like extended producer responsibility (EPR) schemes and can influence the cultural shifts towards the circular economy [15].

Designers can act as super catalysts towards circular transformation. They can achieve this by balancing the ecological restoration along with economic pragmatism and aesthetic excellence. They can weave the very elements of sustainability into the fabric of products, materials and community narratives.

8.2 Conceptual foundations of circular design

To introduce circular design in the fashion and textile industry, the finite and extractive models of regenerative systems should be re-evaluated and converted into businesses that keep materials in a cycle, support resilient economies and reduce textile waste. This whole change needs a paradigm shift from linear design to circular design [16].

8.2.1 Evolution from linear to circular thinking

Since the Industrial Revolution, the product manufacturing industries shifted towards a linear model consisting of take–make–waste in these models the resources are extracted, utilised and then discarded. This model of economy created a devastating environmental situation. The textile and fashion sector alone discards 92 million tonnes of waste while utilising immense amount of resources and consuming nearly 80 trillion litres of water every year [2].

On the other hand, the circular economy offers a great alternative consisting of three principles: (1) keep products and materials in use; (2) reduce waste and pollution; (3) help regenerate natural systems. Many countries, including the USA, EU and China, among many others, are converting the circular economy into policy frameworks, which shows a serious initiative towards the circular economy [17, 18].

8.2.2 Principles of circular economy applied to textiles

The Ellen MacArthur Foundation has done a great job of identifying strategic priorities for the textile and fashion sector with respect to the circular economy. These properties include designing for regeneration, new business-model creation and environmentally friendly inputs [19]. Keeping up with these priorities could mean products are made with less toxicity, while emphasising longevity, natural resources are preserved, there are better textile and fashion services and full use of clothing is ensured before dumping.

Recent studies have shown that by utilising the principles of circular economy in fashion and textiles, it would help not only regenerate the natural environment, but also create economic value, durability, reusability and recyclability in products and help achieve a regenerative model [20–22].

8.2.3 Circular design vs. sustainable design: nuanced distinctions

Circular design and sustainable design (both terms are often used alternatively and broadly) have the same aims, i.e., to reduce harm to the environment. The circular economy specifically focuses on closing resource loops, ensuring that material is rotated in circles through biological and technical material cycles or product cycles [15]. The term 'sustainable design' consists of far greater aims and objectives, for example, culture and social equity, meanwhile helping the environment to regenerate [8]. Circular design often involves material compatibility for recycling, creating new business models for resale, rental, leasing, take-back systems, etc., integrating these into product design thinking, designing the products for longevity and modularity [14].

8.2.4 Role of life-cycle thinking in circular design

The introduction of life-cycle thinking into product design plays a crucial role in validating circular design in the fashion and textile sector [23]. Life-cycle thinking helps designers to learn how the product would impact right from the acquisition of materials to the manufacturing processes, the utility of the product and the end of life would burden the environment.

LCA is one of the methodologies for life-cycle thinking in circular design. It offers three different models: (1) cradle to gate, (2) cradle to grave and (3) cradle to cradle [24]. All these involve assessing the environmental impact, which helped the designers to compare different life-cycle management strategies for the products right from the beginning of our product design. These tools are very helpful and help extend the in-life lifetime of the product and are also effective in end-of-life management and reducing GHG emissions, saving resources and energy.

8.3 Upcycling in textiles and fashion

8.3.1 Definition and terminology

Upcycling is a process that transforms waste, either post-industrial or post-consumer, into products of higher or at least equal value, either in the form of a product function or environmental benefit. This is different from recycling because the recycling process breaks down the material chemically or mechanically into lower volume material, such as raw material and mostly degrades the value of a product and hence true recycling process is also equivalent to downcycling [25].

Recently, many studies highlighted that upcycling helps maintain the footprint of materials, improves efficiency and reduces resource and energy consumption as compared to recycling systems. Product marketing often misuses the term and applies it to recycling of products, mostly dead stock or industrial waste, while ignoring the value integration into the product and due to this media strategy, it dissolves the benefits of sustainability [26, 27].

8.3.2 Typologies and models of upcycling

There are various models of upcycling; any of the models can be applied on the basis of various parameters and availability of waste type and technology. A few ways of upcycling are mentioned in Figure 8.5.

8.3.2.1 Pre-consumer upcycling

The pre-consumer waste recycling is also called post-industrial upcycling and it involves upcycling of products from the leftover garment or other textile materials, such as fabric cut pieces, pattern waste, dead stock garments, etc., into new products. The pre-consumer upcycling

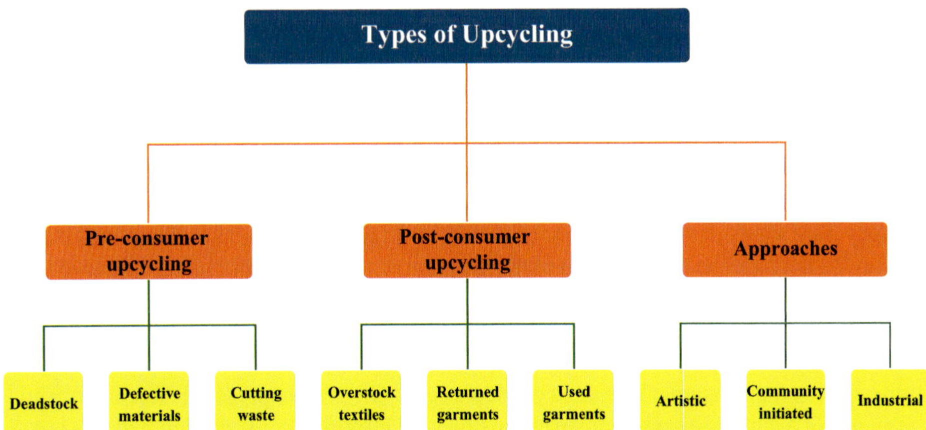

Figure 8.5 Typologies of fashion and textile waste upcycling.

consists of utilising the existing supply of fresh materials within the industry or between different industries without involving the consumer at all [28].

8.3.2.2 *Post-consumer upcycling*

Post-consumer upcycling consists of using second-hand garments, discarded garments, or returned garments. It involves the transformation of fashion garments into new fashion items or repurposing the fashion items into accessories or homeware, etc. [29].

8.3.2.3 *Artistic, industrial and grassroots upcycling approaches*

The approaches of upcycling show a range of approaches, the grass-roots level, the industrial scale or the individual artist. For example, artistic designers can create a new luxury piece from the waste, or an industry or a brand can integrate upcycling strategies on fashion waste materials to manufacture high-value products. Other scale can be community driven initiative, for example using the scrap materials for producing cultural artefacts or using waste for ecological purposes for example Ralli made in Pakistan's Sindh province is one of the fine examples of community driven and cultural initiative [30].

8.4 Circularity and upcycling

8.4.1 Upcycling design in fashion design: a definition

In the fashion and textile sector, the term 'upcycling design' is defined as a process of creatively adding value to the fashion or textile waste, either pre-existing or discarded post-consumer clothing and transforming these into fashion products of superior quality, increased functionality, environmentally friendliness, or epitomising some cultural importance [31]. The upcycling process presents huge development in terms of sustainability in the fashion and textile industry, shifting this sector from a linear model of production, consumption and disposal towards a more circular and regenerative model of economy [32].

Unlike conventional recycling, which typically involves mechanically or chemically breaking down textiles and repurposing them in energy-intensive industrial processes that in most cases degrade their quality, upcycling design focuses primarily on preserving materials and enhancing the original appearance and properties of products or materials, thus avoiding the impact of downcycling processes and adding more value to waste.

Upcycling is more of a cultural or ethical action rather than a mere technical process. The upcycling process challenges the conventional paradigms of mass consumption, cheap fashion and planned obsolescence. Upcycling design reimagines waste as a resource for innovation, storytelling and artistic expression rather than creating pollution [33]. Therefore, in embarking in the upcycling route, designers mainly rely on craftsmanship, material memory and individuality. They do not hide the previous life of the garment or the product but rather preserve the visible textures, prints, wear patterns and the visible traces from the previous life as key elements of the aesthetic and conceptual value of the product. By following this approach, they not only provide aesthetic and conceptual value to the product but also provide a deep emotional connection between the consumer and, hence, encourage them to have personal attachment and use the garment for a longer life with more responsible and informed consumption patterns [34, 35].

The upcycling of fashion and textile products is a part of broader circular economy movements, the product consumerist theory and regenerative design. It advocates for small-scale production, local manufacture that reduces the environmental impact, and it empowers local artists and communities through reusing of discarded products creatively. It helps in enhancing social responsibility, ecological stewardship and economic resilience in communities. The implementation of upcycling design in the fashion and textile industry forces us to rethink the relations between identity, label materiality and repurposing of waste, meanwhile recognising the potential within the material [36, 37].

From centuries-old historical traditions, clothing production has a significant involvement of craftsmanship, resources and manpower and are therefore treated as precious objects in social and domestic economies. [38]. Every culture focused on preserving the value of garments for as long as possible, often constructing the garment with hidden fabric seams that would allow the garment to be resized easily as the human body transforms over time. Furthermore, it could be historically seen that women's clothing had more generous seam allowances to allow alterations of the garments during pregnancy [39]. The garments are often seen to be shared among cousins, siblings and multiple generations. The old garments were adjusted, adapted to new fashion ideas and extended their functionalities and aesthetic appearance over time. This demonstrates that the centuries-old historical value of textiles is expressed through maintenance, repair and alteration, not as a sign of poverty, but as a symbolic respect towards the value of textiles [40, 41].

From a historical perspective, circular practices in the fashion and textile industry circular practices in the fashion and textile industry can be seen as evolving not only from material consciousness but also from necessity and craftsmanship, material awareness but also from necessity and craftsmanship, providing historical context and cultural importance for understanding upcycling design [42]. This reveals that modern valorisation of reuse and adaptation of garments is seen as innovative or disruptive, but in fact it is historically rooted in the traditional materials stewardship, now read and contextualised in new terms of sustainability, environmental challenges and socio-economic concerns [43]. Therefore, it is important to keep in mind that upcycling is not a new technique, but rather a culture, an artistic language and a design philosophy that also demonstrates social, economic and environmental commitment. Upcycling design reframes the use of discarded fashion and textile items not as the end point but the beginning of a new life as a catalyst for resilience, creativity and sustainable and circular transformation in the fashion and textile sector [44].

8.4.2 From necessity to innovation: a brief history of upcycling in fashion

Although the term 'upcycling' seems to be a fancy and new term, in practice the reuse of textiles by creatively transforming them into a new product has deep historical roots within many cultures before the clothing manufacturing industrialisation. Fabric manufacturing was a very labour-intensive and expensive commodity [45]. In pre-industrial societies, garments were carefully cared for and were mended many times or refashioned before being broken down into smaller pieces to be used in children's clothing or utility items as mops, cleaning clothes, etc. Especially during the wars and times of economic depression, material shortages were normal, and upcycling design became a widely diffused response to the scarcity of resources [46]. Official campaigns, like 'make do and mend' during the Second World War, fostered a culture that creatively promoted repairing and reimagining the fashion and textile products rather than disposing of them [47].

After the Industrial Revolution, the mass production of fashion and textile products became a global norm, and the supply chain globalisation and consumer capitalism introduced a disposability-oriented mindset that replaced creative and sustainable approaches like upcycling design [48]. All these gave rise to fast fashion and the consequent generation of an immense quantity of textile and fashion waste, which became an inevitable by-product of industrialisation. The upcycling has now re-emerged not merely as the continuation of historical practice but as a crucial act of sustainability [49].

8.4.3 Early pioneers of contemporary upcycling fashion

Orsola de Castro and Filippo Ricci, in 1997, founded 'From Somewhere' to make upcycling design the brand's core activity [50]. They recognise a huge amount of surplus fabric produced by the fashion industry and try to counter this waste stream from getting incinerated or landfilled. Therefore, they used pre-consumer textile waste production, fabric offcuts, unsold garments, misprinted fabric, or production of cuts, etc. to manufacture highly designed collections. Their work became prominent because of sharp tailoring, intentional exposure of patchwork and playful use of colour in upcycled construction that added aesthetic value to the garment [51]. In 2010, a major project came to light when Speedo requested to From Somewhere to produce a small collection from unsellable LZR Racer swimsuits originally designed for Olympic-level competitions. From somewhere sliced, bound and reconfigured the complex high-performance swimmer suits into jackets, dresses and accessories, and hence they demonstrated that specialised synthetic textiles are not impossible to be recycled or upcycled but could be imaginatively recontextualised [52]. This collaboration opens a room for forward-thinking approaches and enhance the possibilities of raising awareness and mitigate the volume of waste generated, even from niche-specific garments.

In early 2000s, Nin Castle founded Goodone by adopting a scalable and strategic approach towards upcycling. Goodone combined systematic thinking process with aesthetic innovation and established a factory model that sourced and sorted pre-consumer textile waste and rooted it towards structured textile production processes. Instead of relying solely on the unpredictability of the material availability, they developed a design process that could be adapted to any level of the input material, hence creating a commercially viable and visually cohesive collection. Their distinctive approach characterised by colour-blocked patterns, contemporary silhouettes and careful craftsmanship challenged the current limitations of upcycling design. Their collections have reached high-end retailers, won recognition from the media and become recognised as competitive with conventional brands rather than being limited to niche environmentally friendly boutiques [53, 54].

Junkie Styling, founded in 1997 in London by Annika Sanders and Kerry Seager, provided a new level of rebellious and avant-garde energy to the upcycling design movement. They specialised in the disassembly of post-consumer garments, especially menswear and transformed them into gender-fluid designs. Their garments prominently celebrated a combination of new and old, often consisting of recognisable pieces from the old garments to expose the radically altered forms [55]. It was not just any other brand but a participatory practice that invited the customers, through their 'Wardrobe Surgery', to collaborate in the design process and provide them with their old and unwanted clothes. This democratised the upcycling process, showing the urgency for sustainability in fashion.

Other significant designers and brands contributed to the legitimisation of upcycling design. Martin Margiela's designs from the late 1980s and 1990s, for example, used techniques of deconstructing and reconstructing ready-made garments, anticipating what is the contemporary upcycling design. Through reconstructed denim jackets, dresses made from vintage gloves and repurposed military wear, Margiela engaged in an intellectual and critical manner with the value and history of material.

Rei Kawakubo of Comme des Garçons also explored imperfection, asymmetry and repurposing in Japan, in line with those principles that only in the late 2000s will describe upcycling fashion design. Although not explicitly branded as 'upcycling', these early experiments, even though not presented and upcycling design practices, contributed to standardise the reuse and transformation of material thanks to the creation of sophisticated designs.

Recently, large fashion brands and platforms like Patagonia Worn and Wear and Eileen Fisher Renew have formalised upcycling design within corporate-level frameworks, which showed a significant the shift from small-scale practice to a broader acknowledgement at the industrial level [56] demonstrating that with proper strategic vision and infrastructure involvement circular economy can be implemented effectively within the fashion industry [57].

In 2013, Patagonia launched Worn and Wear, introducing the opportunity for customers to their garments repaired, trade them or purchase refurbished ones. The ReCrafted programme created a collection from unrepairable garments by transforming them into completely new pieces and creating unique and durable garments, often including parts from multiple items [56]. This effort not only reduces the carbon footprints and environmental footprints of the company but also helps structure its brand identity, highlighting products' transparency, durability and consumer responsibility, ultimately encouraging the shift towards longevity instead of disposability.

Similarly, the Renew initiative by Eileen Fisher, initiated in 2009, invited customers to return their waste garments in exchange for store credit. The garments, when possible were repaired, cleaned and resold, otherwise upcycled, through their Resewn programme, into completely different designs. Similarly to Patagonia's effort, Elieen Fisher did not claim it as a mere recycling effort but demonstrated it as a voluntary commitment towards ethical consumption and circular economy [57]. They maintained high quality and design standards in their upcycled products and challenged the assumption that upcycling fashion is not always inferior or compromised aesthetically as compared to the new product.

These last two are early examples of institutionalising upcycling design in the corporate sector. Applying a strategic vision and the inclusion of infrastructural investment demonstrates that circular practices can be effectively integrated into mainstream retail models.

8.4.4 The social and cultural implications of upcycling in fashion design

Upcycling is not only about material reuse, but it has far-reaching consequences beyond social and cultural perspectives. Upcycling used clothing and garments conceptually transforms fashion and textile waste in a source of creativity rather than a sign of failure. The use of upcycled fashion encourages the consumer to have a deeper connection with the garment and model the practices of longevity, personalisation and utmost care that are of little relevance in the culture of mass consumption [58]. Upcycling design promotes the respect and valorisation of craftsmanship and labour within the fashion and textile sector. In an industry often associated with exploitative labour practices upcycling design could help in experiencing esteem for the labour and the effort put into manufacturing these items, highlighting techniques such as reconstructing, restitching and patching as source of special meaning and authentic value [59]. By upcycling the textile items, the conservative norms can be challenged by heterogeneity, hybridity and imperfection, rather than sticking to the ideas of perfection and newness. Upcycling design raises awareness about the material memories and irregularities layered within the history embedded within reused textiles. This sense aligns with the culture shift towards an anti-waste approaches and sustainability, in which adaptation and imperfection are increasingly valued.

Upcycling can also serve as a form of activism and resistance by diverting the sector waste, often directed exclusively to landfills and incineration, towards new and possibly political value creation. Today global campaigns like for example Fashion Revolution, co-founded by Orsola de Castro, link upcycling design practices to the demand for systemic change in the fashion and textile industry, while advocating for social justice, accountability and transparency [60]. Upcycling design is now not a matter of technical and aesthetic innovation but represents an ethical shift and involves a broader culture that invites all human beings to consider and reimagine our relationship with environmental resources, clothing and labour.

8.4.5 Contemporary practices and future directions of upcycling design in fashion

8.4.5.1 *The expansion of upcycling design across fashion sectors*

The 21st century transitioned upcycling within the fashion and textile sector into a more diverse and broader field of practice from the historically confined do-it-yourself culture and creative artisanal workshops. Starting in the early 2000s upcycling design gradually gained credibility and legitimacy in the fashion industry, and evolving this approach from the small-scale independent artisans and designers, directly reclaiming used clothing and thus creating unique garments having legitimate and distinct design approaches [61].

Upcycling design was at first adopted by the elite fashion houses, reinterpreting it through couture craftsmanship. These interpretations of upcycling design transformed it into a high fashion form of creativity, rather than presenting it exclusively as an ethical and pragmatic gesture. This approach helped redefine the perception of upcycling design from a sign of necessity and scarcity to a symbol of exclusivity and innovation [62].

Due to user-driven subcultures, upcycling gained great popularity, in which the small and independent designers made clothes that clearly comprised the signs of previous existence, for example, hybrid jackets, patchwork T-shirts and reconstructed denim, which celebrated the anti-mainstream, aesthetic, authenticity and individualism. These concerns aligned upcycling with reclaiming personalised consumer identity and shifting away from homogenised mass production [63]. The do-it-yourself culture, strongly embedded in punk, hip-hop and grunge styles echoed the upcycled design in which the garments became a mode of cultural commentary and self-expression. These efforts made upcycling an appealing idea rather than the mass-market strategies. Some high-end brands introduced upcycling capsule collections, sometimes consisting of collaborations between independent small craftspeople and designers, to show how important sustainability and artisanal products are [64]. These initiatives, different sincerity and depth, signalled that the industrial reuse of materials is no longer marginal or impractical. Upcycling design now spans not only from experimental ideas to urban-influenced styles, luxury pieces of garments and corporate initiatives that try to integrate sustainability and reuse of materials into conventional retail models. Upcycling has introduced new authenticity, artisanal values, scalability, creativity and systemisation, along with commodification, which would continue to shape the evolution of the fashion and textile sectors [65].

8.4.5.2 *Upcycling design in contemporary fashion: key approaches and trends*

Upcycling design in the fashion industry has evolved to incorporate early artisanal crafts or grass-roots activism in today's business practice. It consists of innovation in a dynamic field that combines multiple sectors of the fashion and textile industry, ranging from social enterprises and corporate initiatives to experimental and luxury items [66]. Several case studies can be seen in this field, which show how upcycling has evolved to utilise and reinterpret fashion and textile waste in multiple contexts and communities and how they utilise it through their own economic, cultural and aesthetic implications.

Upcycling design is now a widely accepted approach, and it is often termed as a creative approach with great cultural significance. Some designers have used material regeneration as a core principle of their design process while combining vintage sportswear, reclaimed textiles, deadstock fabrics from industrial processes, and household linens to create hybrid and futuristic styled garments [67]. Despite creating the upcycled garments as an exceptional gesture, they have seamlessly used them as aesthetic and for projecting a sustainable vision for the future, while normalising waste as a resource rather than as refuse. One such example can be seen from the work of Marine Serre, who had adopted upcycling and created it as a source of innovation, crafted storytelling identities and demonstrated the urgency of sustainability for conserving the natural environment [68].

Walid Al Damirji, through its brand By Walid, also used upcycling to produce unique luxury pieces from antique textiles like tapestries and linens from the 18th and 20th centuries. The garments made from aged fabric celebrate the value of aged textile items. Each of his products carries visible traces of the original life, which offers the wearers a connection to history layered into a tangible garment. His approach challenged the traditional aspect of luxury, conventionally being perfect and new, proposing an upcycled product with emotional resonance, irregularity and a patina [69].

Upcycling has gained a lot of attention in the recent decade. For example, a British designer, Bentley Williams, example of a new generation of fashion designers and practitioners, for whom upcycling design is an act of social activists by collaborating with social enterprises and marginalised communities. She collected waste textiles and integrated social narratives directly into her product line. Through projects, such as Breadline line that address food insecurities in the UK and 'Women of Change' that celebrated the stories of incarcerated women [70]. Her work proves that upcycling can be used as a tool to enhance environmental responsibility, help meet social justice issues and eventually broaden the scope of ethics in the fashion and textile industry.

Recently, upcycling design has been institutionalised into corporate culture and frameworks of the industrial sector, which has created a significant change from its origin, i.e., artisanal practices. Now it has been adopted as a part of the fashion industry formally and is now integrated into broader production systems, which demonstrates its cultural, economic and environmental scope. Estonian designer Reet Aus demonstrated exceptional work in this field by collaborating with Bangladeshi manufacturing firm Beximco and by creating a certification system known as Upmade. Upcycling systems mostly rely on manual techniques using small-scale approaches focused on unique items. But Reet Aus took it to a whole different level by creating a systematic process on an industrial scale that reduces the waste generated by the fashion and textile sector. They incorporated the waste into manufacturing processes to prepare new products even before leaving the factory. This process and certification not only reduce the environmental impact but also prove to be a viable framework for manufacturers in the global South, which is the hub of fashion industries' waste production. This work demonstrates how upcycling can be scaled above artisanal and boutique levels, and it challenges the conventional waste-producing textile industrial processes [71].

In a similar effort, the New Zealand Post in collaboration with Space Between counters the niche-specific waste, for example, uniforms and transforms them into new fashion products. Their work demonstrated how old niche-based bulk quantity garments can be used to prepare high-quality and difficult-to-recycle products. They produced reconstructed garments as limited-edition pieces that preserved the culture and the history of the original uniforms. Their approach offered an alternative, for institutions that face great environmental risk, to rethink textile waste as a resource rather than as a grappling liability [72].

RE;CODE, a Korean brand, founded by Kolon Industries, that incorporated upcycling strategies in mainstream fashion, transformed the deadstock and outdated garments into new contemporary collections contrasting the issue overproduction in this way. They also collaborated with other designers and automotive companies to present upcycled products of cultural and fashion significance. They proved how upcycling design can have commercial practicality beyond a sustainability niche customer base [73].

All these initiatives collectively show that upcycling is not confined to limited-scale feasibility but can also be implemented in industry on commercial, institutional and corporate scales. This transformation shows the urgency of waste mitigation in the fashion and textile sector and the growing interest in the legitimisation of upcycling as a key production method. Upcycling design has proved to be a contemporary approach which shows significantly broad adaptability and practicality in different settings and across the different geographic locations from New Zealand to Bangladesh and from Global North to Global South, upcycling has been widely adopted and it reflects the broad cultures that celebrate the history, social and environmental awareness, ethical responsibility and resourcefulness of textiles, meanwhile highlighting the solution to challenges in adoption of sustainable materials.

8.4.6 Industry applications and case studies of upcycling

8.4.6.1 Designer brands

The number of case studies can be seen for upcycling examples in the fashion and textile industry, for example, Re/done that transforms old denim into modern style denim pants, Re/done by Sean Barron and Jamie Mazur that upcycles the used textiles into narrative-rich products with the help of social enterprises, or another example can be Marine Serre that we consume the surplus fabric and the manufacturers' ready-made garments [74].

8.4.6.2 Mass-market and SME strategies

An example of emerging labels can be taken from ELO, the export leftover of a Pakistani start-up that resells the dead stock from the local industries and sells it to the community at marginal prices, helping conserve the environment while producing value and providing the public with low-priced garments [75]. Another startup of Pakistan, the Recomp Solutions, converts dead waste from the textile and agricultural sectors into ceiling tiles and wall panels, helping the recycling of waste that would otherwise be incinerated or landfilled, but also makes it possible to manufacture fashion and textile products to meet the net zero waste target.

8.4.6.3 Community-led upcycling and maker culture

Many community-led practices can be observed, for example Visible Mending Project based in the UK, which provides a second life to a garment, while plastic weaving collectives in Dharavi, India, repurpose plastic waste into new products. Upcycling helps achieve sustainability by reducing the use of energy and water, as compared to through cycling. It helps preserve the integrity and the value of the materials, meanwhile empowering the community's designers and the brands to achieve a circular economy and heal the environment by reconstituting the waste as a creativity and not as a burden on the environment [76].

8.5 R strategies for circularity

Strategies of embedding circularity into textile and fashion products can be understood by the circularity framework consisting of 9R strategies. These 9R strategies build upon the three-stage model that explains the strategies from closing the loop towards recovery.

8.5.1 Overview of the 10R framework

The first strategy and the most important one is reduced by design, which is the guiding principle and helps the designers to completely exclude or reduce the linear model intrinsic into the product right from the design phase. The 9Rs of circular economy are visually displayed in Figure 8.6. The 9Rs of circularity and the tenth are often considered altogether and are given below [77].

R1 **Refuse**: Refuse is the first strategy to implement circularity, which involves the elimination of non-environmentally friendly or unnecessary materials from the products at the design stage; meanwhile, it also involves educating consumers to refuse products that are unnecessary or non-environmentally friendly.

R2 **Rethink**: Rethinking means altering the product system or concept to enhance the resource efficiency or the resource utility during product manufacturing.

R3 **Reduce**: Reduce means to limit the use of virgin raw materials or minimise the raw material energy consumption and the intensity or the number of processes.

R4 **Reuse**: Reuse involves circulating the product in its original form multiple times without alteration or repair.

R5 **Repair**: Repair involves fixing the product or enhancing its maintenance to extend its utility life.

R6 **Refurbish**: Refurbish means restoring the full functionality of a product by incorporating upgrades or servicing the components.

R7 **Remanufacturing**: Remanufacturing involves building the product again using the used components or parts, meanwhile achieving the original specifications.

R8 **Repurpose**: Repurpose means the transformation of the product into a product of more value or at least equal value.

R9 **Recycle**: Recycling means recovering the raw materials from the used product so that the reclaimed material can be used to manufacture new products.

R10 **Recover** (or incineration): Sometimes counted as the 10th R strategy, this strategy involves extracting energy from the product of the materials at the end of their life, often by using biological are thermal methods.

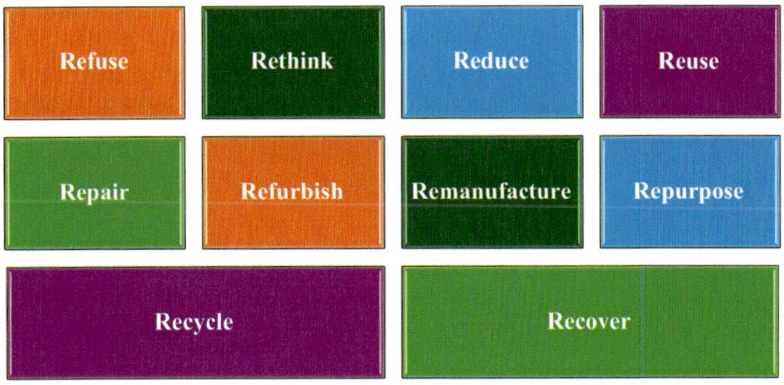

Figure 8.6 9Rs of circular economy.

8.5.2 Integration of R strategies into design briefs and product development cycles

It is a commonly accepted view that the first two strategies have the most potential to improve the environmental impact of the products by eliminating inefficiencies. These strategies help prevent waste while designing the product. For example, designers can incorporate modularity while designing the product and so each component of the product can be recycled individually rather than the whole product [78].

The next two strategies are then focused on the longevity of the product; for example, a designer may think of reuse while creating the garments that can be shared as a service or can be altered in simple steps by incorporating redesign options. The repairability can be embedded into the product by incorporating the modular parts or stretching the fashion product in a way that makes it easier to repair [79]. The aesthetics of the garment can be improved by easily replaceable hardware or parts like closures, buttons, etc., or the parts that are prone to fading or common wear and tear, for example, the cuffs and the collars.

The next strategies are more focused on industrial fashion production or enhancing the durability of the product, for example, providing reconditioning services to the worn-out garment and remaking it to achieve the standards of the original product. Remanufacturing can be conducted, whereas turning the products into another distinct product, like manufacturing bags from the leftover fabric from the apparel manufacturing process. Reclaiming strategy can be used to recover the material, either by chemical or mechanical processes, which is often more relevant to the products having a single material construction [80].

The last strategy, the 10th R, is different from all other strategies, and it is also the least preferred strategy, which usually halts the circular flow of material and fails to generate new products, but rather helps recover the energy from the products that are deemed inefficient or impossible to recycle in the circular life cycle [81].

8.5.3 Mapping R strategies onto fashion and textile use phases

The strategies of the circular economy can be implemented right from the concept, the use of raw material, the production, distribution, use, ownership and the end-of-life management of the product. For example, the reduce by design principle can be implemented to avoid the environmentally harmful raw materials in product development process. Utilising the regenerative or recycled fibre and maximising the yield of the production process as described by R1 and R2 can be effective. Facilitating the garment as a service for integrating the modular design for auxiliaries or closures or including the repair kits with the garments during the manufacturing and distribution phase, this can be helpful for consumer to self-repair their garment during the product utility. Another such example can be including instruction manuals for repairing of product or offering after sale services like maintenance or remanufacturing, all these strategies can be introduced to incorporate the circular economy concepts into product. At the end of life the repurposing of waste fabric or recycling of pre-consumer, post-industrial or post-consumer waste can be helpful to keep the material in loop. If none of this remain viable the last and the least preferable strategy involves the extraction of energy from the material. Energy recovery is all though the least preferable route but it at least helps recover the minimal value from the material or the product [82].

Many examples can be taken from businesses that illustrate the utility of the circular economy in their operations. For example, an accessory brand that transforms the cut pieces into patchwork pieces, a business that remanufactures the returned clothes and resells them by certifying them as refurbished, whereas many businesses recycle cotton fibre from the textile waste and provide it as raw material for new product manufacturing. It is commonly accepted that businesses that implement multiple circular strategies will be more effective, for example, preventing waste, extending the life cycle of the product and preparing the material to be recycled altogether [83].

8.6 Design for disassembly

Designing the fashion and textile products with a clear picture of the end-of-life stage of the product is very important for efficient reuse, repair and recycling. Design for disassembly (DfD) helps product management during the life cycle, especially ensuring that the parts of the product can be taken apart, separated, or replaced, hence improving the repairability throughout the life cycle [84].

8.6.1 Importance of disassembly

A clear picture of disassembly is very important for perfect circular product design. The disassembly helps in facilitating the repair and maintenance of a product. For example, integrating a design that helps fetch zippers or buttons to be replaced when they become faulty or damaged, therefore omitting the need to replace the entire product. Figure 8.7 demonstrates the importance, examples, strategies and tools used to incorporate DfD in fashion and textile products. The disassembly of parts helps in separating different kinds of materials, for example, the separation of cotton fabric parts from polyester or nylon parts, or polyester parts from leather and hence enhances the material recovery at the end of product life [85]. The disassembly can also help in introducing modularity into textile and fashion products so that the garment or any other product can be more adaptable to changing fashion trends and help users to meet their needs for a longer time, making the garment relevant over a large span of time, even with changing fashion trends [86]. Hence, the concept of disassembly aligns very well with circular economic principles, especially in fashion and textile products, so that they can maintain and retain their maximum value over a long useful life [87].

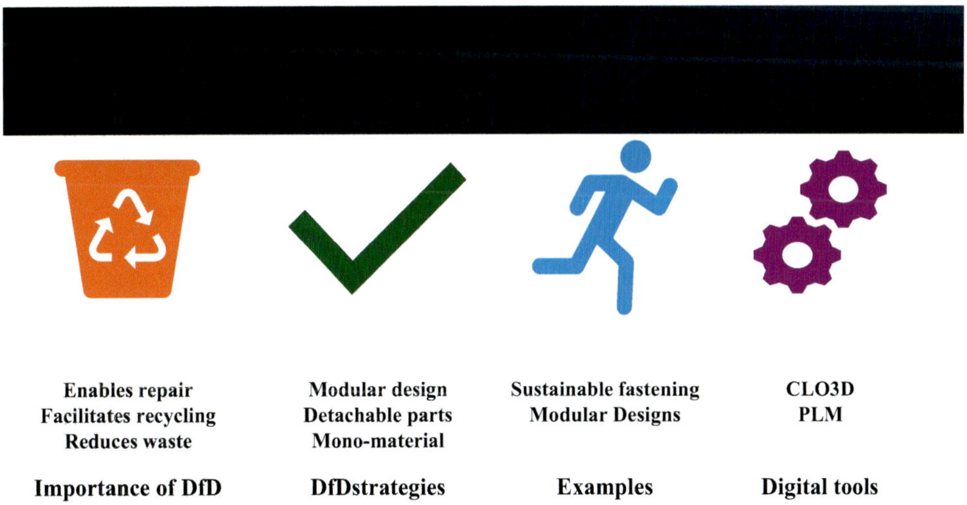

Figure 8.7 Importance, examples, strategies and tools used to incorporate DfD in fashion and textile products.

The design for this assembly has various strategies [85, 86, 88], as given below:

- **Using a mono material composition.** Using only one type of material, for example, single-fibre textile and fashion products, can help in the recovery of material at the end of product life. The same mono materiality concept can also be applied to fabrics with a single colour composition, so that the use of chemicals and bleaches can be avoided during the recycling process.
- **Detachable or seamless product construction.** Using detachable or seamless construction techniques, for example, snap fastening or ultrasonic welding, can help eliminate the need for stitching, which helps in the separation of parts for easy replacement or recovery of materials at the end.
- **Robust fastening systems.** Incorporating the fastening elements designed with the core intention in mind so that they can be removed before recycling without a significant effort can help achieve circularity and enhance the repairability, maintenance, servicing and end-of-life recovery of material, which ultimately helps in achieving the circular life cycle of a product.

8.6.2 Modular designs

Modular design means constructing the fashion industrial products in such a way that their components can be separated and rejoined so that they can become interchangeable, for example, the callers' sleeves or linings that can be detached and easily remade, customised, or repaired, thereby extending the product life cycle and also helping keep up with fashion trends [86].

8.6.3 Digital integration

Designing the software or tools to help designers analyse material complexity, recyclability, assembly, or manufacturing process can also easily help in introducing circularity within material processing or product manufacturing. Some examples are BLM and CLO 3D software. Multiple case studies can be taken as an example of the circular economy in the textile and fashion industry. Many of these are practical and thriving on an industrial scale, for example, Patron's sportswear that joins hood and liners by snap-on with regular jackets that help enhance the lifespan and therefore increase garment efficiency by using modular jacket designs. Another such example is demonstrated by ZKK zippers that have features like easy and fast removal zippers, therefore helps in recycling processes. In another example, it was demonstrated that changing the design of the denim with detachable waistband buttons and reinforcement helps in refurbishing or repairing [89].

8.7 Closed-loop textile systems

Closed-loop fashion textile systems involve bringing the post-industrial and post-consumer products and materials back into the loop, which helps in reducing the reliance on virgin raw materials and maximises the use of waste in the production of new products. Such systems are truly circular and are important because they promote biological and technical material cycles to keep the materials in a loop for as long as possible [90].

8.7.1 Defining closed-loop systems

There are multiple types of systems, for example, biological systems in which natural or biodegradable fibres are used to manufacture a product, so they decompose or degrade at the end of

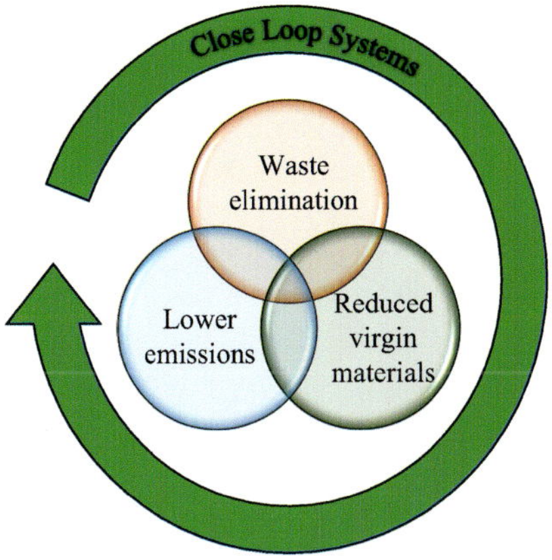

Figure 8.8 How closed-loop systems help align processes and products with the circular economy.

their life and return the nutrients to the ecosystem, as displayed in Figure 8.8. Therefore, these materials help in reducing the pollution and waste problem. Conversely, the technical loops recover the material at the end of the product life. The materials can include coatings, synthetic or natural fibres and metals, etc., that can be reprocessed into new products by chemical or mechanical recycling [91]. The closed-loop systems are not merely just the name of a process, but they use a combination of vast systems that are heavily dependent on materials compatibility and design for end-of-life recovery. For example, using mono material in product construction, the collection and sorting infrastructure that helps in collecting and bringing the material back into the loop, the capital initiatives and policies that help in maintaining a system that forces the brand's manufacturers or other stakeholders to ensure the viability of material recovery operations, either through legislation or consumer-driven approaches [85].

8.7.2 *Industry initiatives and technologies*

There are several initiatives that strive to implement closed-loop manufacturing, some of which are visually displayed in Figure 8.9 and discussed below.

8.7.2.1 *Fibre-to-fibre recycling technologies*

INFINITED FIBRE COMPANY OF FINLAND

Infinited Fibre Company recovers the textile product and produces cellulose from the waste material. They manufacture human-made cellulosic fibres and hence keep the material in a closed loop through a biochemical process. These processes are efficient in maintaining the fibre length and, therefore, the strength of the produced item [92].

EVRNU USA

Another example can be taken from Evrnu USA, which uses various enzymatic treatments to recover polymer and regenerate high-performance fibres, meanwhile maintaining resource optimisation and complete transparency of the process [93].

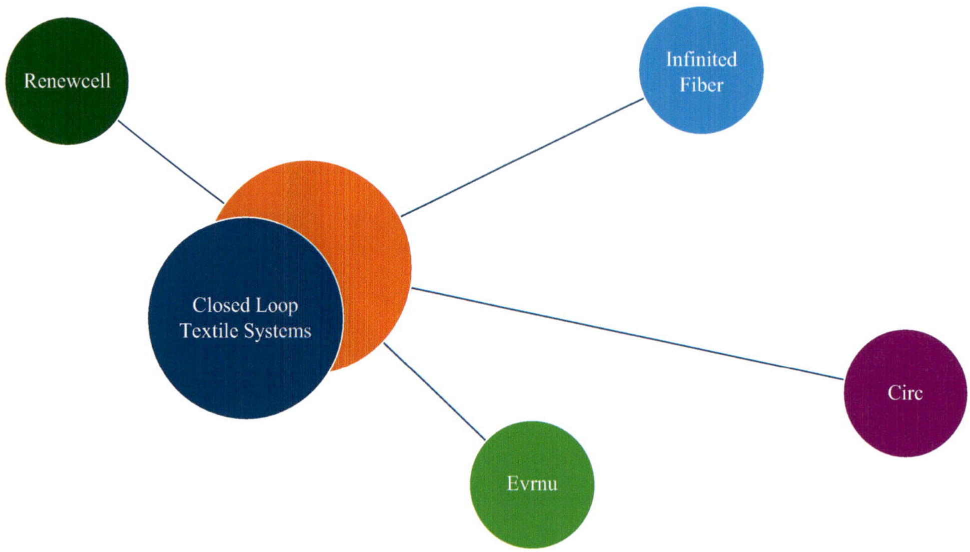

Figure 8.9 Initiatives that work towards closed-loop systems in the fashion and textile sector.

RENEWCELL OF SWEDEN

Renewcell recovers cellulose from cotton waste through a chemical process, which is then used as raw material for viscose rayon production. Their process not only reduces the cotton waste but also adds economic benefit, along with the reduction in deforestation for wood pulp required for viscose production [94].

CIRC NETHERLANDS

Circ Netherlands uses a mechanical recycling process to recover fibres from various blends and prepare fibre mats. The mechanical process not only extracts valuable raw materials but also ensures that no harmful chemicals are used during the process [95].

8.7.2.2 Take-back schemes

Several brands are now offering refurbishing and take-back schemes. For example, Patagonia offers repair and resale services, and these initiatives not only add longevity to the product but also enhance customer loyalty. Their programme, Worn and Wear, helps reduce GHGs by up to 75 per cent as compared to other products, according to a study. Another brand, H&M, has an in-house store-based facility where the customers can drop off their textile and fashion items, which are later recycled mechanically, all used to produce insulation panels. The North Face offers refurbishing services to the damaged product within 90 days of purchasing; their offer increases the revenue and maintains high-quality standards [96–98].

8.7.2.3 Platform-based circular business models

There are multiple circular platforms operational around the world. One such example is Lindström very well established in Denmark, offering a subscription service to provide repair, maintenance and rental services of workwear items. Their services our focused on the professional life cycle of a product, meanwhile helping in maximising the lifespan of the product [99].

8.8 Remanufacturing as a scalable mechanism for circularity in fashion

Remanufacturing is one of the key strategies for the advancement of circularity in the fashion and textile industry. It provides a scalable and practical alternative to the conventional production method. Recycling mostly involves breaking down materials through energy-intensive processes or chemicals and water, but remanufacturing directly reintegrates the waste from the production into new garments without much processing. The remanufacturing approach extends the use of material, reduces the dependence on virgin material inputs it also is compatible with the existing industrial systems and hence making it a viable solution for mass production that helps bridge the gap between manufacturing realities and sustainability objectives. And by reusing the additional material generated during the manufacturing process or diverting the waste from disposal towards the input materials, efficient resource utilisation can be achieved along with greater applicability. Remanufacturing offers a model that is environmentally impactful and operationally feasible at the same time [84].

Reverse Resources can be taken as an example, providing a structural model for remanufacturing in the fashion industry. Their model highlights the importance of reusing the leftovers from the production processes of textile and fashion goods and uses these materials to remanufacture new items, meanwhile reducing the material waste that prevents circularity. Their model relies on three different remanufacturing methods that can be implemented on the same systems with proper efficiency and standardisation [100, 101].

In the first method, they use the leftover fabrics in non-visible garment components like linings, interfacing and other structural supports. This approach allows them to use the waste materials without even showing them on any external parts of the garment. This integration is suitable for high-volume productions as well.

In the second strategy, they use leftover materials for minor features that are often used for contrasting or introducing complementary details. Although these material utilisations are visible on the garment, they do not disturb the overall aesthetics, therefore enabling proper adherence to the original design of the garment.

In the third approach, they use design-led manufacturing by incorporating recycled waste into design processes. In this approach, new surplus material is introduced during the design stage, which enables consistent use of this material while ensuring proper design production scalability and uninterrupted industrial production at the same time.

These remanufacturing methods offer a range of benefits, like reduction in reliance on virgin raw materials and enhancing the longevity of materials that would otherwise have been downcycled without more demands for technology or capital cost. With this approach these materials are pre-segregated within the production facility and re-channelled as material inputs using proper design strategies. Remanufacturing not only provides environmental benefits but also provides an opportunity to have transparency in fashion and textile supply chains, which is one of the key barriers limited by the non-availability of data, non-visibility of production facilities and lack of valuation of waste materials. Implementing remanufacturing can improve inventory management and data sharing can incentivise collaboration and accountability between suppliers and buyers.

8.9 Textile circularity pathways: fibre-to-fibre recycling and industrial symbiosis

The management of post-industrial and post-consumer waste can be conducted by fibre-to-fibre recycling. These processes keep textile fibres within the textile and fashion value chain by converting post-industrial and post-consumer textile waste into raw material to manufacture new products. Reclaiming the fibres can play a crucial role in circular economy implementation, this strategy embeds the reuse of fibres within the original sector and therefore reduce its reliance on new virgin raw materials, thereby mitigating the impacts of this sector on the environment, specifically related to virgin fibre production [102].

Although fibres can be recovered in different ways, mechanical recycling is most commonly used. It involves breaking down the fabrics into fibre, such as wool, cotton, or synthetic, simply by shredding and then reclaiming the fibres step by step. Later, these fibres are used again to make yarn. The reclaimed fibres can be used in the manufacturing of new textile products, but the mechanical fibre reduces the overall length and quality of the textile product. Due to this, the reclaimed fibres are often blended with virgin raw materials to achieve the minimum performance requirement. Other challenges of mechanical recycling involve the removal of dye, prints, or other contaminations; other than these, the most significant challenge to mechanical recycling is processing blended textiles or products made from composite fibre or yarns [103]. To counter these challenges, chemical recycling techniques have been implemented as a complementary system. These processes break down fibres at the polymer or monomer level. The significance of this process is that the output quality matches that of virgin fibres. This method is very relevant and of utmost importance for textile blends, which are very common in textile and fashion products. Chemical recycling can deliver an extended number of recycling cycles meanwhile providing high-quality outputs [104].

The implementation of chemical recycling has significant challenges, including the use of chemical reagents and solvents, energy-intensive processes, the lack of proper infrastructure for pretreatment, sorting and the requirement of continuous waste input according to the given specifications. In chemical recycling, economics play a vital role, being a capital-intensive technology [105].

Both chemical and mechanical recycling can help transition towards a sustainable textile sector by proper management of material within the sector. Simultaneously industrial symbiosis keeps redirecting textile waste into alternative industries that utilise this waste for insulation, structural integrity, or absorption applications. Although this inter-industrial utility reflects broader resource efficiency, it minimises the overall waste generation. A large quantity of textile waste is often used for insulation purposes in the construction industry, in which the fibres are repurposed for acoustic or thermal insulation panels. Textile-based products are also used as non-woven textiles and are converted into composites like seat padding, dashboard insulation and liners. These applications rely on the bulk density of these fibres, which provide both structural and functional value, along with the aesthetics of the components. Other than these, the textile waste can also be used to make cleaning clothes and wipes, which is a low-tech but high-volume application and has applications in the industrial and service industries. Textile fibres can also be used to develop composites and act as reinforcement when combined with the matrix materials. These composites can be used in construction products, furniture and packaging applications [106, 107].

The industrial symbiosis keeps leaching out the precious raw material from the fashion systems, but it significantly helps in the conservation of material and enables these textiles to be used beyond their first phase and therefore contributes to the circular economy. Unfortunately, the material cannot be recovered for further recycling and therefore seems to be a one-directional flow of materials, therefore closed-loop recycling in these cases is not possible, making the fibre-to-fibre recycling more viable in the context of circular textile recycling.

As these promote the use of textile fibres within the textile industry, there is industrial symbiosis that promotes collaboration and optimisation of resources through the redirection of waste materials. Both of these have their own limitations and benefits and both can act as complementary strategies to enhance resilience and sustainability in the broader materials economy. With the advent of technology, industrial practices and policy frameworks, these models can become crucial in the establishment of a circular paradigm for a robust fashion and textile industry.

8.10 Advances in textile recycling technologies

8.10.1 The current status of recycling technologies

Technological requirements and capacity are crucial in recycling textile fibres and advancing in circular economy. How much policy instruments and legislation may be central to circularity, but the implementation of a real closed-loop textile and fashion industry depends significantly on the availability of technologies that can be scaled and provide optimal performances. In the midst of this mounting pressure for conservation of natural resources, minimising climate change and mitigating the environmental impact in harvesting or extracting raw materials, innovation within these technologies has gained a lot of attention [108]. Various techniques are implemented for the recycling of textiles; some renowned techniques are displayed in Figure 8.10.

These technologies are helping to unlock the true potential of textile waste and, therefore, are considering it as a secondary raw material. There are three major approaches to textile recycling being followed: chemical recycling, mechanical recycling and thermo-mechanical recycling. Each of these is implemented on specific fibres and composition of fibres and has its own limitations and opportunities [109]. All these pathways require some key enablers, including a proper and consistent segregation of textile waste, specific commitment from manufacturers and brands to integrate the recycled fibres back into their supply chains even at higher raw material costs and sustained support from legislation. Without these conditions, these technologies cannot be scaled up.

Out of the three options, chemical recycling is most technologically and innovatively advanced textile recycling system. It provides a potential solution for increasing the circularity of both pre-consumer and post-consumer textile waste. It is mostly in the development stage with most of the project in their pilot phase and only few commercial plants. Chemical recycling can be achieved through a diverse set of technologies and can be divided into multiple phases, including the following:

Purification of waste by using reagents to reduce additives and contaminants, meanwhile preserving the original polymer chains, making this technology more suitable for pure waste streams like 100 per cent polyamide or polyester. Although this technique can be applied to the blends as well.

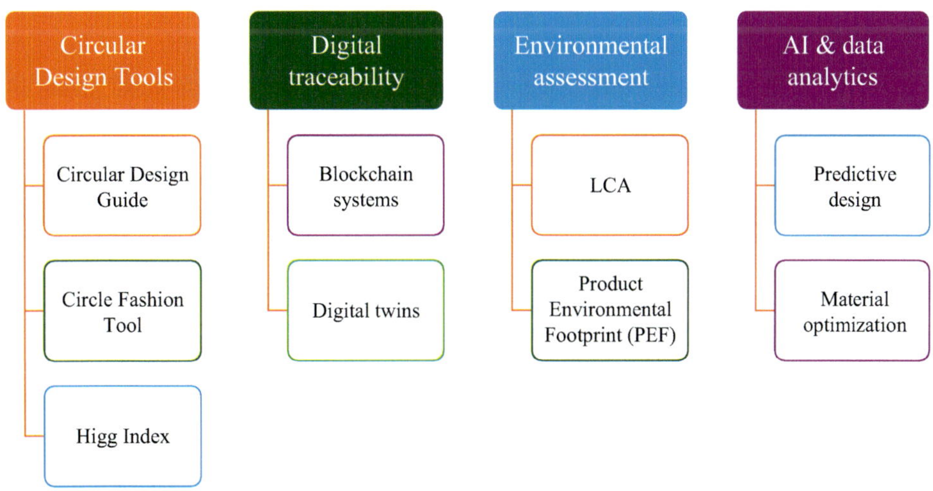

Figure 8.10 Types of recycling methods for fashion and textile products.

Depolymerisation is a process in which the polymer is broken down into its smaller constituents, the oligomers or the monomers. These smaller molecules are then re-polymerised into dope material that is later converted into fibres. The fibres developed by this method provide equivalent properties to those of virgin raw material. The depolymerisation process can target both synthetic fibres and natural fibres and their blends and provide output in the form of synthetic or regenerated cellulosic fibres.

Although chemical recycling has some drawbacks related to the high consumption of energy or water another subcategory is gaining attention: the enzymatic recycling. It uses biological enzymes as catalysts to break down the natural fibres at the molecular level and then indeed polymerising these molecules. Using enzymatic recycling uses less energy, reduces chemical and water consumption, but this process is in very early stages of development, due to which it has problems associated with economic feasibility and throughput rates if applied on a large scale.

Enzymatic recycling has a lot of benefits as compared to the traditional chemical recycling because chemical recycling use high temperatures, high energy, higher solvents, release of harmful GHGs and toxic by-products but enzymatic depolymerisation is a cleaner alternative and although has some limitations in processing of biomaterial products and achieving high scale commercial production but with time and further research this could be possible on large scale as well [110–112].

8.10.2 Mechanical recycling of textiles

Mechanical recycling is one of the most diffused and commercially established methods for converting textile waste into a valuable resource. In this process, the fibres are physically disintegrated by cutting, shredding and grinding. These fibres can subsequently be used for the spinning of yarns. But due to the high mechanical stresses, the integrity of these fibres is often compromised, and therefore, the reclaimed material is often blended with virgin fibres to be used for textile-to-textile applications. This method is mostly suitable for post-industrial textile waste due to the high production of fabric offcuts, where colour and composition consistency of material is uniform and allows efficient processing, sorting and therefore can increase the quality of the recycled output material [113].

8.10.2.1 Feedstock compatibility

For textile-to-textile application mechanical recycling can only handle single waste stream composition, but it cannot deal with the blended fibres. For products having multiple fibre compositions, textile-to-textile mechanical recycling is not viable; in these cases, the recycled material is mostly diverted to low-value applications, including geotextiles, automotive padding, or non-woven applications for installations, etc.

8.10.2.2 Advantages

The mechanical recycling technology is far less capital-intensive as chemical recycling and, therefore, it can be initiated with meagre capital investment, depending on the throughput and the quality of machines.

Recent advancements in mechanical recycling have enabled new computerised machinery to have enhanced fibre length output and therefore, when combined with proper waste quality and sorting, these innovations can help the output material to meet the desired characteristics required for spinning.

8.10.2.3 Limitations

Mechanical processes shorten the fibres; therefore, the quality of recycled material is inferior as compared to the virgin alternative and consequently it restricts its adoption in a wide range of industries and is limited to coarse yarn products, including upholstery, knitwear and denim only.

Due to the reduction in properties of mechanical reclaimed material, the subsequent recycles of the output material can be very difficult and often impossible. Blended textiles can hardly be processed using a mechanical recycling technique because of the lack of fibre separation capabilities and therefore, the blended material process using mechanical recycling is often downgraded and is used in low-value products.

8.10.3 Thermo-mechanical recycling of synthetic fibres

Thermo-mechanical recycling is often used for thermoplastic synthetic materials such as polyester and polyamide. It involves grinding the textile waste then melting it to extract the plastic material and convert it into pellets that can then be spun into filament or staple fibres and ultimately yarn. This method is similar to a plastic recycling system and can be easily integrated into various industries, and the recovered polymer can then be redirected to the textile system [114].

8.10.3.1 Feedstock compatibility

Thermo-mechanical recycling requires homogeneous and uncontaminated waste. It requires the absence of any kind of contamination, including any non-thermoplastic fibres or dyes, or finishing chemicals, that can degrade the polymer quality and hence reduce the yield. Even micro-level contaminations, such as dye stuff or finishes, compromise the structural integrity of polymer molecules.

8.10.3.2 Advantages

One of the key benefits of this method is the well-established infrastructure in plastic recycling that can uptake huge volumes of waste and is normally applied to PET bottled recycling but can also be adopted for textile waste. The thermo-mechanical recycling system can be embedded in polymer-producing chains that can facilitate scaling and resource sharing among various kinds of industries.

8.10.3.3 Limitations

Thermo-mechanical recycling can only process pure mono materials, and any blend, coating, finish, or any other impurity can lower the yield or degrade the output. Due to the thermal and oxidative reactions during each processing cycle, the material becomes contaminated, leading to a lower output and eventually making this process a downcycling process. The recycled material can be successfully converted into fibres, which mostly have inferior mechanical properties and therefore cannot be utilised for high-end products, due to which there is limited use of these materials in high-value circular textile supply chains.

8.10.4 Chemical recycling

Chemical recycling involves molecular-level recovery of the polymer. Chemical recycling processes recover monomers from the fibres and can remove the contaminants, meanwhile enabling high-quality raw material outputs that are highly suitable for producing new fibres. A transformed kind of chemical recycling uses enzymes in the recycling process. These are biologically derived catalysts, rather than chemicals, and this technology is therefore capable of processing the material in moderate environmental conditions [115, 116].

8.10.4.1 Feedstock compatibility

Chemical recycling can handle complex waste streams; for example, it can handle multi-material textiles, printed fabrics, heavily dyed or finished materials and fabrics. Chemical recycling technology can target specific fibres, such as polyamide or polyester, as well as natural polymers like cellulose or PLA.

The enzymatic recycling shows more sustainable and promising results as compared to the chemical recycling technique, but it can only entertain selective natural fibres, and therefore, pre-sorting is required. The presence of synthetic coatings or other contaminants can stop enzymatic activity, which is a big limitation in its adaptability; therefore, this process is limited to low-contaminant or pure materials only.

8.10.4.2 Advantages

The chemical recycling process can provide high-quality outputs especially due to deep polymerisation. The properties of regenerated polymer are similar to those of virgin material, and, therefore, chemical recycling enables multiple recycling cycles without losing the complete integrity of the material, and therefore, offers a more circular solution. Chemical recycling techniques can accommodate different blends and has tolerance towards contamination. Unlike other recycling methods, it can process fibres contaminated with dyes, elastin and other functional finishes, and therefore, it can process larger volumes of mixed textile waste, which cannot be currently utilised by any other processing technique. Chemical recycling can lead to a truly circular and regenerative textile economy.

8.10.4.3 Limitations

The capital cost of chemical recycling set-ups is high, and substantial capital is required to purchase separation units, specially designed reaction chambers and purification systems. Chemical recycling technologies are mostly on a pilot scale, and a fully functional industrial setup capable of processing huge quantities is very limited; therefore, their commercial viability is limited owing to their technical complexities and the severity of the process

Chemical recycling process needs huge energy input, uses a lot of water and hazardous chemicals, and therefore, the ultimate benefit is dependent on the type of technology utilised, along with the recovery of by-products in recycling and subsequent processes. Chemical recycling needs high-quality input materials, although it can handle contaminated materials as well. But the clean and well-sorted raw materials would result in better yields and therefore would need pretreatment, or otherwise, the process will provide lower output.

The chemical recycling technique overcomes the limitations offered by thermo-mechanical and mechanical recycling systems, and therefore, it is very important for achieving complete circular textile value chains, especially when the textile materials are blended fibres [117]. Only chemical recycling can provide closed-loop recycling while providing high-quality material outputs. Despite a lot of challenges in the context of infrastructural barriers, technical limitations and economic limitations, chemical and enzymatic recycling techniques are improving through continuous innovation in the field and are transforming how materials can be recycled from the textile and fashion sector and can support low-impact and regenerative textile and fashion value chains [118].

8.10.4.4 Industrial applications of chemical recycling

Several manufacturers and brands are working towards the development of commercialise-able processes. Some of these are discussed below:

- Circulose: It produces regenerated material from cotton waste. Currently, it is under new management due to Renewcell's bankruptcy [119].
- Syre: A Swedish company that chemically recycles the polyester waste and has aimed to establish a large-scale production facility, including a giga-scale recycling plant in Vietnam [120].
- Aquafil's Econyl: Based in Italy, it utilises a chemical recycling process to convert fishing nets, textile scraps and carpets of nylon 6 waste to produce regenerated nylon, keeping the original properties intact.
- Infinited **Fibre**: A Finnish manufacturer that uses carbamide technology to reduce cell loss and develop Infinna fibre that resembles the properties of virgin cotton [92].
- Gr3n: Based in Switzerland, Gr3n employs a microwave-assisted recycling process that deep polymerises the polyester textiles and converts them into monomers. Their process is energy efficient than conventional chemical recycling and can provide high-quality polyester even from blends [121].
- Worn Again Technologies: Based in the UK; recycles polyester and cellulose fibres from textile waste. It aims to achieve fibre-to-fibre reuse on an industrial scale recovery process [122].
- Ambercycle: Based in the United States and uses molecular regeneration for the conversion of post-consumer polyester into recycled polyester resin that performs like virgin polyester; their output is Cycora [123].
- Carbios: Based in France, a pioneer of enzymatic recycling technology for polyester fibre. They have employed specially engineered enzymes that can first depolymerise the polyester and then re-polymerise it into new material [124].
- BlockTexx: A company based in Australia that has developed a fibre separation process that uses a chemical process to separate and recover cell loss from polyester, which is very common in blended textile waste. It started its operations in 2022 and became the world's first commercial facility for the recycling of polyester-cotton blended textile waste [125].

8.10.5 Positioning recycling within the circular economy framework

The initiatives discussed above show how wide a variety of solutions are currently being studied in chemical recycling technologies. Using different approaches and different technologies, these methods are exploring the opportunities to recover material from uniform and blended fabrics, although they have some differences in their approach, but are offering various solutions related to energy requirements, the purity of output material, the scalability of the technology and therefore have their own level of complexities in achieving the circular economy. The scaling up of these techniques would require systematic coordination among the complete fashion and textile supply chain, the consumer market framework, along technical optimisation [126]. Some design tools that can help achieve sustainability in the fashion and textile sector are visually displayed in Figure 8.11.

Figure 8.11 Digital tools that can help achieve sustainability in the fashion and textile sector.

Mechanical and thermo-mechanical recycling processes, despite their current limitations, are commonly being employed; meanwhile, the chemical and enzymatic technologies are still in development mode. Mechanical recycling remains the fastest employable technique and can divert the textile waste from incineration and landfills immediately. The innovations and the quality protocols are important because these quality requirements and protocols will play a crucial role in the process improvements that will be continually demonstrated throughout all these recycling systems. The selection of proper recycling technology should be decided on the basis of the type of waste to be processed, the life-cycle impacts and the end use of the process material. All these recycling techniques bridge the linear practices and the sustainable and circular economy and therefore have their essential importance in advancing the pathways for the fashion and textile sector for the years ahead.

8.11 Tools, frameworks and digital enablers

Multiple quantitative tools are available to evaluate the circularity of a specific product, for example product environmental footprint (PEF), the material circularity indicator and recycling rate and the input quality tracker.

8.11.1 Product environmental footprint

The PEF helps in my ring the cradle to grave in terms of a product on the environment and therefore helps in quantitative measurement of product impact on land, water and climate change, etc. [127].

8.11.2 The material circularity indicator

The material circularity indicator helps in analysing the amount of virgin and recycled materials used in the product and the potential for end-of-life recycling or reuse of the product [128]. These tools help in maintaining a record of the purity of recycled inputs and the volume of all recycled material, along with the collection efficiency of the product. These metrics and tools help the product designers in the fashion and textile sector to organise their circular designs and hence identify the loopholes such as contamination during the collection sorting or recycling process, or load take back efficiency of the product and therefore help in optimising of circular model.

8.11.3 Circular design tools: Circle fashion tool, Higg Index, circular design guide

Implementation of the circular economy needs a lot of tools that can be based on various digital systems and predefined frameworks; these tools help in making product designs as well as in tracking the material throughout the life cycle [129]. With the increasing use of technology and better understanding of environmental parameters nowadays, multiple circular design tools are available.

8.11.3.1 Circular fashion tool

The circular fashion tool was originally developed to visualise the circular economy. This tool assists the design teams in studying the flow of material throughout the product supply chain and, hence, helps in making decisions during circular product designs and visualises the dependencies across the supply chain and aligns the targets with key performance indicators or key performance indicators necessary for achieving sustainability [130].

8.11.3.2 Circular design guide

The circular design guide provided by IDEO and the Ellen MacArthur Foundation provides the strategies for designing fashion and textile products, product lifetime extension, material looping and a framework for circular business models. This tool includes simple instructions for designers to incorporate circular economic principles in their products [131].

8.11.3.3 Higgs material sustainability index MSI

The Higgs material sustainability indicator helps in quantifying the impact of a material or a product in terms of GHG emissions, acidification, water consumption and eutrophication quantification etc. These matrices help in making informed design decisions [132].

8.11.3.4 Digital twins and blockchain for material tracking and circular verification

DIGITAL TWINS

Digital twins provide a virtual replication of the products by integrating the matrices from various production stages and the lifetime of the product. These digital twins help enable the designers and the manufacturers to observe and simulate real-life circular scenarios, for example, end-of-life recycling, reuse, etc., along with observation of the environmental impacts, all of which happen before the product manufacturing starts [133].

TRACEABILITY PLATFORMS AND BLOCKCHAIN-BASED SOLUTIONS

Several systems exist, for example, textile genesis, provenance and Ever-ledger, enabling the traceability of the material in the form of tokens or batches to verify the sustainability across the supply chain. This ensures the transparency and the implementation of certifications like

Fairtrade or Global Organic Textile Standard (GOTS), etc. These systems and tools help in maintaining circularity and help ensure material flows are reliable and traceable throughout the life cycle [134, 135].

8.11.3.5 Life-cycle assessment and product environmental footprint in circular decision-making

The PEF and LCA are tools that help study the environmental impact of the products throughout their life cycle. LCA methodologies rely on tools like GaBi, SimPro and OpenLCA to measure the impacts of a product throughout its life cycle. These tools help the designers to identify the right materials or processes and compare them to other materials in the design of a product and align it with circular economy principles, for example, comparing the recycled polyester with virgin polyester as a raw material [136, 137].

PEF was built by the European Commission and helps in defining the standards of environmental impact across various sectors. Integration of PEF with LCA can help designers in the evaluation of various scenarios in the context of the circular economy, for example, the identification of loopholes like increased life-cycle service to make informed decisions while product designing [138].

8.11.4 Integrating artificial intelligence and data analytics for circular product design

Nowadays, various data analysis and artificial intelligence (AI)-driven tools are also valuable, which help in providing analytics signs and powerful material traceability. Some applications of AI tools in the fashion and textile industry for circular product design are listed below [139–141].

8.11.4.1 Make predictive analysis

Various machine-learning methods and models help forecast the lifetime of a fashion and textile product's market demand, supply quantity, returns rates etc. and therefore help the brand in take-back logistics, rental-scheme viability, or inventory control, etc., during different stages of the product life cycle.

8.11.4.2 AI-powered material discovery

AI-maintained databases can help brands analyse materials that are conducive to biodegradability or recyclability while maintaining the aesthetic and functional requirements. Such services are already being provided by Resynthetic Analysis and Unspun vast material databases.

8.11.4.3 Generative design tools

Multiple AI techniques are available that can help designers in crafting ideas related to disassembling or modulated designs, thereby helping in complex product designs to promote longevity without losing the options for recyclability.

8.11.4.4 Predictive analysis

To manage returns, product lifetime data, or simply forecast the useful life of a garment, machine learning can be implemented; not only this, but the brands and retailers can also use these models to manage rental programmes, take-back programmes, inventory, or other processes.

8.12 Consumer, engagement and behaviour

The material loops and the application of the circular economy are mainly dependent on the participation of end consumers and users in effectively implementing the sustainability strategy, which must be implemented efficiently, considering the economic, cultural and longevity aspects together to encourage sustainable consumption and mode circular models like reuse, repair and resell, etc. [142]. The barriers to implementing sustainability in the fashion and textile sector, in light of consumer behaviour, consumer engagement concerns, barriers and other variables, are visually demonstrated in Figure 8.12.

8.12.1 Educating consumers about value retention and upcycling

Longevity is one of the key aspects that is mostly neglected by consumers while making product-buying decisions, and, therefore, this is one of the leading causes of overconsumption of products, often due to economic motivations or cultural norms. To address this issue, policymakers and brands must educate them to demonstrate the value of longevity, upcycling and proper certification systems [143].

8.12.1.1 Demonstrate the value by experience

The policymakers and the brands can highlight the importance of upcycling value retention along with longevity by offering upcycling courses, hosting workshops and running campaigns and sharing through education and certifications [144]. Educate the consumers and workers by seeing transparency and storytelling, consisting of supply chain environmental impact and product life cycles. Certifications can provide a stop solution for credibility by offering trust to the consumers. There are various labelling and certifications, including global recycling standards (GRS), GOTS, Oeko-Tex, Fairtrade and Cradle to Cradle. These certifications are a sign of success for consumers who are already aware of the value of retaining and upcycling textile waste and fashion products.

8.12.2 Barriers to circular behaviour (cultural, economic, aesthetic)

There are numerous barriers that consumers may be influenced by number of barriers [145, 146]. These barriers can be perceptual or structural and include the following:

- **Economic trade-offs**: Sustainable products until now have relatively higher prices, and this is one of the core barriers that restrain customers from buying sustainable products.

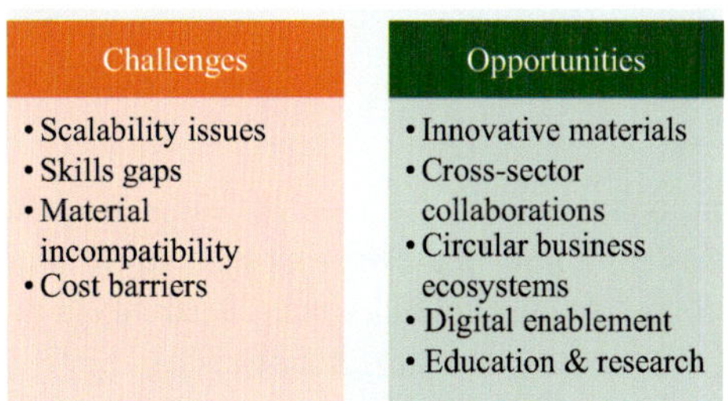

Figure 8.12 Barriers to achieving sustainability in light of consumer behaviour.

- **Aesthetic concerns**: The second-hand items are the upcycled ones, which may often have lower aesthetic design or quality and which may influence the consumer's behaviour to shift towards more attractive products.
- **Cultural norms**: Fast fashion, as compared to slow fashion, mostly follows the trends we change day-to-day and therefore incentivises their disposability of products.
- **Convenience gap**: The repair operations or drop-off and collection for recycling are often time-consuming processes, due to which the consumers may be prone to buying a new product rather than opting for these services.

The proper implementation of sustainability in the fashion and textile sector requires a thorough understanding of these barriers, and these barriers should be overcome by introducing design interventions in such a way that these alterations resonate with the consumers as well.

8.12.3 Nudging strategies to promote use-phase circularity (repair, resale, rental)

The strategies based on human behaviour can provide deep insights to help develop consumer behaviours and habits to promote the circular economy [147–149]. These strategies are explained below.

- **Social proofing**: The share repair and similar campaigns, especially the influencer promotions, again help normalise the circular behaviour as secondary to human nature if properly utilised.
- **Incentives**: Providing loyalty points for repair and maintenance, or serviceability of the product, along with discounts for returns at the end of product life, can promote circular economy practices in end consumers.
- **Pledges and commitments**: Consumers may be replaced or offered commitments to buy long-lasting products, or to buy fewer products and live minimalist lives by promoting mobile applications, or may be provided incentives by keeping up put into their pledges and commitments.
- **Options for repair or rental during product sale**: If consumers are offered or, by default, are provided rental or repairability services during the product's life can help incorporate the circular economy models.
- **Convenience tools**: The mail-in return systems, home-repair cards, drop-off locations, and other similar services can be helpful and would be convenient for consumers to help achieve the circular economy in the product.

By applying these social and behavioural barriers with the incentives, the policymakers or the brands and other start-ups can help make the circular economy and sustainability secondary to human nature more appealing and more accessible.

8.13 Policy and regulatory landscape

Legislation can play an important role in the incorporation of circular-economy practices in the fashion and textile industry. These frameworks can help in implementing the regulations that force the brands to incorporate sustainability and circularity into their product designs and complete supply chains of the product, along with EPR.

8.13.1 EU Green Deal, Ecodesign for Sustainable Products Regulation

The sustainable products revolution and the European Green Deal both have the main aim of integrating the circular economy into product designs systematically. The Ecodesign for Sustainable Products Regulation necessitates [150]:

- enforcing the restrictions on using non-recyclable compositions and the use of hazardous substances and materials in products, mandating designs for longevity, durability, repairability, upgradability and reuse
- introducing digital passports to make supply-chain traceability possible, along with post-consumer waste recycling, reselling, or upcycling.

These frameworks help to play a crucial role in transforming the textile and fashion sector's priorities and binding the manufacturers and brands operating in Europe, specifically.

8.13.2 Extended producer responsibility in fashion

EPR means forcing the brands to sort, collect and recycle the textile and fashion products at the end of their useful lives. Many EU member states, including France and Germany, are now requiring the brands to follow these strategies for the betterment of the planet. The revenue is raised through echo-modulated charges based on repairability, recyclability and other circular economy models that the brand could offer. It is forecasted that these schemes should be able to effectively drive eco-designing of the products, help innovative business models and circular models focused on recyclability, longevity and product durability and ultimately reduce the waste generated by the products [151].

8.13.3 National and global frameworks shaping circular transition

Multiple governments are now shaping the frameworks that enhance the transition from linear to circular economy models. Such frameworks not only position the national interests become the guiding principles for the brands but also become the binding forces that enforce circular economy and sustainability in shipping business practices [152]. Some of these frameworks and legislation are discussed below.

- UK Textiles 2030 road map: This legislation helps to set up voluntary targets for take-back infrastructure set-up, supporting innovation in recycling and ultimately waste reduction [153].
- France's Anti-Gaspillage pour une Économie Circulaire law of 2022: This framework binds the brands and the manufacturers to refrain from disposing of the textile and fashion products and also convinces them to collect and recycle the products at the end of the product's life [154].
- California Government Worker Protection Act: This legislation promotes traceability in the brand's supply chain, incorporates brand liability, social accountability and environmental accountability to better serve its workers [155].
- United Nations Alliance for Sustainable Fashion and the Global Fashion Agenda's Policy Hub: These frameworks help in promoting the sustainable procurement incorporation of circular design initiatives and enhancing standardised regulations in trade [156].

8.13.4 How legislations share design practices

Legislation and policy frameworks facilitate and transform circular design by incorporating the following [157–159]:

- providing transparency and data-management tools, such as life-cycle guides, assessments and product passports to enable accountability through a vast range of supply-chain processes;
- setting up the repairability and recyclability standards that force and push the designers to integrate DfD options within product designs and also designing products with longevity in mind.
- offering financial incentives for moving away from single-use fashion products and encouraging eco-modulation;
- encouraging public-private partnerships to enhance innovation, offer funding to uphold the circular economy and the infrastructure required for enabling the circular business models.

The manufacturers, designers and brand leaders should be able to interpret and forecast the policy and legislation signals, and let's see regulation not as a burden but as an opportunity to make a difference and innovate in circular value creation, along with integrating compliance into their designs.

8.14 Challenges and future opportunities

Several challenges remain unsolved while the implementation and scaling up of the circular economy and the relevant businesses, despite the circular economy being the most promising sustainable economic model in the fashion and textile sector. The following are the constraints in implementing the circular economy, along with some opportunities that may help overcome these challenges.

8.14.1 Material incompatibility and design constraints

Fast fashion, enormous product designs, the trending fashion and ever-changing desires to look distinct have created a huge opportunity for designing for disassembly and closing recycling loops. Design constraints, including the technological ones, play a crucial role in determining the compatibility of the material. For example, the incorporation of stretch fabrics for extended comfort or the incorporation of attachments or accessories for DfD options can be a design constraint. Bi-materiality and complex fibre blends are one of the main constraints faced by the fashion and textile sector because the fibres, especially the polyester-cotton blends, are very difficult to separate without the use of additional hazardous chemicals during the recycling process. These challenges are very crucial and are limited by the material properties themselves, but investing in new materials research and finding alternatives to innovate processes or materials to enhance separability and recyclability without compromising comfort or performance can be very promising and fruitful [160, 161].

8.14.2 Economic and scalability hurdles

The high cost of recycled materials, the capital cost of recycling infrastructure in case of mechanical or chemical recycling and the implementation of the circular economy have become a great challenge and often rely on third-party, government, or legislative compensation. Due to this factor, implementation of the circular economy majorly relies on government investment and legislative support rather than a self-sufficient model. Another big hurdle is due to the limited collection of discarded products, proper sorting and the gaps in the supply chain that limit the volume of raw material and therefore, these fluctuations make the recycled material competitive in availability and prices. That's sustainable fashion, owing to the government and legislative support offering a high financial risk based on the interests of ever-changing governments, which limits investments into circular economy models, especially in cases where the payback periods are longer [162, 163]. Although these are huge challenges, collaborative efforts can help scale these technologies and with the support of investors and green procurement policies, or by building multi-brand consortia, again help distribute these costs and fill the knowledge gaps through collective research and development.

8.14.3 Opportunities for cross-sector collaboration (designers, recyclers, software developers)

The textile and fashion sectors often lack standardisation and data transparency, which makes it very difficult for different stakeholders in the supply chain to efficiently coordinate and manage the flow of materials and analyse circular performance and sustainability. The supply chain of the fashion and textile industry is scattered across the globe, due to which the facilities of waste management and recycling in one part of the world are highly dependent on collection in a distinct part of the world. Furthermore, due to this fragmentation, the designers waste management farms and the fibre recyclers, all of the manufacturers work in isolation, which creates a lack of innovation and therefore becomes a great hurdle in the implementation of the circular economy [164, 165]. To date, a lot of skill gaps exist in the implementation of the circular economy. Competency to implement it practically at managerial and designer levels requires a lot of practice and competency, which is unfortunately not very well established in the industry. A great opportunity to establish platforms and digital tools to enable tracking and transparency across the supply chain and launch interdisciplinary certifications and training programmes for engineers, designers and businessmen to fill these gaps. The challenges and opportunities for the future in the context of the implementation of SDGs are provided in Figure 8.13.

8.14.4 Regulatory and policy gaps

Voluntary guidelines to incorporate sustainability and circularity in the fashion supply chain are often weak and lack enforcement and wider coverage, due to which the mass adoption of these practices is very limited in the fashion and textile sector. The regulatory frameworks, even though defined in some parts of the world, are not uniform globally, owing to the different

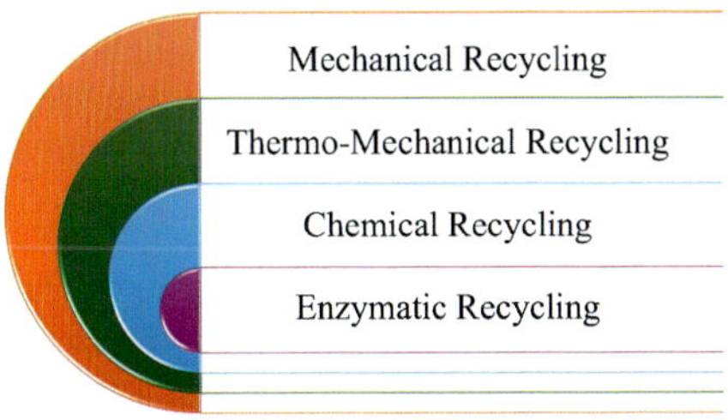

Figure 8.13 The challenges and opportunities for the future in the context of the implementation of SDGs in the fashion and textile sector.

EPR methods, sustainability initiatives and recycling mandates [151]. The implementation of uniform systems is quite impossible until a great intervention is made on a global scale [166]. By promoting and struggling for global-level trading rules and standards, the circular-economy matrices could be improved along with a level playing field for group brands and manufacturers across wide geographies, which would definitely help reduce the misunderstandings during trade and other supply chain operations.

8.14.5 Role of education and research in mainstreaming circular practices

A lot of gaps exist in practical research, especially in socio-economic and socio-cultural integration of circular-economy models and systems, and therefore, a major shift is required in consumer value retention and perception mindset. To date, the educational curriculum often limits the understanding of sustainability just limited to books, and unlimited hands-on training experiences are available, especially in the case of repair and DfD [167, 168]. To support the circular-economy implementation, universities and business incubation centres should introduce fashion-focused institutes that combine prototyping, research and stakeholder engagement to promote the implementation of sustainable and circular economy while training the future professionals of this sector [169].

8.15 Conclusion

This chapter represented a comprehensive discussion on upcycling and circular design applications in the fashion and textile sector. It started with the introduction of circularity in contrast to the linear economy, discussing the circular economy in the context of SDGs. Then we discussed the basic principles and fundamentals of circular economy, the relationship between sustainable and circular design and the relevance of life-cycle thinking while implementing circular economy principles. After that the concepts and typologies related to the industry designer and the consumer on grassroot levels were discussed, the 9Rs of circular economy in context to circular fashion and textile product design were discussed. The chapter then transitioned to DfD strategies along with the discussion on how it can be implemented in fashion and textile products and what challenges and opportunities exist in DfD. After that, closed-loop systems, including business models, technology and industrial applications, were discussed. The digital product passport framework and tools were discussed, which can help managers and designers optimise, evaluate and track the implementation of sustainable practices in the industry. The chapter then discussed behavioural and policy interventions which may be very important in shaping consumer behaviour, along with opportunities for challenges in implementing circular economy and sustainability practices in the fashion and textile industry. Finally, the chapter closed with a discussion of key challenges and upcoming opportunities while suggesting the goals for research collaboration, future innovation and system-wide transition of circular economy practices.

References

[1] K. H. D. Tang, 'State of the Art in Textile Waste Management: A Review', *Textiles*, 3 (4) (December 2023): 454–67, doi: 10.3390/TEXTILES3040027.

[2] K. Bailey, A. Basu and S. Sharma, 'The Environmental Impacts of Fast Fashion on Water Quality: A Systematic Review', *Water*, 14 (7) (March 2022): 1073, doi: 10.3390/W14071073.

[3] Earth.org, '10 Concerning Fast Fashion Waste Statistics', (2023), https://earth.org/statistics-about-fast-fashion-waste.

[4] S. Andreadakis, P. Owusu-Wiredu, S. Andreadakis and P. Owusu-Wiredu, 'Fashion Footprint: How Clothes Are Destroying Our Planet and the Growing Impacts of Fast Fashion', *Global Warming: A Concerning Component of Climate Change*, (July 2023), doi: 10.5772/INTECHOPEN.1002000.

[5] S. Schützenhofer, I. Kovacic, H. Rechberger and S. Mack, 'Improvement of Environmental Sustainability and Circular Economy through Construction Waste Management for Material Reuse', *Sustainability*, 14 (17) (September 2022): 11087, doi: 10.3390/SU141711087.

[6] A. K. Das, M. F. Hossain, B. U. Khan, M. M. Rahman, M. A. Z. Asad and M. Akter, 'Circular Economy: A Sustainable Model for Waste Reduction and Wealth Creation in the Textile Supply Chain', *SPE Polymers*, 6 (1) (January 2025): e10171, doi: 10.1002/PLS2.10171.

[7] Sugam Upadhayay, Omaima Alqassimi, Edmund Khashadourian, Alex Sherm and Dinesh Prajapati, 'Development in the Circular Economy Concept: Systematic Review in Context of an Umbrella Framework', *Sustainability*, 16 (4) (February 2024): 1500, doi: 10.3390/SU16041500.

[8] United Nations, 'The 17 Goals: Sustainable Development', (2015), https://sdgs.un.org/goals.

[9] Ellen MacArthur Foundation, 'Circular Fashion: Circular Economy for the Fashion Industry', (n.d.), www.ellenmacarthurfoundation.org/topics/fashion/overview.

[10] A. Anderluh, R. Hackl, Z. Mesbahi, T. Pellegrini and C. Raffler, 'Consumers' Affinity to the Circular Economy: Empirical Evidence Utilizing the Circular Economy Cube', *J Clean Prod*, 525 (September 2025): 146539, doi: 10.1016/J.JCLEPRO.2025.146539.

[11] N. C. Onat, M. Kucukvar, A. Halog and S. Cloutier, 'Systems Thinking for Life Cycle Sustainability Assessment: A Review of Recent Developments, Applications, and Future Perspectives', *Sustainability*, 9 (5) (April 2017): 706, doi: 10.3390/SU9050706.

[12] M. Ribul, A. Lanot, C. Tommencioni Pisapia, P. Purnell, S. J. McQueen-Mason and S. Baurley, 'Mechanical, Chemical, Biological: Moving towards Closed-Loop Bio-based Recycling in a Circular Economy of Sustainable Textiles', *J Clean Prod*, 326 (December 2021): 129325, doi: 10.1016/J.JCLEPRO.2021.129325.

[13] X. Zhang, A. Le Normand, S. Yan, J. Wood and C. E. Henninger, 'What Is Modular Fashion: Towards a Common Definition', *Resour Conserv Recycl*, 204 (May 2024): 107495, doi: 10.1016/J.RESCONREC.2024.107495.

[14] M. D'Urzo and C. Campagnaro, 'Design-Led Repair and Reuse: An Approach for an Equitable, Bottom-Up, Innovation-Driven Circular Economy', *J Clean Prod*, 387 (February 2023): 135724, doi: 10.1016/J.JCLEPRO.2022.135724.

[15] Raghu Raman, Payel Das, Rimjhim Aggarwal, Rajesh Buch, Balasubramaniam Palanisamy, Tripti Basant, Urvashi Baid, Pozhamkandath Karthiayani Viswanathan, Nava Subramaniam and Prema Nedungadi, 'Circular Economy Transitions in Textile, Apparel, and Fashion: AI-Based Topic Modeling and Sustainable Development Goals Mapping', *Sustainability*, 17 (12) (June 2025): 5342, doi: 10.3390/SU17125342.

[16] C. Klemm and S. Kaufman, 'The Importance of Circular Attributes for Consumer Choice of Fashion and Textile Products in Australia', *Sustain Prod Consum*, 45 (March 2024): 538–50, doi: 10.1016/J.SPC.2024.01.021.

[17] Raimund Bleischwitz, Miying Yang, Beijia Huang, Xiaozhen Xu, Jie Zhou, Will McDowall, Philip Andrews-Speed, Zhe Liu, and Geng Yong, 'The Circular Economy in China: Achievements, Challenges and Potential Implications for Decarbonisation', *Resour Conserv Recycl*, 183 (August 2022): 106350, doi: 10.1016/J.RESCONREC.2022.106350.

[18] K. Nowak-Marchewka, E. Osmólska and M. Stoma, 'Progress and Challenges of Circular Economy in Selected EU Countries', *Sustainability*, 17 (1) (January 2025): 320, doi: 10.3390/SU17010320.

[19] Ellen MacArthur Foundation, 'What We Do', (2025), www.ellenmacarthurfoundation.org/about-us/what-we-do.

[20] United Nations Environment Programme, Sustainability and Circularity in the Textile Value Chain: A Global Roadmap. Nairobi, Kenya: UNEP, (2020), www.unep.org/resources/publication/sustainability-and-circularity-textile-value-chain-global-roadmap.

[21] K. Saha, P. K. Dey and V. Kumar, 'A Comprehensive Review of Circular Economy Research in the Textile and Clothing Industry', *J Clean Prod*, 444 (March 2024): 141252, doi: 10.1016/J.JCLEPRO.2024.141252.

[22] P. Gazzola, E. Pavione, R. Hillebrand, V. Vota and R. Rosa, 'The Circular Economy and the Role of Technology in the Fashion Industry: A Comparison of Empirical Evidence', *Sustainability*, 17 (7) (April 2025): 3104, doi: 10.3390/SU17073104.

[23] A. R. Shafqat and A. Saccavini, 'Eco-design of Textiles', in R. M. Kozlowski and M. Mackiewicz-Talarczyk (eds), *Sustainable Textiles: Production, Processing and Recycling*, Cham: Springer Nature, (2023), pp. 39–58.

[24] M. Hussain and A. R. Shafqat, 'Sustainability and Materials', in K. Shaker and Y. Nawab (eds), *Engineering Materials*, vol. 11, Cham: Springer Nature, (2025), pp. 257–274, doi: 10.1007/978-3-031-72263-9_11.

[25] Banyan Nation, 'Recycling vs Upcycling: What's the Difference?', (April 2025), www.banyannation.com/blog/recycling-vs-upcycling/.

[26] J. Singh, 'The Sustainability Potential of Upcycling', *Sustainability*, 14 (10) (May 2022): 5989, doi: 10.3390/SU14105989.

[27] Por Janah Cycle, 'Recycling vs. Upcycling: Are They Both Sides of the Same Coin?', Tech Barcelona, (August 2024), www.janahcycle.org/post/recycling-vs-upcycling-are-they-both-sides-of-the-same-coin.

[28] Maria Angela Butturi, Alessandro Neri, Francesco Mercalli and Rita Gamberini, 'Sustainability-Oriented Innovation in the Textile Manufacturing Industry: Pre-consumer Waste Recovery and Circular Patterns', *Environments*, 12 (3) (March 2025): 82, doi: 10.3390/ENVIRONMENTS12030082.

[29] R. De Silva, 'Waste Isn't Waste until We Waste It', (n.d.), https://circulareconomy.europa.eu/platform/sites/default/files/2023-10/WearnRepair_casestudy.pdf.

[30] A. Rauf and A. Kamal, 'Preserving and Promoting the Sindhi Ralli Quilts: A Case Study of Pakistan's Artistry', *Journal of Design and Textiles*, 2 (2) (December 2023): 88–113, doi: 10.32350/JDT.22.05.

[31] M. Shamsuzzaman, M. M. R. Khan, T. M. H. R. Jony, M. S. Islam, and M. R. H., 'Fashion Circularity: Potential of Reusing and Recycling Remnant Fabric to Create Sustainable Products', *Sustainability*, 17 (5) (February 2025): 2010, doi: 10.3390/SU17052010.

[32] R. Aus, H. Moora, M. Vihma, R. Unt, M. Kiisa and S. Kapur, 'Designing for Circular Fashion: Integrating Upcycling into Conventional Garment Manufacturing Processes', *Fashion and Textiles*, 8 (1) (November 2021): 34, doi: 10.1186/S40691-021-00262-9.

[33] A. Milad, 'Recycled and Upcycled Materials in Contemporary Architecture and Civil Engineering: Their Applications, Benefits, and Challenges', *Cleaner Waste Systems*, 10 (March 2025), p. 100203, doi: 10.1016/J.CLWAS.2025.100203.

[34] Y. Zhi and Y. Zhi, 'The Upcycling and Reconstruction of Garments and Fabrics', *Art and Design Review*, 10 (1) (January 2022): 72–102, doi: 10.4236/ADR.2022.101007.

[35] M. Burcikova, 'Mundane Fashion: Women, Clothes and Emotional Durability', Ph.D. dissertation, School of Art, Design and Architecture, University of Huddersfield, Huddersfield, (2019), https://eprints.hud.ac.uk/id/eprint/35322/.

[36] A. D. Pranta, Md. R. Repon, A. Haji, M. Rahman and M. M. Rahman, 'Sustainable Merchandising for Recycled Textiles and Apparel in the Western Apparel Industry: Strategies, Challenges and Market Opportunities', *J Eng Fiber Fabr*, 20 (August 2025), doi: 10.1177/15589250251359273.

[37] Neetu Singh, 'Circular Economy in Textile and Apparel Supply Chains', in Syed Abdul Rehman Khan (ed.), *Innovation and Sustainability through Circular Economy in Businesses*, IGI Global Scientific Publishing, (2025), pp. 25–64. https://doi.org/10.4018/979-8-3693-4123-0.ch002.

[38] E. O. Alegbe and T. O. Uthman, 'A Review of History, Properties, Classification, Applications and Challenges of Natural and Synthetic Dyes', *Heliyon*, 10 (13) (July 2024): e33646, doi: 10.1016/J.HELIYON.2024.E33646.

[39] C. Quintero Rodriguez, A. Anisimova, S. Ryan and A. O. Troynikov, 'Conference Paper Critical Design Aspects of Maternity Support-Garments and Its Contemporary Perspective', (2017), doi: 10.18502/keg.v2i2.600.

[40] I. G. Klepp and K. Laitala, 'Shared Use and Owning of Clothes: Borrow, Steal or Inherit', in I. Cruz, R. Ganga and S. Wahlen (eds), *Contemporary Collaborative Consumption*, (2018), pp. 153–77, doi: 10.1007/978-3-658-21346-6_8.

[41] S. Pitman and P. Hohti, 'Remaking Dress History: Applying Reconstruction Methods to Early Modern Textiles and Clothing', *Text Hist*, 54 (1) (2023): 4–9, doi: 10.1080/00404969.2024.2409295.

[42] C. Manolchev, R. Cherrington and G. Crabolu, 'Untangling Circularity: Transitioning to a Regional Circular Economy in the Clothing and Textile Sector in England', *J Clean Prod*, 482 (December 2024): 144188, doi: 10.1016/J.JCLEPRO.2024.144188.

[43] S. Abbate, P. Centobelli, R. Cerchione, S. P. Nadeem and E. Riccio, 'Sustainability Trends and Gaps in the Textile, Apparel and Fashion Industries', *Environment, Development and Sustainability*, 26 (2) (February 2023): 2837–64, doi: 10.1007/S10668-022-02887-2.

[44] A. de A. Hugo, J. de Nadae and R. da S. Lima, 'Can Fashion Be Circular? A Literature Review on Circular Economy Barriers, Drivers, and Practices in the Fashion Industry's Productive Chain', *Sustainability*, 13 (21) (November 2021): 12246, doi: 10.3390/SU132112246.

[45] A. R. Shafqat, M. Hussain, Y. Nawab, M. Ashraf, S. Ahmad, and G. Batool, 'Circularity in Materials: A Review on Polymer Composites Made from Agriculture and Textile Waste', *Int. J. Polym. Sci.*, (1) (2023): 5872605, doi: 10.1155/2023/5872605.

[46] A. Palmsköld, 'Reusing Textiles: On Material and Cultural Wear and Tear', *Culture Unbound*, 7 (1) (March 2015): 31–43, doi: 10.3384/CU.2000.1525.157131.

[47] Meer, '1940s Fashion: Wartime Necessity and Modern Sustainability', (2024), www.meer.com/en/85710-1940s-fashion-wartime-necessity-and-modern-sustainability.

[48] Fashinnovation, 'How Industrial Revolution Changed Fashion', (2022), https://fashinnovation.nyc/fashinnovation-industrial-revolution-fashion/.

[49] P. Y. Donzé and R. Fujioka, 'European Luxury Big Business and Emerging Asian Markets, 1960–2010', *Bus Hist*, 57 (6) (August 2015): 822–40, doi: 10.1080/00076791.2014.982104.

[50] Pearls Magazine, 'Orsola de Castro: 'Upcycling Is the Key to Transform the Industry', (2023), https://pearlsmagazine.com/en/orsola-de-castro-upcycling-is-the-key-to-transform-the-industry/.

[51] Barbara Chiodi, 'Orsola de Castro, the Upcycling Pioneer', 2020, https://eco-a-porter.com/2020/05/29/orsola-de-castro-the-upcycling-pioneer/.

[52] 'Speedo LZR Racer "Upcycle" Student Collaboration: Exhibiting Fashion', (2012), https://fashionexhibitionmaking.arts.ac.uk/speedo-lzr-racer-upcycle-student-collaboration/.

[53] Jessica Wretlind, 'Nin Castle', (2010), https://jessicawretlind.wordpress.com/tag/nin-castle/.

[54] Worn Again, 'Q&A with CAP Member: Nin Castle', (2009), https://wornagain.co.uk/qa-with-cap-member-nin-castle/.

[55] M. Lee, 'Junky Styling: Breathing Life into Last Season's Clothes', The Ecologist, (August 2009), https://theecologist.org/2009/aug/18/junky-styling-breathing-life-last-seasons-clothes.

[56] A. D. Pranta, Md. R. Repon, A. Haji, M. Rahman and M. M. Rahman, 'Sustainable Merchandising for Recycled Textiles and Apparel in the Western Apparel Industry: Strategies, Challenges and Market Opportunities', *J Eng Fiber Fabr*, 20 (August 2025): 1–30, doi: 10.1177/15589250251359273.

[57] H. Seo and B. E. Jin, 'Engaging in Fashion Take-Back Programs: The Role of Loyalty and Perceived Benefits from a Social Exchange Perspective', *Sustainability*, 16 (22) (November 2024): 10031, doi: 10.3390/SU162210031.

[58] Y. Zhi and Y. Zhi, 'The Upcycling and Reconstruction of Garments and Fabrics', *Art and Design Review*, 10 (1) (January 2022): 72–102, doi: 10.4236/ADR.2022.101007.

[59] A. Kozlowski, Duan Wu, Mingyu Zhuang, Xinni Zhang, Yuheng Zhao, 'Towards Circular Fashion: Design for Community-Based Clothing Reuse and Upcycling Services under a Social Innovation Perspective', *Sustainability*, 15 (1) (December 2022): 262, doi: 10.3390/SU15010262.

[60] Fashion Revolution, 'Written Evidence to the Sustainability of the Fashion Industry Inquiry, UK, Environmental Audit Committee', (2018), www.fashionrevolution.org/fashion-revolution-written-evidence-to-the-sustainability-of-the-fashion-industry-inquiry-u-k-environmental-audit-committee/.

[61] P. Pandit, K. Singha, L. Kumar, S. Shrivastava, and V. Yashraj, 'Business Paradigm Shifting', in Pintu Pandit, Shakeel Ahmed, Kunal Singha, and Sanjay Shrivastava (eds), *Recycling from Waste in Fashion and Textiles: A Sustainable and Circular Economic Approach*, (2020), pp. 151–76, Wiley, doi: 10.1002/9781119620532.ch7.

[62] Luxury Tribune, 'Upcycling: Turning Trash into Fashion Treasures', (2021), www.luxurytribune.com/en/upcycling-turning-trash-into-fashion-treasures.

[63] ZoneModa Journal, 'View of Upcycling: The Art Nouveau of Fashion?', (2025), https://zmj.unibo.it/article/view/21931/19926.

[64] J. Yue, 'The Development and Popularity of Hip Hop Culture and Its Clothing: Supreme vs. Louis Vuitton', *Journal of Education, Humanities and Social Sciences*, 43 (December 2024): 16–20, doi: 10.54097/BNTKK071.

[65] L. Zhang and J. Hale, 'Extending the Lifetime of Clothing through Repair and Repurpose: An Investigation of Barriers and Enablers in UK Citizens', *Sustainability*, 14 (17) (September 2022): 10821, doi: 10.3390/SU141710821.

[66] Emily D'Silva, 'Upcycling Is Revolutionising the Fashion Industry', (2021), https://emilydsilva.com/2021/04/20/upcycling-is-revolutionising-the-fashion-industry/.

[67] M. Chidozie Ogwu, E. A. Kosoe and M. C. Ogwu, 'Innovative Approaches to Recycling, Upcycling, and Downcycling for Sustainable Waste Management', *CleanMat*, 2 (3) (September 2025): 242–61, doi: 10.1002/CLEM.70013.

[68] ArtReview, 'Fashion Designer Marine Serre's Dark Art of Upcycling', (2022)(, https://artreview.com/fashion-designer-marine-serre-dark-art-of-upcycling/.

[69] Walid al Damirji, 'One of a Kind: Walid al Damirji's Antique Fashion – The Thread', (2022), https://blog.fabrics-store.com/2022/11/27/one-of-a-kind-walid-al-damirjis-antique-fashion/.

[70] Creative Boom, 'Fashion as a "Force for Change": Bethany Williams Shows Us an Alternative Fashion Industry', (2022), www.creativeboom.com/news/bethany-williams/.

[71] European Circular Economy Stakeholder Platform, 'UPMADE® by Reet Aus Turns Textile Leftovers from Cost into Value', (n.d.), https://circulareconomy.europa.eu/platform/en/good-practices/upmader-reet-aus-turns-textile-leftovers-cost-value.

[72] NZ Post, 'Social Procurement at NZ Post', (n.d.), www.nzpost.co.nz/about-us/sustainability/social-procurement-at-nz-post.

[73] Korea.net, 'Upcycling Fashion Brand Gains Fame for Designing Suits for BTS: The Official Website of the Republic of Korea', (2021), www.korea.net/NewsFocus/HonoraryReporters/view?articleId=204516.

[74] Denimology, 'RE/DONE Spring 2025: A Closer Look', (2025), https://denimology.com/re-done-spring-2025-a-closer-look/.

[75] Peak Point, 'Image Garments Turns Surplus into Sustainable Fashion', (2025), www.peakpoint.pk/en/2025/09/09/image-garments-sustainable-fashion/.

[76] A. Gwilt, 'What Prevents People Repairing Clothes? An Investigation into Community-Based Approaches to Sustainable Product Service Systems for Clothing Repair', *Making Futures Journal*, 3 (2014), https://shura.shu.ac.uk/8125/.

[77] CE Grow Circular, '9R Framework: CE Grow Circular', (2017), https://grow-circular.eu/knowledge-base/9r-framework/.

[78] W. Gu, 'Research on Strategy Optimization of Sustainable Development towards Green Consumption of Eco-Friendly Materials', *J King Saud Univ Sci*, 36 (6) (July 2024): 103190, doi: 10.1016/J.JKSUS.2024.103190.

[79] M. L. Kambanou, J. Matschewsky and A. Carlson, 'Business Models and Product Designs that Prolong the Lifetime of Construction Workwear: Success, Failure and Environmental Impacts', *Resour Conserv Recycl*, 206 (July 2024): 107602, doi: 10.1016/J.RESCONREC.2024.107602.

[80] U. N. Haq and S. M. R. Alam, 'Implementing Circular Economy Principles in the Apparel Production Process: Reusing Pre-Consumer Waste for Sustainability of Environment and Economy', *Cleaner Waste Systems*, 6 (December 2023): 100108, doi: 10.1016/J.CLWAS.2023.100108.

[81] L. Malooly and T. Daphne, 'R-Strategies for a Circular Economy', (2023), www.circularise.com/blogs/r-strategies-for-a-circular-economy.

[82] A. Diaz, T. Reyes and R. J. Baumgartner, 'Implementing Circular Economy Strategies during Product Development', *Resour Conserv Recycl*, 184 (September 2022): 106344, doi: 10.1016/J.RESCONREC.2022.106344.

[83] R. Rathinamoorthy, 'Circular Fashion', in *Circular Economy in Textiles and Apparel: Processing, Manufacturing, and Design*, (2019), pp. 13–48, doi: 10.1016/B978-0-08–102630-4.00002–9.

[84] M. Islam, M. Shamsuzzaman, H. M. R. U. Hasan and M. A. R. Atik, 'Environmental Sustainability of Fashion Product Made from Post-consumer Waste: Impact across the Life Cycle', *Sustainability*, 17 (5) (February 2025): 1917, doi: 10.3390/SU17051917.

[85] Muhammad Babar Ramzan, Muhammad Salman Habib, Muhammad Omair, Jawad Naeem, Hajra Mustafa, Muhammad Waqas Iqbal, Asif Iqbal Malik, 'Role of Design for Disassembly in Educating Consumers for Circular Behavior', *Sustainability*, 15 (21) (October 2023): 15505, doi: 10.3390/SU152115505.

[86] R. Lin, M. Lukáč and M. Leake, 'Refashion: Reconfigurable Garments via Modular Design', (September 2025), pp. 1–18, doi: 10.1145/3746059.3747632.

[87] A. Joy, J. F. Sherry, A. Venkatesh, J. Wang and R. Chan, 'Fast Fashion, Sustainability, and the Ethical Appeal of Luxury Brands', *Fashion Theory: Journal of Dress Body and Culture*, 16 (3) (2012): 273–95, doi: 10.2752/175174112X13340749707123.

[88] M. Wang, Y. Jing, J. Wang, P. Mika, F. Li and Y. Yan, 'Design Strategies for Modular Demountable Building Products Oriented to Design for Manufacturing and Assembly: A Case Study of M-Box1.0', *Buildings*, 15 (18) (September 2025): 3424, doi: 10.3390/BUILDINGS15183424.

[89] A. Pruhs, A. Kusch, J. Woidasky and T. Viere, 'Design for Circularity in Manufacturing Industries: Operationalisation and Decision Support', *Resour Conserv Recycl*, 202 (March 2024): 107376, doi: 10.1016/J.RESCONREC.2023.107376.

[90] G. Burini, J. Xu, M. Pero and E. Sandberg, 'Reverse Supply Chain Configurations in the Fashion and Textile Industry', *Sustain Prod Consum*, 56 (June 2025): 504–18, doi: 10.1016/J.SPC.2025.04.016.

[91] A. Farrukh and A. Sajjad, 'Drivers for and Barriers to Circular Economy Transition in the Textile Industry: A Developing Economy Perspective', *Sustainable Development*, 32 (6) (December 2024): 7309–29, doi: 10.1002/SD.3088.

[92] Infinited Fiber, 'About Us', (n.d.), https://infinitedfiber.com/about-us/.

[93] Evrnu, 'About Us', (2004), www.evrnu.com/about-us.

[94] B. Webb, 'Renewcell Secures a Future', Vogue, (June 2024), www.vogue.com/article/renewcell-secures-a-future-circulose.

[95] Circ, 'About Us', (2025), https://circ.earth/about-us/.

[96] I Love Ski, 'Patagonia Turns Repair into Environmental Commitment: I Love Ski®', (13 March 2025), www.iloveski.org/en/2025/03/13/worn-wear-patagonia/.

[97] Michael Brady, 'How Patagonia's Repairs Program Drives Loyalty and Customer Advocacy: CX Dive', (24 February 2025), www.customerexperiencedive.com/news/patagonia-repairs-program-loyalty-customer-advocacy/740670/.

[98] H. Seo and B. E. Jin, 'Engaging in Fashion Take-Back Programs: The Role of Loyalty and Perceived Benefits from a Social Exchange Perspective', *Sustainability*, 16 (22) (November 2024): 10031, doi: 10.3390/SU162210031.

[99] Lindström Group, 'Simplify Workwear with a Digitally-Supported Solution', (18 February 2025), https://lindstromgroup.com/simplify-workwear-with-a-digitally-supported-solution/.

[100] Redress Design Award, 'Reverse Resources', (18 February 2025), www.redressdesignaward.com/academy/resources/reverse-resources.

[101] Reverse Resources, 'White Paper: Digitally Enhanced Circular Economy Within Global Fashion Supply Chains', (1 February 2025), https://reverseresources.net/white-paper-digitally-enhanced-circular-economy-within-global-fashion-supply-chains/.

[102] S. Z. Haider, Y. Nawab, B. Qadir and M. Hussain, 'Impact of Post-industrial and Post-consumer Waste on the Characteristics of Denim Fabric Made from Recycled Yarns', *Journal of Industrial Textiles*, 55 (January 2025): 1–27, doi: 10.1177/15280837251322951.

[103] Jana Enking, Amrei Becker, Gilbert Schu, Marcel Gausmann, Stefano Cucurachi, Arnold Tukker, and Thomas Gries, 'Recycling Processes of Polyester-Containing Textile Waste: A Review', *Resour Conserv Recycl*, 219 (June 2025): 108256, doi: 10.1016/J.RESCONREC.2025.108256.

[104] D. Peti, J. Dobránsky and P. Michalík, 'Recent Advances in Polymer Recycling: A Review of Chemical and Biological Processes for Sustainable Solutions', *Polymers*, 17 (5) (February 2025): 603, doi: 10.3390/POLYM17050603.

[105] O. O. Alabi, T. O. Akande, O. Joseph Gbadeyan and N. Deenadayalu, 'Advanced Technologies for Plastic Waste Recycling: Examine Recent Developments in Plastic Waste Recycling Technologies', *RSC Adv*, 15 (48) (2025): 40541, doi: 10.1039/D5RA06715D.

[106] T. Samardzioska, M. Jovanoska and V. Grujoska, 'Sustainable Thermal Insulation Derived from Recycled Textile Waste', *Advances in Environmental and Engineering Research*, 4 (1) (January 2023): 1–8, doi: 10.21926/AEER.2301004.

[107] H. Sezgin, M. Kucukali-Ozturk, O. B. Berkalp and I. Yalcin-Enis, 'Design of Composite Insulation Panels Containing 100% Recycled Cotton Fibers and Polyethylene/Polypropylene Packaging Wastes', *J Clean Prod*, 304 (July 2021): 127132, doi: 10.1016/J.JCLEPRO.2021.127132.

[108] Fiona Charnley, Ruth Cherrington, Florian Mueller, Ajay Jain, Cherie Nelson, Saskia Wendland, and Sonia Ventosa, 'Retaining Product Value in Post-consumer Textiles: How to Scale a Closed-Loop System', *Resour Conserv Recycl*, 205 (June 2024): 107542, doi: 10.1016/J.RESCONREC.2024.107542.

[109] Manikant Tripathi, Minaxi Sharma, Saroj Bala, Vijay Kumar Thakur, Anoop Singh, Kavya Dashora, Phil Hart, and Vijai Kumar Gupta, 'Recent Technologies for Transforming Textile Waste into Value-Added Products: A Review', *Curr Res Biotechnol*, 7 (January 2024): 100225, doi: 10.1016/J.CRBIOT.2024.100225.

[110] Sustainability Directory, 'How Does Chemical Recycling Compare to Mechanical Recycling? Question', (11 February 2025), https://fashion.sustainability-directory.com/question/how-does-chemical-recycling-compare-to-mechanical-recycling/.

[111] Kuyichi, 'Mechanical vs Chemical Recycling', (n.d.), https://kuyichi.com/de/blog/mechanical-vs-chemical-recycling/.

[112] Mehrdad Seifali Abbas-Abadi, Brecht Tomme, Bahman Goshayeshi, Oleksii Mynko, Yihan Wang, Sangram Roy, Rohit Kumar, Bhargav Baruah, Karen De Clerck, Steven De Meester, Dagmar R. D'hooge and Kevin M. Van Geem, 'Advancing Textile Waste Recycling: Challenges and Opportunities across Polymer and Non-Polymer Fiber Types', *Polymers*, 17 (5) (March 2025): 628, doi: 10.3390/POLYM17050628.

[113] K. Mishra, M. Islam, R. Yin and A. West, 'A Brief Review of Mechanical Recycling of Textile Waste', *Textiles*, 5 (4) (September 2025): 41, doi: 10.3390/TEXTILES5040041.

[114] Jana Enking, Amrei Katharina Sabine Becker, Gilbert Schu, Marcel Gausmann, Stefano Cucurachi, Arnold Tukker, and Thomas Gries, 'Recycling Processes of Polyester-Containing Textile Waste: A Review', *Resour Conserv Recycl*, 219 (June 2025): 108256, doi: 10.1016/J.RESCONREC.2025.108256.

[115] J. Ghosh, M. R. Repon, N. S. Rupanty, T. R. Asif, M. I. Tamjid and V. Reukov, 'Chemical Valorization of Textile Waste: Advancing Sustainable Recycling for a Circular Economy', *ACS Omega*, 10 (12) (April 2025): 11697–722, doi: 10.1021/ACSOMEGA.4C10616.

[116] Re_fashion, 'Homepage', (2024), https://pro.refashion.fr/.

[117] K. Choudhury, M. Tsianou and P. Alexandridis, 'Recycling of Blended Fabrics for a Circular Economy of Textiles: Separation of Cotton, Polyester, and Elastane Fibers', *Sustainability*, 16 (14) (July 2024): 6206, doi: 10.3390/SU16146206.

[118] Re_fashion, 'Chemical Recycling of Textiles', (19 May 2025), https://pro.refashion.fr/en/news/recyclage/chemical-recycling-of-textiles.

[119] Circulose, 'About', (n.d.), www.circulose.com/en/about/.

[120] Syre, 'About', (n.d.), www.syre.com/about.

[121] GR3N, 'A New Recycling Concept', (n.d.), https://gr3n-recycling.com/about-us/.

[122] Worn Again, 'About Us', (n.d.), https://wornagain.co.uk/about-us/.

[123] Ambercycle, 'Technology', (n.d.), www.ambercycle.com/technology.

[124] Carbios, 'Other Innovations', (n.d.), www.carbios.com/en/other-innovations/.

[125] BlockTexx®, 'About', (n.d.), www.blocktexx.com/about.

[126] Tarun Grover, Lijing Wang, Rajkishore Nayak, Rajiv Padhye and Asimananda Khandual, 'Sustainable Strategies for the Valorization of Cotton and Cotton Blended Waste: Pathways to Circular Economy', *Cellulose*, 32 (7) (May 2025): 4077–100, doi: 10.1007/S10570-025-06524-9.

[127] Worldly, 'Product Environmental Footprint: What It Is and What You Need to Know', (2025), https://worldly.io/resources/product-environmental-footprint-what-it-is-and-what-you-need-to-know/.

[128] 'What Is Material Circularity Indicator? Question', (3 December 2025), https://pollution.sustainability-directory.com/question/what-is-material-circularity-indicator/.

[129] E. Kaynak, I. S. Piri and O. Das, 'Revisiting the Basics of Life Cycle Assessment and Lifecycle Thinking', *Sustainability*, 17 (16) (August 2025): 7444, doi: 10.3390/SU17167444

[130] Circle Economy, 'Circle Fashion Tool: Textiles Programme', (n.d.), https://legacy.circle-economy.com/programmes/textiles/circle-fashion-tool.

[131] Ellen MacArthur Foundation, 'The Circular Design Guide', (23 November 2022), www.ellenmacarthurfoundation.org/circular-design-guide/overview.

[132] Green Story, 'A Comparative Delve into the Higg Material Sustainability Index (MSI) Methodology Versus Green Story's LCA Study', (23 November 2022), www.greenstory.io/blogs/a-comparative-delve-into-the-higg-material-sustainability-index-msi-methodology-versus-green-storys-lca-study.

[133] B. F. Mon, M. Hayajneh, N. A. Ali, F. Ullah, H. Ullah and S. Alkobaisi, 'Digital Twins in the IIoT: Current Practices and Future Directions toward Industry 5.0', *Computers, Materials and Continua*, 83 (3) (May 2025): 3675–712, doi: 10.32604/CMC.2025.061411.

[134] T. K. Agrawal, V. Kumar, R. Pal, L. Wang and Y. Chen, 'Blockchain-Based Framework for Supply Chain Traceability: A Case Example of Textile and Clothing Industry', *Comput Ind Eng*, 154 (April 2021): 107130, doi: 10.1016/J.CIE.2021.107130.

[135] T. K. Agrawal, V. Kumar, R. Pal, L. Wang and Y. Chen, 'Blockchain-Based Framework for Supply Chain Traceability: A Case Example of Textile and Clothing Industry', *Comput Ind Eng*, 154 (April 2021): 107130, doi: 10.1016/J.CIE.2021.107130.

[136] M. Salati, A. A. Costa and J. D. Silvestre, 'A Comprehensive Review of Dynamic Life Cycle Assessment for Buildings: Exploring Key Processes and Methodologies', *Sustainability*, 17 (1) (December 2024): 159, doi: 10.3390/SU17010159.

[137] D. A. Lopes Silva, A. O. Nunes, C. M. Piekarski, V. A. da Silva Moris, L. S. M. de Souza and T. O. Rodrigues, 'Why Using Different Life Cycle Assessment Software Tools Can Generate Different Results for the Same Product System? A Cause–Effect Analysis of the Problem', *Sustain Prod Consum*, 20 (October 2019): 304–15, doi: 10.1016/J.SPC.2019.07.005.

[138] EPLCA, 'European Platform on LCA', (n.d.), https://eplca.jrc.ec.europa.eu/EnvironmentalFootprint.html.

[139] B. I. Oladapo, M. A. Olawumi and F. T. Omigbodun, 'AI-Driven Circular Economy of Enhancing Sustainability and Efficiency in Industrial Operations', Sustainability (Switzerland), 16 (23) (December 2024): 10358.

[140] H. Nisa, R. Van Amber, J. English, S. Islam, G. McCorkill and A. Alavi, 'A Systematic Review of Reimagining Fashion and Textiles Sustainability with AI: A Circular Economy Approach', *Applied Sciences*, 15 (10) (May 2025): 5691, doi: 10.3390/APP15105691.

[141] L. Ramos, F. Rivas-Echeverría, A. G. Pérez and E. Casas, 'Artificial Intelligence and Sustainability in the Fashion Industry: A Review from 2010 to 2022', *SN Applied Sciences*, 5 (12) (December 2023): 387, doi: 10.1007/S42452-023-05587-2.

[142] G. Batool and Y. Nawab, 'Regulation in Recycling and Circularity: Future Prospective', in S. R. Batool, S. Ahmad, Y. Nawab and M. Hussain (eds), *Circularity in Textiles: Textile Science and Clothing Technology*, Cham: Springer, (2023). https://doi.org/10.1007/978-3-031-49479-6_11.

[143] Sustainability Directory, 'Consumer Behavior Shift towards Product Longevity: Scenario', (5 October 2025), https://prism.sustainability-directory.com/scenario/consumer-behavior-shift-towards-product-longevity/.

[144] K. Stasiuk, A. Tischner, S. Teilmann-Lock and D. Beldiman, 'Empirical Perspective: Upcycling with Branded Material through the Eyes of Consumers', in Péter Mezei and Heidi Härkönen (eds), *The Research Handbook on Upcycling and Intellectual Property*, Cambridge: Cambridge University Press, (2025), doi: 10.2139/SSRN.5296925.

[145] V. Nath and R. Agrawal, 'Barriers to Consumer Adoption of Sustainable Products: An Empirical Analysis', *Social Responsibility Journal*, 19 (5) (April 2023): 858–84, doi: 10.1108/SRJ-12–2020–0495.

[146] Ye Sang, Heeseung Yu and Eunkyoung Han, 'Understanding the Barriers to Consumer Purchasing of Zero-Waste Products', *Sustainability*, 14 (24) (December 2022): 16858, doi: 10.3390/SU142416858.

[147] W. C. Munonye, P. S. Munonye, D. I. Munonye and O. A. Akinloye, 'Behavioural Nudges in Circular Economy Transitions: Evidence and Gaps from Urban Consumption Systems', *Academia Environmental Sciences and Sustainability*, 2 (4) (October 2025): 1–21, doi: 10.20935/ACADENVSCI7946.

[148] L. Congiu, E. Botta and M. Zoli, 'Biases and Nudges in the Circular Economy: A Review', *SSRN Electronic Journal*, 22 (5) (October 2024): 583, doi: 10.2139/SSRN.4975692.

[149] B. Amiri, A. Jafarian and Z. Abdi, 'Discover Sustainability Nudging towards Sustainability: A Comprehensive Review of Behavioral Approaches to Eco-friendly Choice', 5 (2024): 444, doi: 10.1007/s43621-024-00618-3.

[150] S. Sala, A. M. Amadei, A. Beylot and F. Ardente, 'The Evolution of Life Cycle Assessment in European Policies over Three Decades', *International Journal of Life Cycle Assessment*, 26 (12) (December 2021): 2295–314, doi: 10.1007/S11367-021-01893-2.

[151] Eco-shaper, 'EU Textile EPR: New Law for Brands and Consumers', (20 September 2025), https://eco-shaper.com/eu-textile-epr-what-the-new-law-means-for-brands-consumers/.

[152] Bluesign, 'EU Textile Waste Rules and Extended Producer Responsibility', (22 September 2025), www.bluesign.com/en/eu-textile-waste-rules-extended-producer-responsibility/.

[153] Sainsbury's, 'Transforming Our Industry for the Planet', (31 March 2022), https://corporate.sainsburys.co.uk/media/o45dq43i/wrap-textiles-2030-circularity-roadmap-20220331.pdf.

[154] Re_fashion, 'Understanding the Anti-waste Law for a Circular Economy (AGEC) in the Clothing, Household Linen and Footwear Sector', (n.d.), https://pro.refashion.fr/en/marketer/understanding-the-agec-law.

[155] TrusTrace, 'The Garment Worker Protection Act: Leading Fashion Supply Chain Traceability Software', (n.d.), https://trustrace.com/knowledge-hub/garment-worker-protection-act.

[156] UNEP, 'International Trade Centre Lead UN Alliance for Sustainable Fashion: SDG Knowledge Hub', (29 January 2019), https://sdg.iisd.org/news/unep-international-trade-centre-lead-un-alliance-for-sustainable-fashion/.

[157] Giovanni Formentini, Joao P. J. Prioli, Junwon Ko, Buddhika Hapuwatte, Vincenzo Ferrero, Fazleena Badurdeen, Jeremy L. Rickli and Devarajan Ramanujan, 'A Review of Disassembly Systems for Circular Product Design', *J Clean Prod*, 506 (May 2025): 145459, doi: 10.1016/J.JCLEPRO.2025.145459.

[158] S. F. Jensen, J. H. Kristensen, A. Christensen and B. V. Waehrens, 'An Ecosystem Orchestration Framework for the Design of Digital Product Passports in a Circular Economy', *Bus Strategy Environ*, 33 (7) (November 2024): 7100–17, doi: 10.1002/BSE.3868.

[159] A. Christensen, V. Stingl, M. Omair and B. V. Wæhrens, 'Digital Product Passport in Support of Data-Driven End-of-Use Strategies: A Systems Design Perspective', *Cleaner Environmental Systems*, 19 (December 2025): 100354, doi: 10.1016/J.CESYS.2025.100354.

[160] D. Moorhouse, 'Making Fashion Sustainable: Waste and Collective Responsibility', *One Earth*, 3 (1) (July 2020): 17, doi: 10.1016/J.ONEEAR.2020.07.002.

[161] Z. Xue and Z. Huang, 'Current Development and Future Prospects of Designing Sustainable Fashion', *Autex Research Journal*, 23 (3) (September 2023): 420–31, doi: 10.2478/AUT-2022-0013/XML.

[162] A. Lanzalonga, Patrizia Gazzola, Enrica Pavione, Rainer Hillebrand, Vincenza Vota, and Rebecca Rosa, 'The Circular Economy and the Role of Technology in the Fashion Industry: A Comparison of Empirical Evidence', *Sustainability*, 17 (7) (April 2025): 3104, doi: 10.3390/SU17073104.

[163] A. T. Rosário, P. Lopes and F. S. Rosário, 'Sustainability and the Circular Economy Business Development', *Sustainability*, 16 (14) (July 2024): 6092, doi: 10.3390/SU16146092.

[164] L. Aakanksha, M. Aravendan, L. Aakanksha and M. Aravendan, 'Impacts of Transparency and Traceability on Fashion Supply Chain System', *Intell Inf Manag*, 15 (3) (April 2023): 100–19, doi: 10.4236/IIM.2023.153006.

[165] A. Abdelmeguid, M. Afy-Shararah and K. Salonitis, 'Towards Circular Fashion: Management Strategies Promoting Circular Behaviour along the Value Chain', *Sustain Prod Consum*, 48 (July 2024): 143–56, doi: 10.1016/J.SPC.2024.05.010.

[166] David Watson, Nikola Kiørboe, David Palm, Haben Tekie, SteveHarris, Tomas Ekvall, Thomas Lindhqvist and Kari-Ann Lyng, 'EPR Systems and New Business Models: Reuse and Recycling of Textiles in the Nordic Region', (2013), https://norden.diva-portal.org/smash/get/diva2:720972/FULLTEXT02.pdf.

[167] UNESCO, 'Education for Sustainable Development', (n.d.), www.unesco.org/en/sustainable-development/education.

[168] S. Z. Qamar, N. Al Maskari, and T. Pervez, 'Integrating Sustainability into Engineering Education: Execution and Obstacles', in *Proceedings of the International Conference on Industrial Engineering and Operations Management*, Muscat, Oman, (2024), paper 25.

[169] D. Andrews, 'The Circular Economy, Design Thinking and Education for Sustainability', *Local Econ*, 30 (3) (May 2015): 305–15, doi: 10.1177/0269094215578226.

Section 3

Business strategies, policy and industry applications

9 Sustainable design for small and medium enterprises

Muzzamal Hussain and Sadia Razzaq

9.1 Introduction

Currently, the application of small and medium-sized enterprises (SMEs) has developed tremendously, and they are mainly supplying to the economic and social areas of the region. The European Parliament's report states that 99 per cent of European Union (EU) enterprises are SMEs [1]. SMEs play an important role in the EU in terms of providing a large number of jobs. During 2015, approximately 23 million SMEs represented a significant contribution of €3.9 trillion in value, as well as provided jobs to 90 million people. These statistics highlight the crucial role of innovations, jobs and business growth [2]. SMEs act as the main supporting pillar in developing regions, showing 70 per cent of jobs and 90 per cent of business growth worldwide. SMEs chiefly impact gross domestic product (GDP), raise local enterprises, promote regional progress, as well as provide major assistance in poverty reduction and economic growth in South Asia and Africa.

The textile sector plays a significant role in global trade. Major textile-manufacturing regions, for example, Bangladesh, Pakistan and India, have SMEs that are greatly connected with global value chains. These enterprises include weaving, dyeing, printing and tailoring. SMEs produce the big profit in the ready-made garments in Bangladesh, which provides the highest export revenue and employment to 4 million individuals [3]. Handloom weaving clusters such as Varanasi are SME-determined corporations in India, also they have succeeded in global gratitude for their skilled, sustainable production practices [4]. These enterprises originate from social and cultural civilisations in which family-run businesses are common. They rely on community customs and training for learning as compared to formal schemes. Their main purpose is to provide cheap goods and guaranteed items, which helps them to practise new processes without systematic planning. This method makes them flexible and stiff in challenging states. They also create some issues, for instance, low efficiency, low quality of products and difficulties in competing with international standards [5].

According to the European Green Deal, the number of legislation and regulations in terms of textile sustainability approaches is considered; there is high pressure to adopt these standards by all the industrialists or stakeholders. Counting these difficulties, SMEs are in an outstanding situation to advance long-term revolution in the textile industry. Growth of green components, circular-economy models and sustainable practices is made easier at these small levels. This helps flexibility and decision-making skills [6]. Different initiatives from the International Labor Organization (ILO) and the United Nations Industrial Development Organization (UNIDO) have initiated to focus on improving SMEs proficiencies in market contact, supply

DOI: 10.4324/9781003638148-12

chain transparency and sustainability. For instance, UNIDO's SWITCH-Asia Programme has helped SMEs in Pakistan as well as Bangladesh to implement resource-competent equipment and lower ecological influences [7]. The World Wildlife Fund (WWF) has commenced different initiatives to facilitate SMEs in textiles. Another initiative has also started in Bangladesh, Pakistan and Cambodia to support the SMEs by the GIZ's FABRIC programme, which assists in developing energy efficiency and waste reduction approaches [8].

Sustainability is a major portion of these enterprises. This not only helps with developing things but also supports the regulations and market transformations. Currently, a high number of individuals are concerned about social concerns and the ecosystem. Adoption of sustainable approaches makes the SMEs more competitive against their competitors. Consumers request eco-friendly things, specifically in fashion, textiles and food sectors. This is reported in some research that younger customers, principally in North America and Europe, are even eager to pay extra charges for these kinds of stuff [9]. Another attribute of the sustainability approach is the long-term endurance in a changing ecosystem. However, SMEs encounter some issues, which happen due to instant incorporation into global value chains, high struggle from international products, universal climate shift and acquiring buyer preferences [10]. These small enterprises also find that the new business demonstrates that they utilise the digital traceability systems, renewable resources, or the circular economy approaches commonly for better market trends. Moreover, this innovation enhances flexibility and positions SMEs as leaders in the development of green markets [11]. Moreover, SMEs in Bangladesh and India that embrace low-impact dyeing instruments or water-recycling procedures are increasingly being selected by global consumers who are involved in traceability and ecological direct persistence [12].

During COVID-19, many countries encountered anxiety, economic downfall and social disturbance, climate change and living expenses. In this situation, SMEs contribute 90 per cent of the whole business, which is due to the existence of small enterprises. The International Trade Centre (ITC) stated that 70 per cent of employment in the textile areas is provided by these enterprises in developing regions, emphasising their participation in labour-intensive production [13]. But SMEs are vulnerable to universal issues as they lack funds, modern technologies and support. These enterprises have a restricted capability to handle big challenges; therefore, the pandemic can affect their work, specifically in the textile industries.

This chapter will outline how textile SMEs can adopt sustainability practices to meet the different challenges. This chapter starts with various kinds of SMEs with their limitations and impacts, then provides the details of low-cost solutions, circular approaches, advancing water and energy utilisation and sustainable resources. It also explains the function of encouraging policies and organisations in accelerating the transition of SMEs. Real-world examples from India, Pakistan, Bangladesh and the UK illustrate how small companies have previously leading sustainable transformation, emphasising their key applications in forming a better sustainable textile and fashion industry.

9.2 Enterprise fundamentals

Recognising SMEs' primary values and classifications in the textile and fashion industries is necessary for constructing sustainability approaches. Despite having a collection of sizes, structures and schemes of operation, these companies all have some prospects and restrictions that shape their route towards sustainable development. This portion outlines key concepts applied to classify SMEs and the major types of textile and fashion SMEs.

Table 9.1 Classification of SMEs depends on their size.

Category	*Employ*	*Expenses*	*Regions*	*Challenges*
Microenterprise	Less than 10 people	Low-cost startup	Developing countries	Limited resources, market access problems
Small enterprise	10–50 people	Low to moderate	Globally	Constrained growth
Medium enterprise	Up to 250 people	Moderate to high	Globally	Higher competition and regulation

9.2.1 Classifications of SMEs based on size and turnover

Globally, the classification of SMEs depends on the number of staff, annual earnings, or investment in plant and machinery. In this chapter, SMEs are categorised by their size (see Table 9.1) and operational scale as follows:

- **Microenterprise**: A microenterprise, also known as a microbusiness, is a small business that employs a few individuals and is established with a small amount of funds. It generally runs with fewer than ten people. Microenterprises can be owned and operated by one person. They are simple and low-cost to start up. To be considered as a microenterprise, they also need a balance sheet or turnover under a certain amount [14].
- **Small enterprise**: A small enterprise is a privately owned business that operates with a limited number of employees, assets, as well as expenses. Although definitions may vary depending on the country, a small enterprise normally employs between 10 to 50 employees, relies on local or regional markets and has simple management structures. These enterprises are important in local economies and supply chains [15].
- **Medium enterprise**: This is defined by its number of employees and annual income, though the exact conditions can fluctuate from region to region. For example, the EU categorises a medium enterprise, having less than 250 workers and an annual income of under €50 million, while the UK's Department for Business sets the turnover level at less than £25.9 million. Rather than these regional differences, medium enterprises generally function at a larger level than small enterprises, and they also contribute to global markets [16].

9.2.2 Types of textile or fashion SMEs

There are different kinds of SMEs in the textile sector. They are different to each other due to their sizes, operational mode and their targeted markets. These enterprises exhibit a crucial role in the textile industry, by making major contributions to society, job development and innovative practices. These kinds of SMEs categorised into three different following types:

- **Manufacturing SMEs**: These enterprises are the main pillar of the textile development practices. They are contributed to different fields of textiles like spinning, weaving, dyeing and finishing units. Commonly, 10 to 250 individuals are working under this category. This enterprise performs under regional supply chains. In many developing countries, such as Pakistan, India and Bangladesh, these types of enterprises help to contribute to the national GDP as well as exports. Small-scale dyeing units and fabric mills in Bangladesh are main examples of this type of enterprise. They serve domestically and internationally. But sometimes they often work with minimum ecological safeguards and restricted access to green approaches

[17, 18]. These are the main reasons that they are facing many issues in the implementation of sustainable practices, specially involving effluent procedures, energy utilisation and utilisation of harmful substances.

- **Artisan/Handicraft SMEs**: This type of enterprise represents a cultural significance of the textile and fashion industry. These SMEs are based on traditional practices, which are labour-intensive and community-based. They mainly work in different fields of the textile sector, such as natural dyeing, weaving, block printing and embroidery. Famous regions for these kinds of SMEs are India, Sri Lanka, Nepal and Bangladesh. For instance, Kanchipuram silk weavers, Chikankari and Banarasi handloom clusters in India display the richness of textile heritage as well as the economic significance of this area [19]. These enterprises not only sustain cultural attributes but also recommend sustainable procedures of low energy utilisation and the use of biodegradable raw constituents. UNESCO stated that artisan SMEs are one of the greatest sources of rural jobs (especially for women) in South Asia [20]. But these SMEs face various problems rather than their ecological values. So, the problems are a lack of market access, a lack of digital tools and training. Government and non-government interventions are concentrating on connecting these enterprises with global value chains via e-commerce policies.
- **Hybrid models**: A combination of traditional craft elements and modern business models is known as a hybrid model, for example, e-commerce boutiques and social enterprises. These kinds of enterprises prioritise transparency as well as circularity. Normally, these SMEs are initiated by young industrialists, focusing on developing eco-sensible alternatives to fast fashion. These SMEs utilises different technologies like Computer Aided Design tools, online platforms, digital product designs, blockchain (transparency) and AI (anticipation).

Rapanui in the UK works on a circular construction model, utilising organic cotton as well as old garments for recycling. Some brands in India, like Okhai and the Indian Ethnic Co. have assumed digital-first advances to sell handcrafted goods straight to consumers [21]. These hybrid demonstrations are encouraging for SMEs because they assist responsive production, reduce inventory waste and reach ecologically aware clients worldwide. Figure 9.1 demonstrates the various kinds of textile and fashion SMEs.

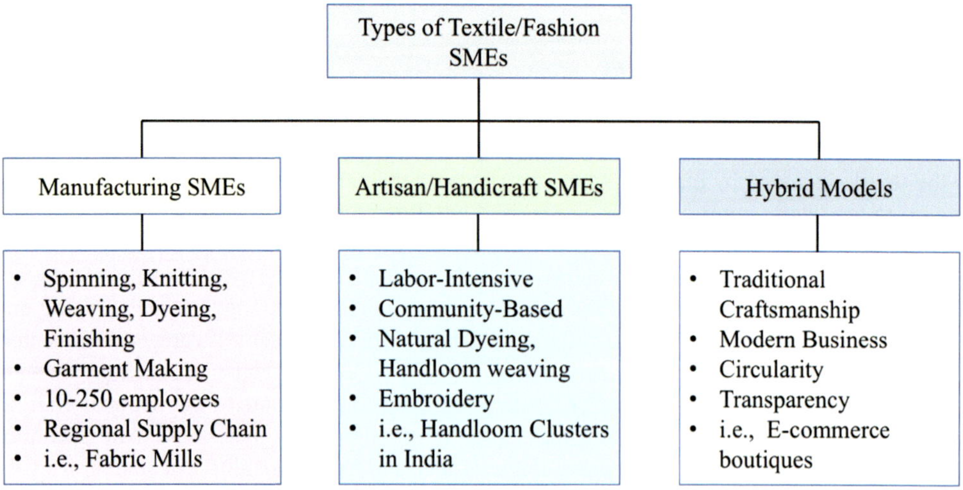

Figure 9.1 Various kinds of textile SMEs.

9.3 Sustainability challenges for SMEs

SMEs play a tremendous role in the textile sectors of different developing regions. But their capability to implement sustainable practices is inhibited by several issues. This section summarises the main obstacles faced during sustainability approaches in SMEs, especially in developing regions that SMEs often face, particularly in developing countries.

9.3.1 Financial barriers (green premium)

Regardless of various initiatives by the State Bank of Pakistan (SBP) to form a methodology to fund local organisations, SMEs have constantly faced obstacles, like BP's prudential regulations and the policy for support of SME expenses, aimed at increasing credit accessibility. Moreover, the employment of the Financial Institutions (Secured Transactions) Act, 2016, has been essential for accelerating the credit approach for SMEs, utilising transferable assets as insurance [22, 23]. Only 6 per cent of SMEs applied for bank loans in 2019. The main challenge for SMEs consists, lack of financial awareness, a lack of certifications and insurances and the implementation of complex loan policies, which inhibit further access to more bank loans.

In addition, some businesses faced various funding problems as banks chose to invest in government protections over proposing an accessible attribute to the private zone [24]. Financial problems are the main economic obstacle preventing SMEs from implementing sustainable technologies. This comprises of restricted approach to reasonable funding, a high upfront expense of green tools and the 'green premium'.

Green premium defines the extra cost related to sustainable opportunities when contrasted to traditional substitutes. This represents the big cost difference in textile and fashion SMEs, especially in developing regions. This cost difference figures as a major obstacle to embracing eco-friendly approaches. Insufficient supply, certification applications and specific manufacturing processes are the primary issues for sustainable methods, but this affects the sustainable items (organic cotton as well as recycled fibres) commonly more expensive than conventional substances [25]. This financial gap delays smaller companies from transitioning to greener components, mostly when profit margins are low. Additionally, green technologies, such as water-efficient dyeing machines or solar energy systems, require considerable funds that many SMEs may find expensive. Finding capital is a major problem for SMEs seeking sustainability, according to an Organisation for Economic Co-operation and Development study. This is because traditional financial institutions are often hesitant to offer money to smaller enterprises without insurance or a strong credit history [26]. The lack of knowledge among SMEs concerning green finance alternatives and the existence of numerous investors make this trouble worse. In some places, financial aid programmes or green financing options may be available.

9.3.2 Knowledge gaps (linear economy vs circular economy)

The conventional linear model of textile-manufacturing is take, make, use and dispose. First of all, fibres are converted into yarn in this model, which is then transform into fabric. After the material is made, this is turned into finished textile items for use (Figure 9.2). Geissdoerfer and colleagues stated that our industrial system still adheres to the linear economic model, where end-products are disposed of, leading to environmental harm [27]. Lenzing shows that 63 per cent fibres come from petrochemicals, which cause carbon-dioxide (CO_2) secretions. The remaining 37 per cent mainly involves cotton, which requires a high quantity of H_2O and dangerous pesticides during agriculture farming, leading to water deficiency as well as toxic pesticide

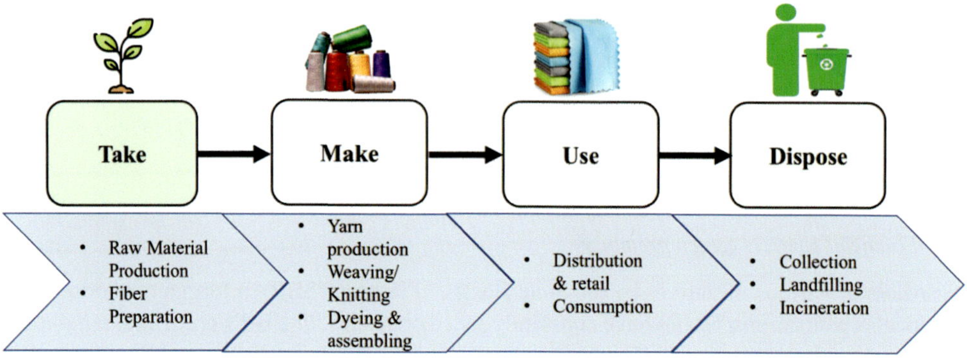

Figure 9.2 Textile linear economy model.

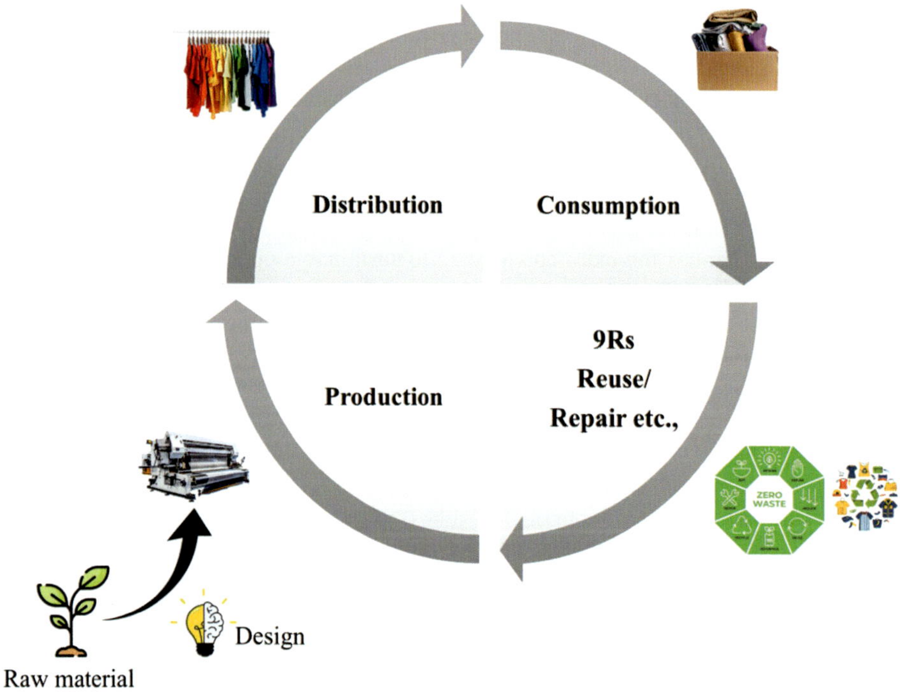

Figure 9.3 Textile circular economy model.

contamination [28]. This practice leads to an unnecessary decline in natural resources as well as waste production, causing environmental and social loss.

The circular simulation advises a potential solution to wasteful resource utilisation (Figure 9.3). As comparison to the linear system, it helps to enhance sustainability as well as eliminate waste components by developing items which are stable, reusable, as well as repairable. Despite potential advantages, most textile SMEs are unaware of the CE standards or relate them to high costs and obstacles [29].

As a conclusion, sustainability is frequently referred to as a secondary problem due to direct effective and financial problems. SMEs also need access to technical proficiency that could allow them to recognise and integrate CE into their enterprise standards. The lack of educational outreach, industry-specific assistance and knowledge-sharing systems increases this difference. This design is a common practice to finish the loop in product cycles. It focuses on designing stuffs for stability, repairability, reusability, as well as recyclability. The target is to decrease waste, expand product life and facilitate items to re-enter the manufacturing cycle, thereby dropping the demand for virgin resources and decreasing the global ecological effect [30].

9.3.3 Regional snapshot (informal sector)

This sector belongs to economic activities that are governed outside of government restrictions, registration, as well as taxation. Many SMEs present in Pakistan and Bangladesh are classified under this category. These enterprises propose mandatory income and employment chances, mainly in rural as well as urban areas. Around 72 per cent of jobs are in the informal sector of Pakistan, but many of the SMEs are unregistered [31]. These enterprises normally have a limited approach to clean machinery, have restricted awareness of environmental regulations and are not exposed to regular assessments. This is central to large processes such as high resource utilisation, bad labour conditions and untreated waste emission. The Competition Commission of Pakistan (2020) reported that 48.5 per cent of SMEs are based on informal funding resources, which delays the industries from adopting modern practices [32].

In Bangladesh, there are limited funding and technical resources, which make the informal textile units in urban regions not able to accomplish their duties. The failure to meet the water pollution standards has a bad impact on the closure of some small dyeing units and the supply chain, according to the World Bank (2021) [33]. Even though numerous small enterprises converted into big units or were replaced by formal companies, this resulted in a loss of stable revenue (especially for women).

9.4 Practical sustainable design strategies for SMEs

In history, textiles were the first producing areas in many states transforming towards industrial development. They served as a basic economic driver for advancement from UK to the USA, assisting industrialisation as well as global product distribution. Now, these sectors contribute considerably to the economic growth of developing regions [34].

Practical sustainable design strategies for SMEs emphasis integrating environmental, social and economic attentions into product development and operations without compromising business feasibility. Energy is important for achieving industrialisation aims, fostering technological innovation, improving productivity and confirming steady economic growth industrialised sector globally [35]. Beyond industrialisation, energy is important for daily activities essential for economic growth via supporting lighting, heating, education, research, transportation, as well as scientific activities [36]. In the garment industry, cutting, sewing and inspection, as well as shipping systems use energy [37]. Commonly, textile manufacturing consumes 40–5 per cent of power for manufacturing yarn as well as fabric and 35–60 per cent of energy in the wet-processing unit [38]. Textile SMEs must implement sustainable (cost-effective) processes to overcome these issues and high regulatory demands. So, the details of the following three key focus areas are discussed below.

9.4.1 Resource proficiency and cleaner production

This strategy provides a vital way to reduce the ecological effects as well as enhance the process functions. The United Nations Environment Programme reported that the cleaner production process contains the consistent role of a combined ecological policy to systems as well as assistance [39]. This is applied to improve productivity as well as to reduce the harmful effects on humans as well as ecosystem. This resulted in the utilisation of incremental modification to achieve significant benefits without large funding for SMEs. The utilisation of these cleaner production approaches mainly enhances the operational functioning by expanding flexibility as well as capacity. This approach makes the product quality better and enhances awareness, decreases emissions and encourages worker safety. Adoption of this practice makes the operational productivity high with the help of optimised time, processes, as well as management policies [40].

Renewable energy resources like water and solar, are proposed to source all the world's energy by 2030, attaining complete support on renewables by 2050 [41]. Now, solar panels/cells have gained consideration because of the depletion of energy sources (which causes global warming). Industrial sectors have a big opportunity to apply solar energies because these are unlimited, eco-friendly and free [42].

9.4.1.1 Energy efficiency

The improvement of this terminology in the textiles industry proposes low greenhouse gas (GHG) production into the environment. It is very important to identify the main areas where energy can be saved by improving the energy efficiency in textile units. This involves sustainable technologies, efficient practices, innovative solutions and a stable energy supply.

This approach can be employed in various systems, for example, compressors, electrical systems, waste heat recovery and lighting. This technology is very significant because it not only saves energy but also decreases pollution and water and chemical utilisation, as well as increases productivity [43]. The following are some technologies which can help reduce energy utilisation.

- **LED lighting**: This technology, which can significantly reduce electricity consumption in textile industries, is very simple and efficient. LED lights utilise up to 75–80 per cent less electricity as compared to conventional bulbs. LED lights have a long functional lifespan, which helps in reducing the bills and costs [44, 45]. Moreover, these lights release less heat, which makes the internal environment cooler, facilitating to minimise of the requirement of air conditioning during the summer season. Many SMEs in India have reconstructed their units with LED lights to support the energy-efficiency programme launched by the Bureau of Energy Efficiency (BEE) [46]. These interventions have confirmed drops in lighting-associated energy source of 30–40 per cent, with restoration on investing usually succeeding within a year. Indian weaving SME also reduces the electricity bills by 30 per cent, which happened due to the upgradation of looms and the unit's lighting consumption [47].
- **Solar panels**: Solar panels are the rooftop systems that facilitate enterprises with satisfactory solutions to save energy consumption along with reduced costs and accessibility to government policies. This strategy becomes a vital source for small enterprises. SMEs in Pakistan, Bangladesh and India are exploiting solar energy to overcome the shortage of electricity and help to maintain the operational expenses. An important example is the medium enterprise weaving unit in Gujarat, a solar system of 30kW installed on the rooftop, resulting in an

Table 9.2 Encouraging influence of different technologies.

Effective technologies	*Positive impact*	*Cost recovery duration*
LED lighting	Less electricity usage, longer lifespan and lower heat emission	6–12 months
Solar panels	Energy independence, carbon reduction and higher savings	3–5 years
Heat recovery	Less thermal energy usage and lower bills	1–2 years

annual profit of 5–6 lakhs rupees. This is also helping in decreasing the carbon footprint [48]. Moreover, solar-powered dyeing as well as finishing units in Bangladesh established by the Infrastructure Development Company Limited have seen up to 50 per cent drops in conventional energy utilisation [49].

- **Heat recovery system**: This system is very important for increasing energy efficiency, particularly in thermal operations, for example, dyeing and finishing. Waste heat from boilers, hot wastewater is utilised by this system, which helps to transform it into beneficial thermal energies for different processes. Moreover, stenter machines, which are extensively employed in textile finishing units, emit substantial heat throughout the procedure. The fixing of a heat recovery system on this equipment can lower fuel intake by 15–25 per cent, with a standard payback cycle of less than two years [33]. A small textile enterprise in Dhaka applied an essential heat exchanger to retrieve heat from effluent and stated a 25 per cent decrease in natural gas intake, subsequent in considerable investments and improved compliance with environmental regulations. The influence of some energy-competent technologies is explained in Table 9.2.

9.4.1.2 Water conservation

This is the vital portion of sustainable manufacturing textile units, particularly for those SMEs that are operating with less water. The textile sector is considered the most water-intensive unit in different processes, like printing and finishing, worldwide [50]. But in Pakistan and Bangladesh, numerous small enterprises are shut down due to the violence of environmental laws, which cause bad effects to the entire region [51]. The utilisation of counter-current cleaning and automatic water flow control systems is a main interest. The employment of wastewater and recycling treatments can be largely valuable [52].

Textile SMEs of Narayanganj in Bangladesh have initiated via digital water meters for real-time intake examination, resulting in enhanced resource administration [53]. Hence, achieving effective water-saving methods can lead to significant ecological and economic benefits for SMEs. The main approaches of water harvesting are discussed below.

- **Rainwater harvesting**: This process involves different processes, like the collection and storage of wastewater. This can be further utilised for industrial applications to decrease municipal waste. In SMEs, this water can be utilised in different applications such as cleaning, cooling and filtration. In India, various SMEs have installed this system to overcome the shortage of water. This system helps in many ways, like reducing water purchasing expenses and improving resilience to droughts [54]. A case analysis advertised by GIZ stated that a textile SME in Coimbatore protected over 25 per cent of its yearly water consumption during the installation of a rainwater harvesting as well as storage system [55].

- **Wastewater recycling**: This strategy is another reasonable solution for textile SMEs. Various processes are used to make water treatment and reutilisation during washing and dyeing; these processes include membrane filtration, biological treatment and reverse osmosis. Even though improved treatment procedures can be expensive, low-cost substitutes like sedimentation reservoirs as well as slow speed sand filters have been recognised efficient for smaller companies. The World Bank indicated that numerous SMEs in Bangladesh's Narayanganj units implemented compacted wastewater recycling schemes, permitting them to use again up to 40 per cent of their route water [33]. A small dyeing organisation assembled a primitive recycling scheme utilising sedimentation reservoirs and sand filters in Pakistan. The approach assisted the reprocess of almost 60 per cent of the course water, lowering water costs plus avoiding governing fines associated with untreated discharge [33].
- **Counter-current rinsing**: This technique is used to make the water consumption at optimised level during dyeing and washing applications in the textile sector. As an alternative of employing fresh water in every washing phase, water runs in the reverse path of the cloth movement, commencing from the cleanest solution to the dirtiest. The maximum effective water is utilised at the end of washing, where it is highest operative and is used again in prior phases where extreme purity is not as critical. This approach primarily decreases water utilisation by expanding the reuse of washing water throughout several developmental stages. It also reduces chemical consumption and decreases effluent volumes. A finding by the UK based organization 'Waste and Resources Action Programme (WRAP)' regulated that SMEs embracing counter-current rinsing in their wet processing companies reduced water consumption by 30–50 per cent compared to conventional approaches [56]. A small textile processing company in Tirupur, India, established counter-current rinsing in its dyeing region and stated a 40 per cent decrease in water utilisation without conceding the property of the finishing product. The impact of water conservation strategies for SMEs has been depicted in Table 9.3.

9.4.1.3 Waste reduction

Waste reduction is a main component of sustainable design, specifically for SMEs in the textile and fashion sectors that often operate under resource limitations. Two widely recognised approaches are zero-waste pattern design and upcycling, facilitating minimising waste and contributing to circular production systems.

- **Zero-waste pattern design**: It is a creative and technical method that reduces textile waste at the design development level by ensuring the whole fabric is used in the final product. Traditional garment production costs up to 15–20 per cent of the fabric during cutting [57]. An SME in Jaipur, India, handcrafted garments to build an impressive impact. The company

Table 9.3 Impact of water conservation strategies for SMEs.

Areas	*Description*
Cost reduction	Lower water bills, reduced operational costs via decreased water and wastewater treatment requirements
Regulatory compliance	Improved adherence to environmental laws
Resource security	Increase resilience in water-scarce regions by reducing the dependence on unreliable external water sources

operated digital pattern-making tools and cut-and-sew optimisation software to rebuild garment patterns and exclude leftover cloth. This idea led to a 20 per cent drop in textile waste and a 15 per cent cost reduction on raw components [58]. Fashion brands are integrating modular design theories in Europe, allowing apparel to be refurbished, reused without producing waste [59]. Rissanen and McQuillan establish that the zero-waste model can notably decrease pre-consumer leftovers while promoting innovation in garment aesthetics. Their research implies that such methods are exceptionally favourable for small-scale inventors who seek to distinguish themselves via ethical and artisanal methods [60].

- **Upcycling**: This process involves redesigning waste components, defective materials into higher-value goods. This technology achieves high-value products with lower costs and makes SMEs more accessible as compared to recycling, which requires complex processing.

Jung and Jin identified that the upcycling process can lower the ecological effects by increasing product lifespans and reducing the necessity for virgin materials. This technique also raises customer engagement by stimulating distinctiveness and innovation in design. Their survey showed that users are progressively willing to buy upcycled goods, remarkably when transparency as well as appealing demand are carried efficiently [61]. A cluster of SMEs in Bangladesh initiated employing textile offcuts to generate patchwork garments as well as home decorations. This idea managed to achieve a 30 per cent decline in textile landfill and started a perception for export-grade handmade stuff in eco-conscious markets [62].

9.4.2 Pollution prevention

An industrial region in Pakistan, mostly fuelled via coal, it is the major energy user, as well as contributes to over 49 per cent of the country's CO_2 releases [63]. Twenty years ago, industrial CO_2 secretions have expanded rapidly from 17.21 to 95.2 million tonnes during 2019 [64]. Lin and Long indicated that the expanding energy need for commercial development is a key problem, generating pollution [65]. SMEs are recognised as vital establishments of the market due to their considerable responsibility in poverty deterioration, economic progress, as well as job formation. In Pakistan, SMEs account for around 92 per cent of all business firms, contribute 40 per cent of the country's GDP and create around 26 per cent of the manufacturing sector's trades. It yields 79 per cent of the country's industrial occupation. SMEs frequently struggle to progress sustainable attitudes, as their initial focus is on extending profits in comparison to ecological concerns [66].

But the pollution generated by leftovers remains a significant issue for sustainability. To address this issue, a practical method that emphasises avoiding pollution from occurring in the first place is considered the highest effective solution. This involves the establishment of new techniques of proficiency [67]. Identifying the production as well as the decrease of waste improves the reduction of trash and safeguards the environment. A pollution prevention (PP) approach is an imperative element of sustainability [68]. PP strategies aim to decrease the quantity of pollution formed at its source [69]. The first stage of the policy is defined by the natural resource-based view, as a PP scheme. Ibrahim and Mahmood reported in their studies that declining waste goods from production methods increase efficiency by reducing compliance and responsibility costs, reducing the need for raw materials and reducing operations [66].

- **Chemical substitution**: Chemical substitution is a practical PP approach in the textile and fashion industry, where hazardous chemicals used in dyeing, finishing and printing are replaced with more eco-friendly substitutes. For SMEs, this approach is specifically favourable as it not only decreases the poisons in wastewater but also confirms compliance with the universal ecological and consumer guidelines. Some examples are heavy metals and azo dyes, as well as formaldehyde, being replaced by non-toxic compounds. In Bangladesh, some SMEs utilise low-salt dyes, which helps to minimise the salt utilisation to 40 per cent. This also enhances the quality of waste materials as well as reduces the ecological expenses [70]. This process also contributed to better working conditions and started new marketplaces and international brands with firm sustainability conditions.

- **Caustic recovery**: This process refers to the procedure of recovering as well as reprocessing caustic soda, which is extensively employed in textile units such as scouring as well as mercerisation. This is particularly crucial for SMEs seeking to decrease both chemical expenses and ecological contamination. In a usual caustic-recovery scheme, the caustic solution is accumulated and filtered for reuse, dropping the requirement for new chemical contribution and decreasing the discharge of harmful components into wastewater. In India, SMEs that employed small-scale caustic recovery organisations stated a decrease of up to 60 per cent in fresh caustic solution consumption. This resulted in annual funds of around 1.2 million Indian rupees per SME [44]. Caustic reclamation not only influences cleaner manufacturing but also improves the long-term cost-effectiveness and facilitates SMEs to gain sustainability certification as well as market approach.

9.4.3 Sustainable product innovations

This technology is becoming an essential component for SMEs in the textile as well as fashion segments. These revolutions converge on reducing ecological effects through the product lifespan, from raw component selection to end-of-life removal, while preserving or upgrading product implementation as well as aesthetics.

- **Eco-friendly fibres**: These fibres seem more widespread due to their decreased ecological footprint than conventional compounds. The examples of these fibres are hemp, organic cotton, banana, linen, bamboo, as well as regenerated cellulose. These fibres frequently need considerably less pesticides, water and energy, while granting more effective biodegradability. Organic cotton cultivation utilises up to 91 per cent less water and emits 46 per cent less GHG than standard cotton [71]. Moreover, bamboo and hemp are fast-growing, pest-resistant crops that require lower agricultural efforts, making them perfect for sustainable agriculture. SMEs, particularly in South Asia, have commenced incorporating these fibres into their supply chains to meet the requirements of ecologically deliberate buyers.
- Natural **dyes**: The adaptation from synthetic to natural dyes is another vital factor that promotes sustainable textile attempts. Synthetic dyes, which grant vibrant colours as well as consistency, are connected to the main ecological problems. The main issues are water contamination, toxic waste and industrial hazards. In comparison, natural dyes derived from plants (turmeric), as well as minerals, are biodegradable, non-toxic and considerably less dangerous to marine and human health [72]. SMEs in Bangladesh, India and Sri Lanka are retrieving conventional dyeing practices using indigenous plants and herbs. The Kalamkari artisans in India employ natural dyes from pomegranate peels, iron acetate and fruit to generate eco-friendly printed textiles. According to a study by Samanta and Konar, the implementation of natural dyeing not only develops sustainability but also improves product distinction as well as export probability, mainly in high-value niche marketplaces that are worth chemical-free inventions [73].

- Circular **design**: Many SMEs are trying to utilise circular design principles, which facilitates increasing their business competitiveness to control the resource deficiency, effluent regulations and customer's demands. Moreover, several fashion enterprises in the Netherlands and the UK are providing fabrics with detachable parts, permitting customers to upgrade their outfits without discarding the whole fabric [74]. Furthermore, some ideas from product-service systems, like rental fabrics, have been created by numerous SMEs, which helps to enhance the use of fabrics and decrease the waste components. Niinimäki also proposes that the implementation of circular design practices during product manufacturing tremendously increases the customers' interests, flexibility and minimise the ecological effects [75].

9.4.4 Social sustainability

A significant component of sustainable development is known as social sustainability, which focuses on the equity and resilience of customers. Some textile SMEs are mainly performing in labour-intensive regions in South Asia, promoting social sustainability. This practice is necessary not only for moral analyses but also for long-lasting business feasibility.

- **Worker safety**: Several textile SMEs are facing some issues, which are caused due to the bad working environment, mainly in contracted and informal activities. Some workplace challenges are exposure to harmful chemicals, insufficient lighting, a bad ventilation system and lack of awareness to handle dangerous equipment [76]. The ILO (2019) emphasises that financial help in working safety enhances productivity, which is considerably beneficial for SMEs with restricted labour resources [77].
- **Fair wages**: Fair wages are a basic principle of sustainability. SMEs, mostly those affiliated with international value chains, frequently struggle to meet wage values due to low margins and cost competition. However, research studies that propose reasonable salary impacts worker maintenance, skill upgrades, as well as better quality of goods [78]. The Fair Wear Foundation offer standards and support for SMEs to execute living wage outlines [79]. Numerous SMEs have executed efficiency-linked bonus programmes as well as worker enterprises to balance business affordability with fair payment [80].
- **Community engagement**: Numerous SMEs in fashion sectors usually run near local regions and frequently retain their labour force from these areas. Community involvement is crucial in advancing mutual trust, obtaining labour supplies and maintaining a social licence to operate. Case studies from rural India and Bangladesh highlight SMEs that have collaborated with NGOs to provide training on financial literacy and health awareness to workers and their families, contributing to broader community development [81]. See Table 9.4 for the benefits of social sustainability for SMEs.

9.4.5 Visual tool: strategy toolbox/checklist (cost-impact matrix)

To facilitate the practical application of sustainable design procedures between SMEs, particularly in resource-constrained areas, it is essential to present a structured decision-making tool. A cost-impact matrix is an effective tool that categorises sustainability strategies based on their relative cost of implementation and the level of environmental or social impact they produce. This matrix lets SMEs arrange activities that give the maximum returns, both economically and environmentally. This matrix enables SMEs to emphasise especially low-cost, high-impact methods, like changing traditional bulbs with LED lights, accomplishing simple zero-waste

Table 9.4 Benefits of social sustainability for SMEs.

Social strategies	*Benefits*
Worker safety	Fewer accidents, enhanced worker morale
Fair wages	Good retention, better productivity and higher product quality
Community engagement	Better local relationships and reputational gains

Table 9.5 Checklist format for action planning.

Factors	*Actions to be done*	*Cost level*	*Impact level*
Economic	Adoption of zero-waste patterns	Low	High
Environmental	Installation of LED lights	Low	High
Social	Make sure of fair wages, worker safety measures	Low	Medium

cutting samples, or targeting energy and water measurements. These involvements are carried out by the results of the United Nations Industrial Development Organization (UNIDO) [44], which confirms that SMEs are embracing more effective production techniques with modest investments. Moreover, some high-impact technologies, such as solar systems and grey-water treatments, can be utilised with the passage of time; this will require the assistance of government and international funding processes (IFC). This toolbox is specifically appropriate for textile and fashion SMEs functional in developing markets, where investment limitations are established. By mapping available strategies against cost and impact, entrepreneurs and factory managers can make informed decisions about their business size, local conditions and strategic objectives [82]. These checklists frequently cover fields such as energy and water conservation, waste production, supply-chain transparency and worker welfare, supporting SMEs with a standard for upgrading and agreement with regulatory standards [83]. A simple checklist format may be adopted for the action planning. It depends on the following three factors of sustainability, with aspects of cost and impact levels Table 9.5.

Such structured formats enhance the effectiveness of SMEs and provide valuable tools for training, capacity-building and grant applications. In addition, the consumption of online key performance indicators (KPIs), control panels and software has become progressively important for tracing the sustainability system of measurement in real time. SimaPro and Ecochain present a powerful programme for accompanying life-cycle assessments (LCAs), carbon footprint analyses and ecological impact tracing.

9.5 Implementation framework for SMEs

9.5.1 Step 1: Baseline assessment and benchmarking

The foundation of sustainable transformation starts with a baseline assessment to assess the current environmental, social and operational implementation. SMEs should analyse resource utilisation, waste production, carbon footprint and compliance with current guidelines. Benchmarking against industry models and competitors facilitates recognising performance gaps [84]. LCA and material flow assessment are effective at this level. Recognising current methods, environmental footprint and sustainability knowledge are the major goals of this stage. The following are the actions that should be executed at this phase.

Activities: (1) Sustainability audits, (2) stakeholder mapping, (3) awareness workshops/seminars, (4) baseline performance indicators.

9.5.2 Step 2: Strategic goal setting

SMEs should plan specific, measurable, achievable, relevant and time-bound (SMART) goals linked with their enterprise aims as well as Sustainable Development Goals (SDGs). This requires setting focuses for energy saving, sustainable sourcing, waste minimisation, or emissions management [27]. Integration into the corporation methodology ensures that stakeholder involvement as well as operational feasibility are achieved.

Activities: (1) Identify short, medium, as well as long-term sustainability aims; (2) line up objectives with previous schemes; (3) incorporate SDGs and CE policies.

9.5.3 Step 3: Implementation phases

The execution of sustainable design methods in SMEs should be built on viability, cost and source availability. This attempt permits businesses to supervise hazards and associate sustainability efforts with their operational capability. Sudden actions genuinely involve cheap, high-impact techniques that can be implemented with less disruption, such as shifting to LED lighting, lowering packing components and digitising governmental activities, for example, invoicing. SMEs can highlight more approaches at medium level, for instance, the performance of sustainable raw supplies, strengthening fabrication of equipment for extensive productivity and proposing staff supervision on eco-design standards. Likewise, long-term actions concern modifications like reordering supply chains for substantial precision as well as sustainability, restructuring outcomes to favour CE standards. This stepwise incorporation ensures that sustainability goals stay valid and practical over time, operating SMEs progress firmly [85].

9.5.4 Step 4: Monitoring, measurement and continuous improvement

Constant examining is essential to observe improvement and assess attempts. KPIs related with supply handling, releases, or sustainability are determined and evaluated constantly (see Table 9.6).

Equipment such as sustainability dashboards and green accounting supports data-driven management. Constant upgrade cycles (Plan–Do–Check–Act) are urged [86]. Below are the actions that should be executed at this phase.

Actions: (1) KPI tracing, (2) utilisation of digital dashboards for clarity.

9.5.5 Step 5: Scaling and certifications

Effective programmes can be scaled throughout departments, for example, Oeko-Tex (textile safety), ISO 14001 (environmental management) and Global Organic Textile Standard (GOTS) (organic fabrics), improving reliability, market retrieve and monitoring agreement [87]. Succeeding these specifications exhibits a company's dedication to sustainability. Scales successful initiatives and communicates outcomes are the main objectives of this stage.

Actions: (1) Continuous improvement loop, (2) stakeholder reporting (customers, investors, regulators), (3) certification pursuit (GOTS, Oeko-Tex). The implementation framework for SMEs is shown in Figure 9.4.

Table 9.6 Summary of the implementation framework for sustainable design in SMEs.

Stages	*Aims*	*Key activities*	*Outcomes*	*References*
Baseline assessment and benchmarking	Identifying current practices Ecological footprint Sustainability awareness	Sustainability audits Stakeholder mapping Awareness workshops/ seminars Baseline performance indicators	Baseline data on environmental/ social impact	[84]
Strategic goal-setting (SMART goals)	Alignment of sustainability goals with the business model	Short, medium and long-term sustainability objectives Align goals with existing strategies SDGs and circular economy principles	Clear sustainability roadmap	[27]
Implementation phase	Execute sustainable initiatives	Launch pilot projects Adopting sustainable materials Supply chain transformation product redesign for circularity	Effective and measurable performance	[86]
Monitoring, measurement and continuous improvement	Monitor performance	KPI tracking (energy use, emissions, waste reduction) Use of digital dashboards for transparency	Effective performance	[86]
Scaling and certifications	Scale successful strategies Communicate progress	Stakeholder feedback Apply for certifications	Credibility	[87]

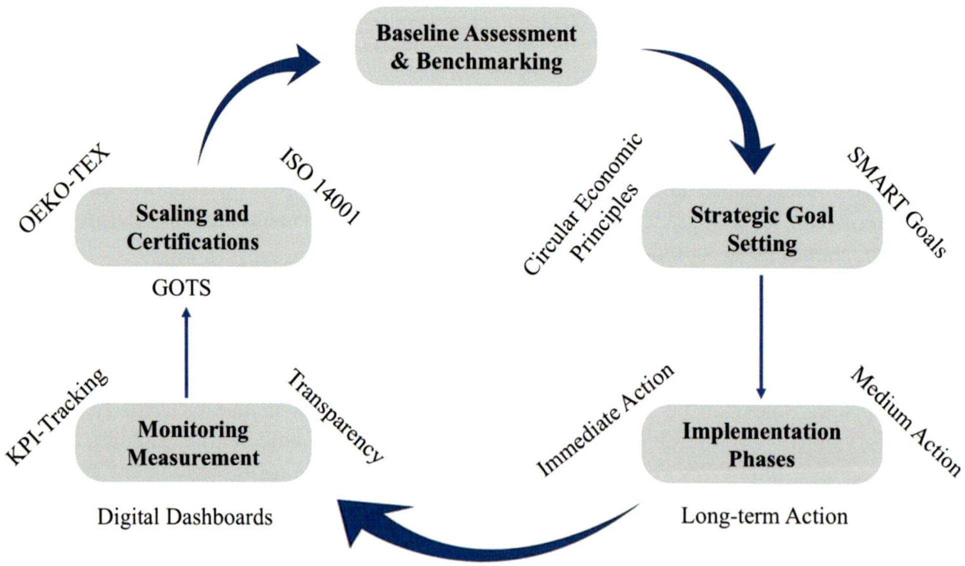

Figure 9.4 Implementation framework for SMEs (cycle of sustainability integration).

9.6 Policy support and institutional frameworks

9.6.1 Role of government policies and incentives

Government involvement is vital in supporting sustainability in SMEs, especially in resource-intensive areas such as textiles and fashion. Although the financial, technical and human supplies of SMEs, well-designed plan contexts and institutional support systems are fundamental to reducing obstacles and persuading the implementation of sustainable design attempts. This assistance is often characterised as financial incentives as well as technical support, both operate together to support measurable adaptations across the value chain.

9.6.1.1 Financial incentives

Financial incentives are the most active and efficient form of government support. These incorporate grants, loans, tax exceptions and approach to green funding methods, which decrease the cost of sustainable funds. In many emerging nations, governments grant investment support to SMEs to assume cleaner tools, upgrade energy productivity, or participate in sustainable raw components. These processes decline upfront investment budgets and advance the financial viability of transitioning to more sustainable business patterns [88].

Furthermore, green financing observes economic outcomes particularly designed to support eco-friendly performance, is progressively available through public banks and advanced financial organisations. Green bonds, sustainability-linked loans and committed credit links for energy efficiency developments have been established to assist investment in sustainability-related strategies. In addition, the Green Climate Fund facilitates SME green enterprises in developing countries via concessional financing and combined finance forms. Additionally, the European Investment Bank (EIB) has trained SMEs with sustainability-linked loan sets to aid in achieving EU Green Deal goals [89].

9.6.1.2 Technical assistance and advisory services

Instead of economic support, technical assistance, as well as advisory services, are important factors in enabling SMEs to adopt sustainable principles of design. Such services assist in addressing the knowledge as well as capacity gaps that often prevent SMEs from understanding or employing sustainability regulations. Agencies at the government level, often in association with NGOs, educational institutions, or international organisations, offer training plans, remote toolkits, workshops and one-on-one consultancy skills that guide SMEs via sustainability planning and its execution. Programmes like the UNIDO Resource Efficient and Cleaner Production initiative, as well as the ITC's Green to Compete programme, offer SMEs practical tools to improve ecological performance while raising production and competition [90]. Laws like the EU's Corporate Sustainability Due Diligence Directive expect corporations to examine as well as decrease risks to people and the conditions in their supply chains, which includes SMEs [91]. Governments have also started sector-specific advisory arrangement plans. For example, India's Ministry of Micro, Small and Medium Enterprises has joined with organisations such as Textile Sector Skill Council for approving sustainable tactics through technical supervision, industrial roadmaps, as well as policy manuals tailored to SMEs. Furthermore, the UK's Business Support Helpline, as well as programmes under Innovate UK, offer financial support and technical help for SMEs accepting the principles of the circular economy [92]. The efficacy of such services is centred on their capability to change complex sustainability aims into actionable, context-specific approaches, permitting SMEs to overcome informational as well as technical constraints. Such services are also a link between sustainability policy and accomplishment, confirming that economic incentives are employed successfully as well as effectively.

9.6.2 Facilitating SME sustainability clusters and shared facilities

Policies of government and institutional frameworks are critical in encouraging sustainability among SMEs through assisting the formation of sustainability clusters and granting mutual facilities. Such groups assist as cooperative networks where SMEs can cooperatively address environmental issues, share resources and improve their competitive benefit. Sustainability groups are globally concentrated clusters of interconnected enterprises and associated institutions in a specific area. These clusters enable SMEs to gain an advantage from shared information, infrastructure, as well as facilities, which might be unachievable individually due to resource restrictions [93]. Additionally, the UNIDO improves the development of the groups by offering technological aid and raising cooperation among firms and helping institutions. This strategy assists SMEs in overcoming restrictions and seizing collective opportunities, strengthening their influence beyond individual capabilities [94, 95].

Shared services within these groups, such as coaching facilities, universal testing labs, as well as waste controlling systems, cause SMEs competent of lowering operational expenditures as well as improving value. By utilising sources, SMEs can consider better technologies as well as expertise that would be cost-effective. This joint model not only promotes sustainable practices but also upgrades the total manufacturing and advanced competence of the supporting enterprises. Moreover, clusters can facilitate SMEs in routing complicated sustainability requirements and conforming with principles. Such clusters help capacity-building and equipment to aid businesses in integrating sustainability into their approaches and strategies. By knowledge-sharing and tailored advisory expertise, groups aid SMEs in improving their marketing as well as compliance with advancing environmental rules [96].

9.6.3 Recommendations for policymakers to empower SME sustainability

The following recommendations aim to guide policymakers in promoting SME sustainability:

9.6.3.1 Expand access to tailored green finance instruments

Policymakers should increase and enhance financial goods design for SMEs, such as green loans, sustainability-associated credit lines and guidelines. These tools can ease the financial effects linked with sustainable funds. The World Bank highlights the consequences of such tailored financial outcomes to support SMEs in their sustainability attempts [97].

9.6.3.2 Simplify sustainability reporting frameworks

Governments should cooperate with international groups to establish smooth and balanced sustainability reporting principles. The Organisation for Economic Co-operation and Development (OECD) promotes simple reporting structures that consider SMEs as regulated sources, permitting them to incorporate into sustainable supply chains [98].

9.6.3.3 Establish SME sustainability clusters and shared facilities

The development of SMEs clusters and services helps gain resources and best approaches to waste management, as well as training institutes. These kinds of collaborative inputs increase the efficiency and innovations, which also enable the sustainable approaches for SMEs. Such collaborative environments can enhance innovation and efficiency, enabling sustainable practices for SMEs.

9.6.3.4 Invest in capacity-building and technical assistance

There should be different training programmes, advisory committees and toolkits at the governmental level to develop sustainable SMEs. The OECD highlights the requirement for non-financial support, incorporating guidance and knowledge, to help SMEs in their sustainability process [99].

9.6.3.5 Integrate SMEs into national sustainability strategies

SMEs should be identified as key collaborators in national sustainability and environmental action strategies. SMEs should be recognised as key partners in national sustainability and environmental action strategies. The enclosure of their requirements and obstacles in policy structures guarantees that sustainability programmes are comprehensive and efficient. The OECD point out the value of concerning SMEs in the green shift to attain wider ecological objectives [100].

9.7 Case studies: SMEs driving sustainability

9.7.1 Case study 1: Rangotri (India)

Rangotri is located at Sanganer, a small township located on the border of Jaipur, India. Sanganer is the home of hand block printing. Block prints form sensational prints with vibrant colours, nonetheless, this old art form has started to fade out due to the introduction of modern and mechanised printing processes. Rangotri has a state-of-the-art production unit/workshop where the processing, dyeing, printing and stitching of fabric (and the packaging of finished goods) are all executed in-house to ensure high-quality standards. The efficient hands of skilled craftsmen, the printer, the block maker and the embroiderer, make every completed product of Rangotri both classic and prestigious. It has been accepted for its unique designs, colour schemes, commitment to quality and environmental problems [101].

Rangotri, a textile SME reveals how small businesses can contribute to sustainability in response to the challenges. Initially facing environmental and regulatory pressures due to traditional dyeing procedures, Rangotri adopted sustainable methods to regenerate its operations. The SWITCH-Asia Sustainable Textiles significantly influenced the company's journey through the Sustainable Development (SUSTEX) initiative, which provided a framework for adopting eco-friendly practices. Rangotri lowered its environmental footprint and increased its market competitiveness by using natural dyes and enhancing waste management. Such example entails the ability of SMEs to convert environmental issues into occasions for sustainable growth.

9.7.2 Case study 2: Living Blue (Bangladesh)

Living Blue is a community-based artisan in Rangpur, Bangladesh, that depicts the combination of traditional craftsmanship with sustainable methods. The Cooperative for Assistance and Relief Everywhere (CARE)Foundation began a programme in 2006 to renew the agriculture and manipulation of natural indigo, a procedure with deep traditional roots in the area. Both the environment as well as economy were strengthened effectively via a closed-loop system, established by Living Blue. Different conventional techniques, such as Kantha quilting and shibori, are used by the cooperatives' artisans, like women, for making a high-quality handmade textile. These things have been recognised globally, utilising partnerships with worldwide vendors and designers. Living Blue's representation not only upholds traditional cultures but also offers reasonable salaries and empowers artisans, contributing to the growth and strength of the community [102].

9.7.3 Case study 3: Rapanui (UK)

Rapanui is a sustainable brand of fashion that was established in the Isle of Wight, UK. This shows how small-scale production can efficiently accomplish a CE model. Mart and Rob Drake-Knight founded this brand in 2008. Rapanui's attempt is the utilisation of GOTS certified organic cotton, a basic source of organic cotton. This option is not only for cotton farming without harmful chemicals, but also for the quality of farming people. The company's dedication to renewable energy is illustrated in its production developments, which are driven by solar and wind energy. Therefore, Rapanui retains closed-loop water systems in its dyeing methods, significantly lowering wastewater and ecological impact.

Rapanui has made on-demand manufacturing tools to address global overproduction issues in the fashion industry. This method promises that garments are generated only after placing an order, thus lowering the waste and unreported sources. Each product has a QR code which permits purchasers to track the journey of the product from raw materials to the final product. Furthermore, Rapanui's take-back policy supports clients in returning damaged goods, which are then recycled into new ones, efficiently closing the manufacturing loop.

Rapanui presented Tee mill, an open-access setup that allows companies and individuals to make their clothing lines through Rapanui's circular supply chain. Such a programme has sanctioned over 10,000 brands to employ sustainable procedures, raising the development towards a more sustainable fashion industry [103]. Advanced efforts of Rapanui have gained considerable appreciation, including the Queen's Award for Innovation, identifying the dedication of the company to enhancing the invention models of the fashion industry. Despite a few problems, like as higher expenses associated with ethical practices, Rapanui remains to establish that integrating CE principles into small-scale manufacturing is not only conceivable but also helpful for the ecosystem, consumers and the business itself [104].

9.7.4 Case study 4: Kamal Limited (Pakistan)

Mercerisation is a common process that improves fabric lustre as well as strength by using sodium hydroxide in the textile processing industry. Therefore, this method generates significant amounts of diluted sodium hydroxide as waste, posing environmental issues and raising operational expenditures [105]. A dyeing SME, Kamal Limited, Faisalabad, gives an interesting case of how low-price technical intervention can bring both economic and environmental assistance. Working within the traditional mercerisation technique, which raises shine and strength of fabric using caustic soda, Kamal Limited handled multiple issues from the production of diluted caustic waste. The firm addressed this by setting up a caustic recovery plant equipped with a three-stage evaporator, allowing the utilisation of concentrated soda. This start-up resulted in the annual recovery of about 300 tonnes of caustic soda, annual savings of about 14 million Pakistani rupees [106].

9.7.5 Lessons learned and practical takeaways from case studies

The following are the lessons learned from different case studies:

Table 9.7 Practical opportunities for SMEs.

Areas	*Proposed strategies*
Leadership commitment	Strengthen a sustainability-focused approach and incorporate it into the company's core values
Community engagement	Develop business models that enable local communities to preserve traditional crafts
Technological integration	Utilise technologies that reduce the amount of waste and promote circular production
Environmental practices	Implement efficient production methods to reduce environmental pollution and operational costs

9.7.5.1 Leadership and values drive sustainability

Rangotri, a textile SME in Rajasthan, India, shows how leadership devotion and values can lead to sustainability. The owner of the company employed environmentally friendly approaches and joined programmes such as SWITCH-Asia's SUSTEX to integrate sustainable approaches into operations. This features the mandatory role of leadership in improving SMEs' environmental accountability.

9.7.5.2 Community-centric business model

In Bangladesh, Living Blue reveals the efficacy of community-based patterns. By combining conventional crafts, i.e., indigo dyeing and quilting, with fair-trade principles, the enterprise encourages local artisans, specifically women, guaranteeing profits are reinvested into their communities [107]. This strategy not only maintains traditional heritage but also improves social and economic development [108].

9.7.5.3 Technological innovation enhances sustainability

Rapanui, a UK-based sustainable fashion brand, shows how technological innovations can strengthen sustainability. Through on-demand manufacturing as well as the Teemill platform, Rapanui lowers waste and improves circular fashion practices. Their model shows the potential of technology in establishing environmentally friendly solutions for business.

9.7.5.4 Adoption of cleaner production technologies

Kamal Limited in Faisalabad, Pakistan, employed a caustic recovery system to reclaim and reuse chemicals generated by the mercerisation method. This programme not only reduced environmental pollution but also led to considerable cost savings, showing that these sustainable strategies can be economically valuable for SMEs (Table 9.7).

9.8 Conclusion and way forward

This chapter entails the crucial role of SMEs in improving sustainability within the fashion and textile industry. SMEs act as major contributors, experts at driving innovation, adapting to rising demands and enhancing sustainability on both local and global levels. Multiple key views develop in SMEs aiming for sustainability; Firstly, sustainable design must be cost-efficient and high-impact strategies (such as resource effectiveness, zero-waste design and cleaner production) can offer available initial points. Secondly, the development of internal expertise through knowledge development, training of skills and baseline sustainability assessments capable SMEs to accomplish sustainability. Furthermore, collaboration across value chains and contributions in clusters can notably lower expenditures and enhance scalability. Strategic directions are compulsory to strengthen this evolution. To raise this shift, SMEs need to insert sustainability into their key approaches by setting SMART goals, involving continuous enhancement, as well as following circular economy principles. policymakers should improve the strategy for green financing, streamline sustainability reporting and integrate SMEs into broader sustainability structures. Stakeholders must guarantee that individuals can maintain supply chains, receive mentorship and access digital tools. SMEs are not marginal players; instead, they are crucial for establishing a sustainable and resilient future for textiles. Collective action among policy, industry, as well as society is essential for training SMEs with the resources required for sector transformation.

References

[1] Eurostat, 'Micro and Small Businesses Make Up 99 Per Cent of Enterprises in the EU-News Articles', (2024), https://ec.europa.eu/eurostat/web/products-eurostat-news/w/ddn-20241025-1.
[2] Anne Ploeger, European Parliament 'SMEs', (2020), www.europarl.europa.eu/factsheets/en/sheet/63/a-kis-es-kozepvallalkozasok.
[3] M. A. R. A. A. M. Saiful Islam, 'Ready-Made Garments Sector of Bangladesh: Its Contribution and Challenges towards Development', *J Asian Dev Stud*, 5 (2) (2016): 50–61.
[4] S. Paul, R. Biswas and R. Bhattacharya, 'The Handloom Weavers of Varanasi: A Study on Their Opportunities and Challenges', *Int J Creat Res Thoughts*, 12 (4) (2024): K680–K688.
[5] L. Zare, M. Ben Ali, E. Rauch and D. T. Matt, 'Navigating Challenges of Small and Medium-Sized Enterprises in the Era of Industry 5.0', *Results Eng*, 27 (2025), doi: 10.1016/j.rineng.2025.106457.
[6] H. Smith, R. Discetti, M. Bellucci and D. Acuti, 'SMEs' Engagement with the Sustainable Development Goals: A Power Perspective', *J Bus Res*, 149 (2022): 112–22, doi: 10.1016/J.JBUSRES.2022.05.021.
[7] Funds for NGOs, 'SWITCH Asia: Promoting Sustainable Consumption and Production', (2023), www.2.fundsforngos.org/health/switch-asia-promoting-sustainable-consumption-and-production/.
[8] GIZ, 'Promoting Energy Efficiency in Indian MSMEs', *Deutsche Gesellschaft für Internationale Zusammenarbeit (GIZ) GmbH*, (2022).
[9] Nielsen, 'Was the Year of the Sustainable Consumer', (2018), https://csnews.com/2018-was-year-sustainable-consumer.
[10] G. Gonda, E. Gorgenyi-Hegyes, R. J. Nathan and M. Fekete-Farkas, 'Competitive Factors of Fashion Retail Sector with Special Focus on SMEs', *Economies*, 8 (4) (2020): 1–18, doi: 10.3390/ECONOMIES8040095.
[11] S. Wang and H. Zhang, 'Enhancing SMEs Sustainable Innovation and Performance through Digital Transformation: Insights from Strategic Technology, Organizational Dynamics, and Environmental Adaptation', *Socioecon Plann Sci*, 98 (2025): 102124, doi: 10.1016/J.SEPS.2024.102124.
[12] UN Conference on Trade and Development, 'Better Trade for Sustainable Development: The Role of Voluntary Sustainability Standards', (2021), https://unctad.org/publication/better-trade-sustainable-development-role-voluntary-sustainability-standards.
[13] International Trade Centre, 'SME Competitiveness Outlook', (2023), www.intracen.org.
[14] European Commission, 'SME Definition', (2020), https://single-market-economy.ec.europa.eu/smes/sme-fundamentals/sme-definition_en.
[15] Organisation for Economic Co-operation and Development, *Enhancing the Contributions of SMES in a Global and Digitalised Economy*, Paris: OECD Publishing, (2017).

[16] Department for Business, Energy & Industrial Strategy, *Business Population Estimates for the UK and Regions 2021*. London: UK Government, (2021), www.gov.uk/government/statistics/business-population-estimates-2021.

[17] F. Khatun, M. Kamal, F. Al Kabir, N. Nawrin, M. Syful and H. Securing, 'Securing Green Transition of the Textile and Readymade Garments Sector in Bangladesh', CPD Report 43, Centre for Policy Dialogue (CPD), (November 2023), doi: 10.13140/RG.2.2.32733.18405.

[18] World Bank, 'Greening the Textile Sector in Bangladesh: A Roadmap for Clean and Competitive Growth', (2021), www.worldbank.org/.

[19] A. Kawlra, 'Kanchipuram Sari: Design for Auspiciousness', *Des Issues*, 21 (4) (2005): 54–67, doi: 10.1162/074793605774597451.

[20] UNESCO, 'Sustaining Traditional Knowledge and Livelihoods through Craft Industries', (2022), www.unesco.org.

[21] Ellen MacArthur Foundation, 'Circular Business Models: Redesigning Fashion for the Future', (2021), www.ellenmacarthurfoundation.org/fashion-business-models/overview.

[22] N. Siddiqui, and R. Iqbal, '*SME Development in Pakistan: Analyzing the Constraints to Growth*, PIDE Working Paper No. 2008:47, Pakistan Institute of Development Economics, Islamabad, Pakistan, (2008).

[23] Government of Pakistan, The Financial Institutions (Secured Transactions) Act, (2016).

[24] M. Amin, '*Credit Constraints and Firm Productivity in Pakistan: An Analysis of SME Financing Gaps*, World Bank Policy Research Working Paper, Washington, DC, USA: World Bank, (2019), https://openknowledge.worldbank.org/.

[25] M. T. Rahaman, A. D. Pranta, M. R. Repon, M. S. Ahmed and T. Islam, 'Green Production and Consumption of Textiles and Apparel: Importance, Fabrication, Challenges and Future Prospects', *J Open Innov Technol Mark Complex*, 10 (2) (2024): 100280, doi: 10.1016/J.JOITMC.2024.100280.

[26] Organisation for Economic Co-operation and Development (OECD), *Financing SMEs and Entrepreneurs 2022*, OECD Publishing, Paris, France, (2022), doi: 10.1787/e9073a0f-en. 2022.

[27] M. Geissdoerfer, P. Savaget, N. M. P. Bocken and E. J. Hultink, 'The Circular Economy: A New Sustainability Paradigm?', *J Clean Prod*, 143 (2017): 757–68, doi: 10.1016/J.JCLEPRO.2016.12.048.

[28] Lenzing, 'The Global Fiber Market in 2016', Lenzing com, (2018). www.lenzing.com/newsroom/news-events/lenzing-generates-substantial-revenue-and-earnings-growth-in-2016.

[29] V. Rizos, A. Behrens, W. van der Gaast, E. Hofman, A. Ioannou, T. Kafyeke, A. Flamos, R. Rinaldi, S. Papadelis, M. Hirschnitz-Garbers, and C. Topi. 'Implementation of Circular Economy Business Models by Small and Medium-Sized Enterprises (SMEs): Barriers and Enablers', *Sustainability*, 8 (11) (2016): 1212. doi: 10.3390/SU8111212.

[30] Ellen MacArthur Foundation, 'A New Textiles Economy: Redesigning Fashion's Future', (2017), https://ellenmacarthurfoundation.org/a-new-textiles-economy.

[31] I. Ahmed, A. Usman, M. S. Nazir and M. Z. Shaukat, 'Safety Practices in Informal Industrial Segment of Pakistan', *Saf Sci*, 110 (2018): 83–91, doi: 10.1016/J.SSCI.2018.07.017.

[32] Competition Commission of Pakistan, 'State of Competition in the Banking Sector of Pakistan', (2020), www.cc.gov.pk/assets/images/Downloads/research_and_publications/state_of_competition_report_2020.pdf.

[33] World Bank, *Sustainable Textile Production: Lessons from South Asia*, Washington, DC: World Bank Publishing, (2021).

[34] S. Haque and I. A. Thaku, 'Development of Textile Industry of India and China: Flying Geese Model Revisited', *Pacific Business Review International*, 8 (2) (2015): 68–75.

[35] M. Hasanuzzaman, N. A. Rahim, R. Saidur and S. N. Kazi, 'Energy Savings and Emissions Reductions for Rewinding and Replacement of Industrial Motor', *Energy*, 36 (1) (2011): 233–40, doi: 10.1016/J.ENERGY.2010.10.046.

[36] M. Çetin and N. Eğrican, 'Employment Impacts of Solar Energy in Turkey', *Energy Policy*, 39 (11) (2011): 7184–90, doi: 10.1016/J.ENPOL.2011.08.039.

[37] A. K. Sharma, C. Sharma, S. C. Mullick and T. C. Kandpal, 'GHG Mitigation Potential of Solar Industrial Process Heating in Producing Cotton Based Textiles in India', *J Clean Prod*, 145 (2017): 74–84, doi: 10.1016/J.JCLEPRO.2016.12.161.

[38] S. J. Shah DS, 'Unconventional Techniques for Energy Conservation in Textile Wet Processing', (2013).
[39] United Nations Environment Programme, 'Environmental Agreements and Cleaner Production', (2007), www.unep.org/resources/report/environmental-agreements-and-cleaner-production.
[40] G. Barbiroli and A. Raggi, 'A Method for Evaluating the Overall Technical and Economic Performance of Environmental Innovations in Production Cycles', *J Clean Prod*, 11 (4) (2003): 365–74, doi: 10.1016/S0959–6526(02)00058-6.
[41] United Nations Framework Convention on Climate Change (UNFCCC), *Outcomes Report of the Global Climate Action Agenda at COP30*, Bonn, Germany: UN Climate Change (UNFCCC), (2025), https://unfccc.int/documents/655037.
[42] Kaniz Farhana, K. Kadirgama, M. M. Rahman, D. Ramasamy, M. M. Noor, G. Najafi, M. Samykano and A. S. F. Mahamude, 'Improvement in the Performance of Solar Collectors with Nanofluids. a State-of-the-Art Review', *Nano-Structures & Nano-Objects*, 18 (2019): 100276, doi: 10.1016/J.NANOSO.2019.100276.
[43] K. Farhana, K. Kadirgama, H. A. Mohammed, D. Ramasamy, M. Samykano and R. Saidur, 'Analysis of Efficiency Enhancement of Flat Plate Solar Collector Using Crystal Nano-Cellulose (CNC) Nanofluids', *Sustain Energy Technol Assessments*, 45 (2021): 101049, doi: 10.1016/j.seta.2021.101049.
[44] United Nations Industrial Development Organization, 'Cleaner and Resource-Efficient Production in SMEs: Good Practice Examples' Vienna, Austria: UNIDO, (2021), www.unido.org.
[45] Organisation for Economic Co-operation and Development, *Environmental Policy Toolkit for SME Greening*, Paris: OECD Publishing, (2019).
[46] Risecorp, 'Surat: India's Top Textile City – Which City Leads the Industry?', (2025), https://risecorp.co.in/surat-india-s-top-textile-city-which-city-leads-the-industry.
[47] GIZ, 'Energy Efficiency: Making Economic Growth in India Sustainable', (2022), www.giz.de/en/worldwide/68984.html.
[48] Arshad Group, 'Weaving', (2025), https://arshadgroup.com/weaving.html.
[49] International Finance Corporation, 'Sustainable Energy Solutions for the Textile Sector in Bangladesh, Washington, DC, (2021), www.ifc.org.
[50] R. Kant, 'Textile Dyeing Industry: An Environmental Hazard', *Nat Sci*, 4 (1) (2012): 22–6, doi: 10.4236/NS.2012.41004.
[51] World Bank, *Greening Industries in South Asia: Accelerating Climate-Resilient and Low-Carbon Industrial Development*, Washington, DC: World Bank, (2021).
[52] A. C. Horne, E. L. O'Donnell, A. J. Loch, D. C. Adamson, B. Hart and J. Freebairn, 'Environmental Water Efficiency: Maximizing Benefits and Minimizing Costs of Environmental Water Use and Management', *Wiley Interdiscip Rev Water*, 5 (4) (2018), doi: 10.1002/WAT2.1285.
[53] A. H. M. Noman, M. A. Mia, H. Banna, M. S. Rana, A. S. A. F. Alam, S.-G. Chan, C. R. Isa, and A. C. Er, 'City Profile: Narayanganj, Bangladesh', *Cities*, 59 (2016): 8–19, doi: 10.1016/J.CITIES.2016.05.020.
[54] The Energy and Resources Institute (TERI), *Sustainable Water Management in Indian SMEs*, New Delhi: TER, (2020), www.teriin.org.
[55] GIZ, 'Water Efficiency in Indian Textile Industry: Best Practice Cases', *Deutsche Gesellschaft für Internationale Zusammenarbeit*, (2021).
[56] WRAP (Waste and Resources Action Programme), *Water Stewardship in the Textile Industry: Case Studies and Practical Guidance*, Banbury, UK: WRAP, (2018), https://wrap.ngo.
[57] K. Niinimäki, 'Sustainable Fashion in a Circular Economy,' in *Sustainable Fashion in a Circular Economy*, Aalto ARTS Books, Espoo, Finland, (2018), pp. 12–41.
[58] Textile Value Chain, 'Zero-Waste Fashion: The Future of Sustainable Clothing', (2021), https://textilevaluechain.in/.
[59] K. Fletcher and M. Tham, *Earth Logic: Fashion Action Research Plan*, London: The JJ Charitable Trust, (2019).

[59] K. Fletcher and M. Tham, *Earth Logic: Fashion Action Research Plan*, London: The JJ Charitable Trust, (2019).

[60] T. Rissanen and H. McQuillan, *Zero Waste Fashion Design*, 2nd edn, London: Bloomsbury Visual Arts, (2023).

[61] S. Jung and B. Jin, 'From Quantity to Quality: Understanding Slow Fashion Consumers for Sustainability and Consumer Education', *Int J Consum Stud*, 40 (4) (2016): 410–21, doi: 10.1111/IJCS.12276.

[62] Intern International Labour Organization (ILO), *Sustainable Practices in Bangladesh's Ready-Made Garment Industry*, Geneva: International Labour Organization, (2020).

[63] Hydrocarbon Development Institute of Pakistan, *Pakistan Energy Yearbook*, (2021), https://hdip.com.pk/pakistan-energy-yearbook.

[64] Emissions Database for Global Atmospheric Research, 'Fossil CO_2 Emissions of All World Countries (Report)', (2020), https://edgar.jrc.ec.europa.eu/overview.php?v=booklet2020.

[65] B. Lin and H. Long, 'Emissions Reduction in China's Chemical Industry: Based on LMDI', *Renew Sustain Energy Rev*, 53 (2016): 1348–55, doi: 10.1016/J.RSER.2015.09.045.

[66] M. Ibrahim and R. Mahmood, 'Proactive Environmental Strategy and Environmental Performance of the Manufacturing SMEs of Karachi City in Pakistan: Role of Green Mindfulness as a DCV', *Sustainability*, 14 (19) (2022): 12431, doi: 10.3390/SU141912431.

[67] B. Elleuch, F. Bouhamed, M. Elloussaief and M. Jaghbir, 'Environmental Sustainability and Pollution Prevention', *Environ Sci Pollut Res*, 25 (19) (2018): 18223–5.

[68] U. Awan, A. Kraslawski, and J. Huiskonen, 'Progress from Blue to the Green World: Multilevel Governance for Pollution Prevention Planning and Sustainability,' in C. Mustansar Hussain (ed.), *Handbook of Environmental Materials Management*, Cham: Springer, (2020), pp. 1–22, doi: 10.1007/978-3-319-58538-3_177-1.

[69] J. S. Appannan, R. Mohd Said, T. S. Ong and R. Senik, 'Promoting Sustainable Development through Strategies, Environmental Management Accounting and Environmental Performance', *Bus Strateg Environ*, 32 (4) (2023): 1914–30, doi: 10.1002/BSE.3227.

[70] A. M. Nuru, 'Review of an Alternative Modification for Salt-Free Reactive Dyeing of Cotton Fabric', *Curr Trends Fash Technol Text Eng*, 9 (2) (2024).

[71] International Finance Corporation (IFC), *Sustainable Energy Solutions for the Textile Sector in Bangladesh*, Washington, DC: World Bank Group, (2021), www.ifc.org.

[72] T. Bechtold and R. Mussak (eds.), *Handbook of Natural Colorants*, Hoboken, NJ, USA: John Wiley & Sons, (2009), doi: 10.1002/9780470744970.

[73] A. K. Kumar and A. Konar, 'Dyeing of Textiles with Natural Dyes', *Natural Dyes*, (2011), https://doi.org/10.5772/21341.

[74] The UK Reporter, 'The Rising Fashion Industry in the UK: A New Era of Innovation and Sustainability', (2024), https://theukreporter.com/the-rising-fashion-industry-in-the-uk-a-new-era-of-innovation-and-sustainability/.

[75] K. Niinimäki, G. Peters, H. Dahlbo, P. Perry, T. Rissanen and A. Gwilt, 'The Environmental Price of Fast Fashion', *Nat Rev Earth Environ*, 1 (4) (April 2020): 189–200, doi: 10.1038/S43017-020-0039-9.

[76] International Labour Organization (ILO), *Safety and Health at the Heart of the Future of Work: Building on 100 Years of Experience*, Geneva, Switzerland: International Labour Office, (2019), www.ilo.org/publications/safety-and-health-heart-future-work-building-100-years-experience.

[77] International Labour Organization, 'Annual Report 2019', (27 March 2020), www.ilo.org/publications/annual-report-2019.

[78] S. Barrientos, G. Gereffi and A. Rossi, 'Economic and Social Upgrading in Global Production Networks: A New Paradigm for a Changing World', *Int Labour Rev*, 150 (3–4) (2011): 319–40, doi: 10.1111/J.1564–913X.2011.00119.X.

[79] Fair Wear Foundation, 'Fact Sheet', (2024). https://standardsmap.org/en/factsheet/19/resources.

[80] Clean Clothes Campaign (CCC), *Tailored Wages: The State of Pay in the Global Garment Industry*, Amsterdam: Clean Clothes Campaign, (2020), https://cleanclothes.org/resources/publications/tailored-wages-2019-the-state-of-pay-in-the-global-garment-industry.

[81] United Nations Development Programme (UNDP), *Accelerating Sustainable Development through Collective Action in the Textile Sector*, New York, USA: UNDP, (2021), www.undp.org/acceleratorlabs/collective-intelligence-sustainable-development-getting-smarter-together.

[82] KPMG Global, *The ESG Journey for SMEs: Practical Guidance for Value Creation*, London: KPMG International, (2019), https://assets.kpmg/content/dam/kpmg/xx/pdf/2019/12/the-esg-journey-for-smes.pdf.

[83] L. Johnstone, 'A Systematic Analysis of Environmental Management Systems in SMEs: Possible Research Directions from a Management Accounting and Control Stance', *J Clean Prod*, 244 (2020): 118802, doi: 10.1016/J.JCLEPRO.2019.118802.

[84] H. Walker, L. Di Sisto and D. McBain, 'Drivers and Barriers to Environmental Supply Chain Management Practices: Lessons from the Public and Private Sectors', *J Purch Supply Manag*, 14 (1) (2008): 69–85, doi: 10.1016/J.PURSUP.2008.01.007.

[85] C. J. C. Jabbour, A. B. L. De Sousa Jabbour, K. Govindan, A. A. Teixeira and W. R. De Souza Freitas, 'Environmental Management and Operational Performance in Automotive Companies in Brazil: The Role of Human Resource Management and Lean Manufacturing', *J Clean Prod*, 47 (2013): 129–40, doi: 10.1016/J.JCLEPRO.2012.07.010.

[86] A. Chiarini, 'Relationships between Total Quality Management and Six Sigma Inside European Manufacturing Companies: A Dedicated Survey', *Int J Product Qual Manag*, 11 (2) (2013): 179–94.

[87] M. A. Delmas and M. W. Toffel, 'Organizational Responses to Environmental Demands: Opening the Black Box', *Strateg Manag J*, 29 (10) (2008): 1027–55, doi: 10.1002/SMJ.701.

[88] Organisation for Economic Co-operation and Development (OECD), *SMEs in a Green Recovery: Challenges and Opportunities*, Paris, France: OECD Publishing, (2020).

[89] European Investment Bank (EIB), *Financing the Green Transition: Supporting SMEs for a Sustainable Future*, Luxembourg: European Investment Bank, (2021), www.eib.org/en/publications/financing-the-green-transition-supporting-smes-for-a-sustainable-future.

[90] United Nations Industrial Development Organization (UNIDO), *Green Industrial Policy: Concept, Policies, Country Experiences*, Geneva, Switzerland: UN Environment Programme (UNEP) & German Development Institute (DIE), (2017).

[91] European Commission, 'Corporate Sustainability Due Diligence', (2022). https://commission.europa.eu/business-economy-euro/doing-business-eu/sustainability-due-diligence-responsible-business/corporate-sustainability-due-diligence_en.

[92] International Trade Centre (ITC), *SME Competitiveness Outlook 2020: COVID-19-The Great Lockdown and Its Impact on Small Business*, Geneva, Switzerland: ITC, (2020), www.intracen.org/SMEOutlook/.

[93] M. J. Abdin, 'Establishing SME Clusters for Sustainable Development', *SSRN Electronic Journal*, (2015), https://doi.org/10.2139/ssrn.2647842.

[94] W. Jordão Filho, *Clusters and Clustering: A Strategic Framework for Innovation, Sustainability and Regional Development*, (2024), http://dx.doi.org/10.13140/RG.2.2.28343.33442.

[95] United Nations Industrial Development Organization (UNIDO), *Inclusive and Sustainable Cluster Development: Key Principles and Project Success Stories*, Vienna, Austria: UNIDO, (2024), https://hub.unido.org/sites/default/files/publications/05062024_Online_UNIDO_CLUSTERS_2024_%28003%29.pdf.

[96] R. D. de Andrade, V. C. Benfica, H. V. E. de Oliveira and N. Suchek, 'Investigating Green Jobs and Sustainability in SMEs: Beyond Business Operations', *J Clean Prod*, 486 (2025): 144477, doi: 10.1016/J.JCLEPRO.2024.144477.

[97] World Bank, 'World Bank and Private Sector', (2025). www.worldbank.org/en/about/partners/the-world-bank-group-and-private-sector/smes.

[98] Organisation for Economic Co-operation and Development, 'Global Dialogue on SME Sustainability Reporting', (2025), www.oecd.org/en/about/programmes/oecd-platform-on-financing-smes-for-sustainability/oecd-global-dialogue-on-sme-sustainability-reporting.html.

[99] Organisation for Economic Co-operation and Development, 'Platform on Financing SMEs for Sustainability', (2025), www.oecd.org/en/about/programmes/oecd-platform-on-financing-smes-for-sustainability.html.
[100] Organisation for Economic Co-operation and Development, 'Greening SMEs', (2025), www.oecd.org/en/topics/greening-smes.html.
[101] Rangotri, 'Manufacturer from Sanganer, Jaipur, India', (2025), www.indiamart.com/rangotri-jaipur/aboutus.html.
[102] The Textile Atlas, 'Living Blue, Bangladesh. The Textile Atlas', (2025), www.thetextileatlas.com/craft-stories/living-blue-bangladesh.
[103] The Business Times, 'Rapanui: Leading Sustainable Fashion Brand with Circular Innovation and Ethical Practices', (2024), https://thebusinesstimes.uk/lifestyle/rapanui-leading-sustainable-fashion-brand-with-circular-innovation-and-ethical-practices.
[104] County Press (Isle of Wight), 'Isle of Wight Eco Clothing Company Rapanui Wins Queen's Award for Innovation for Embracing Circular Economy, (2019), www.countypress.co.uk/news/17657867.isle-wight-eco-clothing-company-rapanui-wins-queens-award-innovation-embracing-circular-economy/.
[105] M. Riding, "pH control for a caustic recovery plant in Pakistan," Process Instruments Ltd., (2025), www.processinstruments.net/news/ph-control-for-a-caustic-recovery-plant-in-pakistan.
[106] Process Instruments Ltd., 'Caustic Recovery Plant (CRP): PDF technical overview', (2025), www.processinstruments.net/news/ph-control-for-a-caustic-recovery-plant-in-pakistan.
[107] The Textile Atlas, 'Living Blue, Bangladesh', (2025), www.thetextileatlas.com/craft-stories/living-blue-bangladesh.
[108] THE KINDCRAFT, 'Living Blue: A Bangladesh Social Enterprise that Handcrafts Indigo-dyed Quilted Textiles', (2025).

10 Corporate responsibility, environmental and social governance, and sustainable business models

Mehvish Sultan and Ghazia Batool

10.1 Overview of global sustainability challenges (climate change, inequality, resource depletion)

The twenty-first century has seen the challenges in world sustainability reach critical levels, and it has become quite apparent that the old conceptualisations of business are going to have to change. The pace of climate change is rising, which leads to an increase in the proportion of extreme weather, sea levels and the destabilisation of ecosystems. The Intergovernmental Panel on Climate Change (IPCC) has suggested that global greenhouse gas (GHG) emissions should be reduced by 43 per cent by 2030 to curb global warming to 1.5 °C [1]. This is a desperate schedule that puts a lot of pressure on governments, civil society, and, more so, the corporate world to become more sustainable.

In addition to environmental concerns, social inequalities are also deeply rooted, as the income difference is growing within and across boundaries. According to the World Inequality Report (2022), the richest 10 per cent of the global population make more than half of all the income, with the poorest half of the world making only 8.5 per cent. Besides, gender discrimination, unequal access to resources and labour exploitation, among others, only exacerbate social instability. In the meantime, the unsustainable consumption and the fast industrialisation process have caused natural resources to disappear at a terrifying pace [2]. The Global Footprint Network (2022) cautions that human beings are using up natural resources at a rate that would have taken 1.75 Earths to feed the yearly population [3].

Considering these dilemmas, it is the role of the corporate sector that is becoming increasingly questionable. Business companies are no longer perceived only as drivers of economic development but also as custodians of environmental, social and ethical responsibility. This transition has triggered the emergence of corporate social responsibility (CSR) and environmental, social and governance (ESG) models, which highlight ethical and moral behaviour, environmental sustainability and social justice. Both CSR and ESG are meant to redefine business responsibilities, although they vary in extent and usage. CSR is usually concerned with voluntary activities of companies to advance social good, whereas ESG is a more defined and investor-focused set of measures of corporate sustainability and ethical activities [4].

The growing importance of CSR and ESG is also connected with changing stakeholder expectations. Current customers expect more transparency and ethical conduct from the brands they follow. Workers would like to work in organisations whose values match their own. Investors are considering ESG in decision-making to determine risk prosperity in the long term. Governments are also becoming stricter and requiring more accountability. Here, CSR and ESG can no longer be seen as add-on elements of the business strategy but rather as a vital part of strategic business planning [5].

DOI: 10.4324/9781003638148-13

This chapter discusses the interplay between CSR, ESG and sustainable business models. Even though each of the concepts in question has evolved separately, they are closely interconnected in their practical terms. The combination of them can be seen as the integrated strategy of recreating the concept of business success in the twenty-first century. CSR is the source of an ethical framework; ESG has accountability frameworks that are measurable; and sustainable business models operationalise them into long-term strategies that generate shared value to both business and society.

With the inclusion of this triad as part of the larger discourse of sustainability, this chapter will attempt to offer a holistic perspective on how contemporary businesses can negotiate and play their role in enhancing the world through global issues. It talks about the ways companies can change their traditional profit-making models to new models that ensure the expansion of the company and provide purpose. Based on practical examples, scholarly sources and practitioner experience, the chapter highlights the strategic value of applying CSR, ESG and sustainability to the very essence of corporate decision-making [6].

10.2 Corporate social responsibility: evolution and modern context

CSR is the ethical aspect of a company making a positive contribution to society that is not part of its business operations. However, the idea of CSR has been developing throughout the years, and it is no longer a marginal activity of doing philanthropy work but has become a pivotal part of the business strategy. Early CSR was even confused with charity on the part of the corporation or community investment, in which business organisations gave donations to causes or local projects as a form of giving back to the community. This has, however, changed drastically in the past few decades as stakeholder expectations, such as those of the consumers, employees, investors and regulators, have risen [7].

The idea of CSR dates to the Industrial Revolution, when business tycoons such as Andrew Carnegie and John D. Rockefeller used their own fortune to advance social causes. These attempts were, however, much detached from the business practices. The contemporary understanding of CSR had its momentum in the mid-twentieth century. A seminal work attributed to Howard R. Bowen, *Social Responsibilities of the Businessman* (1953), is said to have been the basis of the theoretical foundation of the current CSR by positing that businesses should be aware of the social implications of their efforts. Gradually, CSR was institutionalised in the process of corporate governance, especially with increased globalisation and the proliferation of multinational corporations [8].

10.2.1 From philanthropy to strategic CSR

The shift to strategic CSR signifies a pivotal point in the transformation of the concept of traditional philanthropy. Strategic CSR is the integration of company social activities with business interests, to establish shared value – a concept that has gained popularity [9]. Through this strategy, CSR does not serve as a secondary operation but contributes to competitive advantage. Companies recognise social issues that cut across their operations and formulate strategies that are good for both society and the bottom line. Indicatively, the Nestlé Creating Shared Value programme has entailed the integration of nutrition, water sustainability and rural development as part of the core business strategy of the company.

Strategic CSR initiatives are numerous in terms of examples. The Sustainable Living Plan introduced by Unilever in 2010 was designed to have the company grow without having an effect on the environment and become a better contributor to society. This involved activities like cutting down emissions of GHGs, hygiene and health promotion in developing nations and bettering livelihoods in its supply chain [10]. In the same way, Starbucks has also been oriented towards ethical sourcing of coffee beans by its Coffee and Farmer Equity programme that monitors fair labour and sustainable farming.

CSR is also important in engagement with stakeholders and brand reputation. Contemporary consumers are more socially responsible and demand that the companies they patronise show ethical conduct as well as social responsibility. A report by Cone Communications (2020) indicates that 86 per cent of consumers believe that businesses ought to take a position on social matters, and 78 per cent of consumers stated that they will be more likely to purchase from a company that is dedicated to making the world a better place [11]. This highlights the role of CSR activities in increasing customer loyalty and brand differentiation in saturated markets.

The posture of CSR in a company also affects employees. Specifically, millennials and Generation Z employees would like to work for an employer that is socially and environmentally responsible. CSR activities involving workplace diversity, employee welfare and community support may enhance job satisfaction, turnover and recruit the best employees.

With respect to investments, CSR may indicate sustainability in the long term and risk management. Companies that practise responsibility are perceived to be higher and less vulnerable to regulatory scrutiny, reputational risks and changing market demands [12]. In this regard, most institutional investors today consider CSR operating in the decision-making process when making investments (notwithstanding no official ESG indicators).

Overall, CSR has been radically transformed from an individual act of charity to being a strategy-based approach to conducting business. It is such a broad topic today, which includes numerous programmes aimed at not only solving social problems but also promoting businesses. The contemporary CSR environment has been defined in terms of alignment to corporate objectives, incorporation of CSR in governance frameworks and orientation to shared value creation. It improves the relationships with stakeholders, develops brand equity and long-term competitive advantage. CSR is not a mere moral imperative or a strategy in this age of sustainability and stakeholder capitalism [13].

10.2.2 Examples of CSR initiatives in fashion and textiles

CSR has emerged as a pillar that fashion and textile corporations have turned to in the process of trying to strike a balance between profitability and purpose. CSR activities are multidimensional and tend to touch on social, environmental and ethical concerns that are entrenched in the global supply chains of industry. Community development, ethical sourcing, fair labour practice and environmental stewardship are some of the well-known examples of CSR initiatives [14].

Community development programmes to improve the living standards of workers and artisans are one of the frequently mentioned CSR efforts. Some of the brands, such as Pukka Herbs and the People Tree, engage communities, particularly rural ones that are disadvantaged by offering them education, health care and training in skills. People Tree is partnered with fair-trade organisations in Asia and Africa so that not only do artisans receive fair pay but also they are assisted by social programmes such as access to clean water and women's empowerment training [15].

Ethical sourcing is another important issue of CSR in the fashion industry. This would entail the process of tracing the source of raw materials to make sure they are sourced in an environmentally friendly and socially responsible way. Such companies as Patagonia and Eileen Fisher are considered the leaders in this direction. Patagonia seems to have a transparent tool, the Footprint Chronicles, which enables consumers to trace the environmental and social impact of any product. On the same note, Eileen Fisher has also dedicated itself to organic cotton, wool that is responsibly sourced, recycled fibres and fair wages and safe working conditions in its supply chain [16].

Also, CSR in fashion revolves around fair labour practices. The Bangladesh Rana Plaza disaster of 2013 opened the ugly side of the situation of the garment workers, which led the world to demand better labour standards. To counter this, numerous fashion companies, including Levi Strauss & Co., have reformulated their labour policies, had a thorough audit in their factories and become members of such global alliances as the Ethical Trading Initiative to promote the rights of workers [17].

CSR strategies also constitute environmental initiatives. Major companies are making investments in closed-loop manufacturing, waste management and carbon net neutrality. Stella McCartney, as an example, has been the first luxury fashion brand to become sustainable through the removal of animal leather and fur, as well as by using circular production methods. Equally, H&M Conscious Collection also incorporates recycled materials and organic fabrics and its in-store recycling programmes that promote sending customers with old clothes that need a second use [18].

Case study 10.1 People Tree – empowering communities through fair trade

People Tree is a fashion business that was founded in the UK and Japan and is known to have instigated sustainable and fair-trade fashion [15].

CSR initiatives

- Collaborates with more than fifty fair-trade manufacturers in developing nations.
- Gives artisan communities a stable income, capacity-building and welfare services (e.g., health insurance and education).
- Makes use of organic cotton certified and natural dyes to reduce the environmental impact.

Impact

- High standards of living among the rural workers, especially the women.
- Open supply chain enhanced trust among the ethical consumers.
- Assisted in creating a loyal customer base that is in line with conscious consumerism.

People Tree is an example of how CSR can go beyond marketing and actually change the lives of the producers and solidify the sustainability ethos of the brand.

10.2.3 Role of CSR in stakeholder engagement and brand reputation

CSR is a strategic tool of stakeholder engagement and brand-building as well as a moral imperative. Stakeholders such as consumers, employees, suppliers, NGOs and regulators are more conscious of business ethics and sustainability concerns in the contemporary business landscape. Organisations that do not satisfy the expectations of the stakeholders tend to suffer a loss of reputation, government inspection and loss of customers [19].

The process of stakeholder involvement by means of CSR starts with transparency and bilateral communication. Fashion companies that publish their sustainability reports and supply chain information and address the interests of the populace have a higher chance of establishing a sense of trust and loyalty. In one case, Nike, which was once accused of labour violations, has managed to revamp its image with the help of a more open CSR policy involving its sustainability reports and annual supplier audits and forums to hear the complaints of its stakeholders [20].

CSR is also very critical in attracting and retaining talent. The modern worker, more so millennials and Generation Z, would like to be associated with a company that matches their values. Fashion companies that have solid CSR engagement tend to have a higher level of employee satisfaction, employee retention and innovation. Employee engagement and corporate dedication are solidified by programmes such as volunteering opportunities, ethical training and inclusive workplace cultures [21].

As a consumer, CSR initiatives tend to build a distinction in a saturated market. Companies that have proven to be socially conscious and sustainable have higher chances of customers supporting them. In the study conducted by KindLink [22], it was found that two-thirds of consumers globally are ready to pay more for sustainable brands. Fashion brands, which embrace ethical supply chains, non-cruelty products, or carbon-neutral shipping, enjoy the advantages of improved brand equity, customer confidence and brand loyalty over time.

In addition, CSR improves the attractiveness of a company to investors. The ethical and sustainable approaches are becoming the apparent signs of long-term stability and the minimisation of risks. Shareholders match portfolios with ESG measures and fashion companies that incorporate CSR as a central business concept tend to find better investment.

10.3 Understanding ESG: environmental, social and governance dimensions

Over the past few years, ESG models have become a major trend in strategic management and as an investment screening mechanism. ESG in the fashion and textile industry is a multidimensional perspective of the long-term value of a firm that goes beyond the normal financial indicators. Through its analysis of the way an organisation deals with its environmental footprint, societal responsibility and organisational frameworks, it has become an effective instrument in sustainability, risk reduction and corporate branding. In this section, we discuss the main elements of ESG and how they are used to develop ethical and sustainable business models [23].

10.3.1 Environmental dimension: managing ecological impact

The environmental component of ESG deals with the way a company conducts its relations and influences the natural environment. The dimension is especially important to the fashion and textile industry, an industry that is traditionally linked to pollution, resource depletion and the production of waste. The industry is the second-largest water-user worldwide, making it a major emitter of GHGs and microplastic pollution [24].

The most important environmental indicators are climate risk, carbon footprint, resource management and protection of biodiversity. As an example, fashion outlets are already supposed to decrease the usage of fossil-based fibres (such as polyester) and reduce the utilisation of water during textile dyeing, as well as to be actively engaged in decarbonising the supply chain. Measurement of the carbon footprint has turned into the key part of the environmental strategy, and numerous companies have established science-based goals to achieve net-zero emissions [25].

Resource management is the use of raw materials and energy by companies in the course of their operations. The strategies aimed at minimising material waste and saving energy include sustainable design (circular production models, upcycling and low-impact dyeing techniques) and sustainable design. Also, firms are investing in biodiversity programmes to conserve ecosystems where cotton is grown, processed to make leather, or chemical dumping [26].

Pioneering brands such as Stella McCartney and Patagonia have been at the forefront in this field, and they serve as examples of how environmental stewardship can be incorporated into the product design and supply chain processes, as well as the branding message. These initiatives are not only environmentally friendly, but they also make sense to the environmentally friendly consumer and investor.

10.3.2 Social dimension: human rights, welfare and inclusion

The social aspect of ESG is devoted to the way a company deals with relationships with people, its employees, customers, workers within the supply chain and communities. This is particularly essential in fashion and textiles as the industry has highly complex supply chains, which in most cases include the use of low-wage labour in underdeveloped countries [27].

Social responsibility is linked to the welfare of the employees, human rights, diversity, equity and inclusion (DEI), and community engagement. Brands are becoming responsible because their products are manufactured. The problem of forced labour, child labour and unsafe working environments, as well as gender discrimination have put the industry under intense public scrutiny [28]. This disaster was the Rana Plaza in Bangladesh in 2013, which triggered a challenge in the world to enact labour reforms and enhance corporate responsibility.

Employee welfare also incorporates remuneration, work safety, medical facilities and career advancement. Inside companies, they should also pay attention to DEI practices, i.e., all people, irrespective of their gender, race, or background, should be equally provided with opportunities, representation and advancement. Anti-discrimination policies, inclusive hiring practices and sensitivity training have become part of the socially responsible corporate culture.

Besides that, community engagement programmes are essential. Local artisans, indigenous people, or marginalised communities partner with many brands as a way of economic empowerment and cultural heritage. People Tree and Maiyet, among others, have made ethical manufacturing and crafts partnerships to be part of their business strategies, which makes them socially responsible [29].

By showing genuineness in the social practices of the companies, they develop a sense of trust, consumers are loyal to them and even talent that concurs with their values is attracted. The concept of social impact is no longer a peripheral consideration but an element of brand strategy.

10.3.3 Governance dimension: ethical leadership and compliance

The governance component of ESG deals with internal systems, leadership practices and policies that inform decision-making of a company. Good governance guarantees ethical business operations, transparency and business operations that are in line with societal and legal expectations. Powerful governance plays a critical role in ensuring accountability in the fashion and textile industry, where scandals of labour rights violations, greenwashing and corruption have occasionally come into the picture [30].

Some of the aspects of governance entail ethical leadership, diversity in boards, regulatory compliance and transparency. Firms that have a diverse and inclusive board tend to be stronger and more innovative since they introduce different points of view in strategic decision-making. Similarly, open dialogue with stakeholders, such as sustainability reports, ESG reporting, or third-party audits indicate integrity and builds stakeholder trust.

Ethical leadership goes beyond the legal traps; it is about leading the company in a moral direction. This entails creating a concise code of conduct, whistleblower policies, the elimination of conflicts of interest and leadership accountability. Indicatively, numerous international fashion companies have established the position of chief sustainability officer and incorporated sustainability performance in the executive compensation metrics.

Adherence to international frameworks, i.e., the UN Global Compact, the OECD Guidelines on Multinational Enterprises and the Sustainable Development Goals, also turns out to be a norm. These frameworks offer the criteria of responsible business practices and assist companies in aligning with the global sustainability objectives.

10.3.4 ESG as a non-financial performance indicator and investment tool

Especially in addition to being a moral or reputational issue, ESG has emerged as a non-financial measure of performance that is utilised by investors, regulators and stakeholders to assess risk and sustainability in the long term. The use of ESG metrics in the process of investment screening is also becoming a widespread phenomenon with the use of ESG ratings, sustainability indexes and responsible investment funds.

Investors have come to appreciate the fact that firms that have low ESG scores face high risks, which include legal claims, fines and penalties, negative publicity and disruption of operations. On the other hand, companies that have good ESG scores tend to have superior risk management, innovation capability and length of stay. This has seen a boom in ESG-oriented investment vehicles such as green bonds, ESG ETFs and sustainable mutual funds [31].

The ESG-focused brands in the fashion industry can enjoy competitive benefits, such as access to capital, brand equity and customer retention. There are also regulatory requirements, particularly in the European Union (EU), that make it a standard of business to have mandatory ESG reporting and due diligence and are further enshrining ESG as a standard business need.

To summarise, ESG is ceasing to be a niche issue and an emergent outline of twenty-first-century business leadership. In the case of the fashion and textile industry, the adoption of ESG would imply the redesign of operations, supply chain and corporate culture to address the needs of a conscious customer base on a global level. In solving environmental issues, practising social justice and ensuring good governance, companies not only create resilience but also play a significant role in trying to make the world a more just and sustainable place.

10.4 Sustainable business models

Traditional business models are no longer applicable in the global economy, with their growing environmental degradation, social inequality and financial turmoil, which require a more complex approach. The fashion and textile industry, especially, is being scrutinised highly because it is affecting the depletion of resources, labour rights and pollution. It is on this background that the concept of sustainable business models (SBMs) is becoming an important paradigm that assists firms to balance the three aspects of profitability, social and environmental stewardship [32].

SBMs go past the short-term financial benefits to encompass the long-term plans that will bring value to all stakeholders, including the consumers, communities, employees, investors and planet. They are based on the understanding that economic prosperity cannot exist without environmental sustainability and social health.

10.4.1 Definitions and features of SBMs

A business model is said to be sustainable when it is a structure or strategy that can generate, deliver and/or capture value in a manner that is economically viable, socially beneficial and environmentally appropriate. Contrary to the traditional models that tend to focus mostly on the shareholder returns as the key factor of success, SBMs are designed on the premise of the principles of ethical responsibility, systems thinking and long-term impact [33].

10.4.2 Principal characteristics of SBMs

SBMs are not an abstract concept, and they are becoming a reality, with progressive firms in the fashion industry coming to realise that long-term viability is required in an already resource-starved world.

- **Stakeholder orientation**: SBMs weigh the interests of a wider spectrum of stakeholders, such as employees, suppliers, communities and the environment, instead of only focusing on the stakeholders who are shareholders [34].
- **Systems thinking**: Sustainable models acknowledge the interrelatedness of the environmental, social and economic frameworks. This implies that business decisions are considered regarding their more ecological and societal overall impacts [35].
- **Life-cycle approach**: These models evaluate the complete life cycle of products – sourcing of raw materials through production, use and end of life. Life durability, recyclability and waste minimisation are put into focus [36].
- **Purpose-driven strategy**: SBMs can also be consistent with an explicit mission or purpose, other than profit-making, reduction of carbon, promotion of ethical labour, or safeguarding biodiversity [37].
- **Innovation for sustainability**: Firms that use SBMs invest in business models, products and process innovations to achieve sustainability objectives. This is sustainable materials, low-impact technologies and new models of consumption [38].

10.5 Circular economy: closing the loop in fashion

The component of SBMs is the circular economy, a regenerative system that is focused on the removal of waste and the extension of product and material use time to the maximum. Compared to the traditional linear economy, where the mode of action is take–make–dispose, the circular model focuses on the three pillars: reuse, repair, remanufacture and recycling [39].

The principles of the circular economy are currently being implemented in the fashion and textile sector to address such problems as fast fashion, waste of textiles and excessive production. Products are currently being designed in circularity by brands, including mono-material, more readily recyclable, biodegradable fibres and modular garment designs that can be repaired and customised [40].

The main strategic approaches to the fashion industry in the circular economy are:

- rental and resale clothing sites (e.g., Rent the Runway, ThredUp), which prolong the life of the product;
- take-back actions on garments (e.g., garment collection programme of H&M);

- using old garments as a source of new fibres (e.g., Evrnu, Worn Again Technologies) in closed-loop recycling;
- design to be disassembled so that it can be more easily broken down into components that can be disassembled or recycled.

Not only do circular business models create more revenue and value, but they also contribute to the reduction of environmental impact, in addition to providing conscious consumers with the opportunity to find sustainable alternatives to consumption.

10.5.1 Shared value creation: integrating profit and purpose

Shared value creation is a term that was coined by Michael Porter and Mark Kramer, and it pertains to the business practices that create economic value in such a manner that also creates value to society by meeting its needs and challenges. In contrast to philanthropy or conventional CSR, shared value is integrated into the essence of a business, and the advancement of society is the force that pushes economic prosperity [41].

Shared value creation may be in many forms in the fashion industry:

- investing in the local supply chains to enhance the economic opportunities of the artisans and workers;
- enhancing the working environments in factories, aiming at increasing productivity and curtailing turnover;
- creation of environmentally friendly products that satisfy the sustainability trend among consumers;
- training the consumers to create fashionable behaviour by teaching them how to make sustainable fashion decisions;

As an example, Levi Strauss & Co. created its Worker Well-being programme to improve the lives of factory workers by educating them on health, offering financial literacy and gender-equality courses. This enhances the social life of the production communities as well as increases the retention of workers, productivity and brand loyalty. Shared value assists in the reconciliation of the business objectives and the expectations of society, hence it is a robust component of SBMs.

10.5.2 Triple bottom line: people, planet, profit

The other concept of sustainable business is the triple bottom line (TBL), which was proposed by John Elkington. It broadens the classic financial bottom line, adding two more dimensions, namely people (social equity) and planet (environmental health). TBL is a performance measurement system that encourages businesses to review their performance regarding these three dependent dimensions [42].

People

The social aspect dwells on the community, supplier and employee well-being. In fashion, it encompasses decent salary, health and safety, equality with gender and community development. Businesses that are interested in TBL also make sure that the business they run does not take advantage of the labour force or discriminate against vulnerable populations.

Planet

The environmental component deals with the ecological footprint of the business: carbon emission, water consumption and pollution. Fashion companies that include TBL invest in sustainable fabrics (e.g., organic cotton, Tencel), non-toxic dyes, energy-efficiency and low-waste production techniques.

Profit

The economic aspect is not overlooked but is sought in ethical, responsible and transparent ways. The objectives of companies in achieving long-term profitability entail creating resilient supply chains, mitigating risks and enhancing brand identity because of sustainability credentials.

Examples of TBL brands include Patagonia, Veja and Reformation. They are always focused on environmental ethics and social justice and, at the same time, ready to be commercially successful and loyal to the brand. The people, planet and profit balance make TBL-oriented companies more flexible to changes in the market, more appealing to ESG investors and prepared to grow in the future.

A sustainable business model is a paradigm shift that can be used by businesses to survive in a dynamic and fast world. Considering the principles of the circular economy, shared value creation and the TBL, the fashion and textile industry businesses will be able to approach systemic issues and generate long-term value. These models spin the traditional focus on the maximisation of profits in the short-term to long-term sustainability, responsible leadership and regeneration. With the increasing global demand to act on climate change, to be fair and to take responsibility, the adoption and scaling of these models is not a strategic decision but an inevitability in the future of fashion and the world [43].

10.6 Case studies in sustainable business and ESG excellence

Sustainability is no longer a buzzword in the business world, but an incredible force that fosters innovation, strength and competitiveness over the long run. In all industries, firms are remodelling to include the ESG and the circular economy. This part includes three case studies that are interesting and persuasive: Patagonia, Unilever and Interface Carpets, all of them show how sustainability can be a strategic resource and a source of global good [44].

Case study 10.2 Patagonia – transparency and environmental stewardship as a brand strategy

Patagonia is an American outdoor clothing company that was established in 1973 and has a high ethical and environmental policy.

CSR initiatives

- Developed the Footprint Chronicles, an interactive platform that enables the consumer to trace the impact of a product.
- Sources organic cotton, recycled materials and good working conditions.
- Gives back 1 per cent of profits to the environment and acts as an activist in the so-called Action Works platform.

Impact

- Well-established consumer confidence and sustainability loyalty.
- It has always been one of the most recognised brands in the world.
- Gets environmentally friendly investors and employees who share in its mission.

Patagonia shows that environmental responsibility and absolute transparency may be transformed into a brand image, competitiveness and future development.

Case study 10.3 Unilever – ESG-driven growth and brand value

- **Industry**: Consumer goods (FMCG)
- **Founded**: 1929
- **Headquarters**: London and Rotterdam

Unilever, the parent of more than 400 brands such as Dove, Ben & Jerry's and Lifebuoy, has become one of the most powerful users of ESG in the business. The Sustainable Living Plan incorporates the social, environmental and ethical factors into the value chain [10].

ESG integration

Unilever is a performance and investment index that employs executive compensation for sustainability objectives. It also implements transparent supply chain reporting, follows zero-deforestation policies and encourages sustainable sourcing and sourcing of palm oil and tea.

In the social areas, the firm has initiated DEI efforts to enhance gender balance and racial equity throughout its operations. In addition, its brands engage in proactive positions concerning social problems. The Real Beauty campaign by Dove breaks all the norms of beauty and, at the same time, brings body positivity and mental health.

Environmental leadership

The climate action of Unilever involves being carbon-neutral in its operations, transitioning to renewable energy and aiming to achieve net-zero by 2039. It has undertaken to ensure that 100 per cent of its plastic packaging is reusable, recyclable, or compostable by the year 2025.

Impact

- More than €1 billion of 'sustainable living brands' sales (e.g., Hellmann's, Seventh Generation, etc.).
- 67 per cent decrease in GHG emissions (per tonne of production) since 2008.
- Considered one of the best ESG companies in the world.

Case study 10.4 Interface – innovative path to zero-carbon goals

- **Industry**: Commercial Flooring (Textiles)
- **Founded**: 1973
- **Headquarters**: Atlanta, Georgia, USA

The manufacturer of Interface Inc. is a commercial flooring company that has transformed its business by overturning the concept of sustainability as its leader. Its founder, Ray Anderson, said his move to green his company was his so-called Mid-Course Correction, which made Interface one of the most desired green companies in the world [45].

Mission Zero and climate take-back

The Mission Zero programme of Interface was to reduce the negative environmental impact to zero by the year 2020. They also achieved much in the way of GHG emission, water and waste reduction through design, sourcing and manufacturing innovations.
Their second ambitious step is the Climate Take-Back, which is a commitment to turn the world warmer back by creating carbon-negative carpet tiles and donating to building materials that contribute to climate positivity.

Circular business model

The Interface design relies on the modular design strategy in which individual carpet tiles can be substituted by the customers instead of substituting entire floorings. Their ReEntry programme enables the old carpet tiles to be returned, recycled and used in making new products. Bioplastics and recycled nylon are some of the materials that reduce the environmental footprint without interfering with quality.

Governance and transparency

Full annual reports on ESG are published by Interface, science-based targets are applied, and it has third-party certifications, including Cradle to Cradle, GreenCircle and Leadership in Energy and Environmental Design (LEED).

Impact

- Cut down the carbon footprint of carpet tiles by 74 per cent.
- Recycled millions of pounds of waste.
- Introduced the first carbon-negative carpet tile in the world in 2020.

These case studies underline a variety of yet complementary ways of incorporating ESG and SBMs. Patagonia is an example of fashion circularity and activism. Unilever shows how ESG can create value in the brand and corporate change on a large scale. The example of Interface Carpets demonstrates the possibility to transform fantasy dreams of carbon-neutrality into practical results.

Every business demonstrates that sustainability is not a trade-off: it is an innovation, strength and leadership approach. With the growing demand by global consumers and investors to be responsible, those businesses that consider embedding ESG in their DNA will become future-proof leaders.

10.7 Integration of CSR and ESG into business strategy

Sustainability ceased to be a marginal issue in the contemporary business environment a long time ago, as it has become the key element of strategic decision-making, risk management and value creation over the long term. CSR and ESG have become two major models of guiding ethical practices, sustainable growth and responsible stakeholders. Although the terms CSR and ESG are used interchangeably, they play different roles in corporate strategy, yet they can be combined in a manner that complements one another and leads to organisational performance [46].

10.7.1 CSR and ESG: different origins, complementary functions

CSR is a value-driven approach that encourages businesses to contribute positively to society. It emphasises ethical behaviour, philanthropy, volunteerism, community involvement and the moral obligation of companies to do good [8]. CSR often operates as a self-regulatory model where companies go beyond compliance to align business practices with societal expectations.

ESG, in contrast, is a performance-based, data-driven framework used by investors and stakeholders to assess a company's sustainability risks and impacts. ESG provides measurable indicators of corporate performance related to climate impact, labour conditions, diversity, governance structures and ethical conduct [5]. Unlike CSR, which is often qualitative, ESG integrates non-financial metrics into investment and strategic planning.

While CSR tends to focus on 'doing good', ESG emphasises 'doing well by doing good'. The key distinction lies in voluntary values versus measurable metrics. However, the two frameworks are deeply interconnected. CSR sets the ethical foundation and stakeholder expectations while ESG operationalises those values into accountable and reportable strategies. For businesses aiming for long-term sustainability, integrating CSR principles into ESG practices provides both social legitimacy and investment credibility [4].

10.7.2 ESG reporting frameworks: from voluntary disclosure to regulatory mandates

With the increasing institutionalisation of ESG practices, standardised reporting systems have emerged to assist firms in measuring, managing and reporting their sustainability performance. The frameworks inform organisations to report material issues and ensure comparability, transparency and accountability.

Global Reporting Initiative

GRI Standards have become the most popular sustainability reporting standard in the global arena. The Global Reporting Initiative (GRI) addresses the effects of an organisation on the people, the environment and the economy, and is very detailed on such issues as human rights, anti-corruption and emissions [47]. GRI is also stakeholder-focused and assists companies in aligning their CSR efforts with reportable ESG objectives.

Sustainability Accounting Standards Board

The Sustainability Accounting Standards Board (SASB) offers industry-related standards which are specific to assisting companies in reporting financial material problems in sustainability. In comparison to GRI, SASB is more concerned with the interests of investors and capital markets. In the case of the fashion industry, SASB standards are used to draw attention to such important issues as labour practices, sourcing of raw materials, water consumption and waste disposal [48].

Task Force on Climate-related Financial Disclosures

The Task Force on Climate-related Financial Disclosures (TCFD) offers suggestions for reporting climate-related financial risks and opportunities. It focuses on governance, strategy, risk management and metrics and assists companies in aligning with climate resilience objectives [49]. Fashion brands such as Burberry and H&M have started reporting in a manner that is aligned with the TCFD to be ready to regulate climate and be put under investigation by investors.

Corporate Sustainability Reporting Directive

In 2023, the EU introduced the Corporate Sustainability Reporting Directive (CSRD), which requires thousands of companies that trade with or operate in the EU to engage in ESG reporting. It needs digital, audited and standardised sustainability disclosures founded on the European Sustainability Reporting Standards. This guidance represents a change from voluntary to mandatory ESG reporting, which strengthens the idea of strategic inclusion of sustainability in business governance [50].

Collectively, these frameworks enable companies to frame their ESG strategies and harmonise them with the larger CSR commitments, regulatory compliance and expectations of stakeholders.

10.7.3 Role of leadership in embedding responsibility and sustainability

The issue of leadership is the key to the effective implementation of CSR and ESG in business strategy. In the absence of a dedicated and dedicated vision at the top, sustainability initiatives tend to be cosmetic or unrelated to the main business activities. The ethical standards, culture of accountability and entrenching sustainability into the structure of governance, operations and incentives are aspects of responsible leadership [51].

Board members and CEOs need to set an example, with a clear statement of purpose that is not about profit, an investment in sustainability innovation and executive compensation tied to ESG performance. Unilever, as an example, connects executive bonuses to improvements in such sustainability targets as plastic reduction, emissions and DEI [10].

In addition, cross-functional integration should be encouraged by leadership. Sustainability ought not to be kept behind the CSR departments but to percolate into the decision-making processes in supply-chain management, product design, marketing, finance and human resources. As an illustration, the leadership of Patagonia institutionalised environmental ethics throughout its business model, which means that it is connected to procurement, product development and consumer experience. Such internal coherence and external legitimacy are achieved through this type of embedded leadership.

Lastly, leadership has the mandate of cultivating an inclusive culture among stakeholders. Relations with suppliers, customers, employees, regulators and investors promote resilience and legitimacy, particularly in such industries as fashion, where labour rights, transparency and environmental impact are key concerns.

10.7.4 Strategic tools for integrating CSR and ESG

Some of the tools and frameworks that businesses can use to operationalise CSR as well as ESG strategies are as follows.

ESG scorecards

An ESG scorecard is a performance management tool that helps in monitoring sustainability key performance indicators at departmental and supply-chain levels. This can be through emission cuts, energy saving, diversity ratio, waste disposal, or supplier audit. Scorecards can be used to make teams focus on specific quantifiable objectives and to facilitate periodic monitoring and reporting internally [4].

Sustainability audits

Sustainability audits entail the evaluation of business operations to determine the business gaps, risks and opportunities for improving business operations in areas like labour conditions, environmental compliance and waste management. Audits can be third-party or internal and are commonly employed to monitor the progress in compliance with the environmental and social development (ESD) organizational frameworks, guidelines and management systems like ISO 14001 (Environmental Management) or SA8000 (Social Accountability), etc.

Stakeholder mapping and engagement

Stakeholder mapping is critical in identifying people or groups that are impacted or have an influence on the business. This involves customers, employees, suppliers, non-governmental organisations, regulators and investors. When the stakeholders are known, the companies can consult them using surveys, forums, focus groups and partnerships to know the expectations and develop sustainability strategies together [52].

Materiality assessments

Materiality assessments assist the companies in knowing what ESG issues are of most interest to the stakeholders and their financial performance. These evaluations focus resources and report on the greatest issues, say, the water used in textile manufacture, or the rights of the labourer in the garment supply chain.

Integrated reporting

Integrated reporting is reporting that incorporates both financial and non-financial information into one unified report. This method promotes long-term thinking and plays a role in showing the influence of ESG factors on the value creation in the long term. It can be applied particularly well in terms of reporting messages about sustainability strategies to investors and regulators [53].

The inclusion of CSR and ESG in business strategy is a paradigm shift in the way companies view and deal with their duties towards the people, planet and profit. CSR is the moral compass whereas ESG is the performance measure, so that the values become quantifiable results. The combination of the two reinforces a sustainable approach to business that creates resilience, trust and long-term value.

The integrated approach to being employed to achieve success in this integration is a multi-stakeholder, cross-functional and leader-oriented approach, which is backed by effective tools and global reporting standards. This integration is the competitive edge as well as the requirement in the industries where supply-chain transparency, labour ethics and environmental impact are gaining a new standard of scrutiny, such as the fashion and textile industries.

As sustainability is no longer a differentiator but an expectation, the companies that completely incorporate the values of CSR and ESG metrics as part of their DNA will become the leaders of performance and purpose.

10.8 Benefits and business case for ESG and sustainable models

In the current globalised and open market, companies are slowly feeling that it is not only morally right and beneficial but also a strategic competence to incorporate the ESG elements in their business models. Resilience, innovation and value creation are motivated by the ESG and sustainability-oriented models. The benefits are multidimensional and can be seen in the reduction of risks to lure in investment and in increasing brand loyalty, as well as in the area of better interaction with the workforce.

10.8.1 Risk management: the climate, reputational and legal insurance

Risk mitigation is among the short-term effects of ESG integration. Companies that actively deal with environmental and social issues are in a better position to pre-empt and mitigate any form of disruption, liability and regulatory punishment.

Supply chains, manufacturing and distribution are directly affected by climate risks, including the increase in temperatures, extreme weather and a lack of resources, particularly in the spheres of fashion and textiles [54]. As an illustration, cotton production is extremely susceptible to droughts and floods and, therefore, climate-aware supply-chain strategies turn out to be not only sustainable but also necessary.

Moreover, reputational risk in the form of poor working conditions, environmental damage, or failure to show transparency can easily lead to consumer distrust. Negative publicity is highly disseminated in the virtual world, both in brand equity and sales. The Rana Plaza collapse in Bangladesh in 2013, which revealed unsafe work practices in the fashion industry, resulted in global backlash and made many brands take more rigorous ESG procedures and open sourcing policies [55].

The level of legal risks is also growing with the tightening of environmental regulations by governments. An example of this is the EU's CSRD, which requires companies to report the impact on ESG. Companies that plan and follow such regulations have higher chances of avoiding punishment and have a chance of remaining open to global markets.

10.8.2 Long-term profitability and investor interest

SBMs and ESG integration do not harm profits as opposed to old ideals; they optimise them in the long-term perspective. The ability of sustainable companies to have high financial performance is usually because of efficiency, innovation and reduced cost of operation.

As an example, replacing production with energy-efficient technologies, switching to renewable energy and cutting down on the waste of materials not only decreases carbon footprints but also leads to considerable cost-cutting. On the same note, the models of circular economy, such as product repair and resale and recycling, create additional revenue sources and eliminate reliance on raw materials.

Notably, companies that are ESG-oriented increase the interest of investors. BlackRock and Vanguard are the institutional investors that now take ESG scores into account prior to making funding decisions. The report by Morningstar (2021) indicates that sustainable funds had net flows of more than $51 billion and were also performing better than traditional funds [56].

Some ESG measures like the Dow Jones Sustainability Index and FTSE4Good have exceeded such standard indices, demonstrating that ESG is not only an ethical issue but also a prudent investment choice.

10.8.3 Talent attraction and retention

The modern labour force, and especially the younger generations, do not want to receive a salary alone: they require a sense of meaning and an opportunity to have their values reflected in the work environment. Business organisations, which are well informed on the use of ESG in their operations have a higher chance of attracting, motivating and retaining the best talent.

In a 2022 Deloitte Global Generation Z and Millennial Survey, almost three-quarters of the interviewees indicated that they are better inclined to work in a company that demonstrates their environmental and social concerns. Besides, ESG-conscious companies have a higher rate of employee satisfaction, engagement and loyalty [57].

Whether through the implementation of DEI programmes or through the support of mental health, fair wages and sustainable working conditions, the businesses that consider the social factor in the ESG are more noticeable in the competitive labour markets. Also, transparency, fairness and accountability in ethical governance, the employees will develop trust and long-term commitment within the teams. Conversely, opaque leadership systems in toxic work environments have high employee turnover and morale.

10.8.4 Consumer trust and brand loyalty

People are now more socially and environmentally conscious consumers than ever before. Sustainability is no longer a niche issue; it is a mainstream requirement. IBM's (2021) consumer report indicates that almost 60 per cent of consumers are willing to switch their shopping patterns to minimise their environmental footprint, and 70 per cent of consumers are willing to spend more on sustainable brands [58].

Brands in the fashion and textile sector that adopt the ESG principles have an advantage in having more consumer trust, loyalty and brand differentiation. Indicatively, Patagonia, Stella McCartney and Eileen Fisher are among the companies whose reputation as ethical leaders was based on the focus on circularity, fair labour and minimal ecology. Their value-based marketing and openness contribute to high levels of emotional attachment with the consumers, making the brand promoters long-time brand supporters.

On the contrary, businesses that do not ensure sustainability are at risk of being boycotted and criticised. Consumers would find it easy to hold the brands responsible and make better choices with easy access to supply chain information and ethical ratings (through platforms such as Good on You).

10.8.5 ESG and sustainability: competitive necessities, not just ideals

ESG and sustainability are converging into the core of long-term business success due to the combination of social expectations, environmental crises and financial markets. With these models, organisations can:

- prepare and deal with challenges ahead;
- enhance stakeholder trust;
- become innovative and constantly improving;
- develop a strong and meaningful corporate culture.

Quite on the contrary, ESG is a facilitator of efficiency, flexibility and expansion. It marks a new understanding of value by business not alone in the ways that it can be quantified in shareholder returns but in the sense that it affects stakeholders.

With climate change, inequality and scarcity of resources becoming increasingly global issues, firms that incorporate ESG values in their DNA will be more successful in being ahead of the pack, inspiring and winning.

10.9 Challenges in implementing ESG and responsible models

Although the benefits of including ESG frameworks and a responsible business model are considerable, their implementation is not a smooth process. The integration of sustainability has been hindered by practical, financial and organisational barriers in many companies; it is particularly in the fashion and textile industries that companies have tried to integrate sustainability into their core strategies. One of the most urgent issues includes the issue of greenwashing, the absence of global ESG standardisation, high transition costs and resistance from old stakeholders. These challenges are critical to develop authentic, transparent and impactful ESG [59].

10.9.1 Greenwashing and lack of standardisation

A major issue in the implementation of ESG is the threat of greenwashing companies inflating or falsifying their efforts in the environmental or social areas to make them seem more sustainable than they are. Without universal standards, most businesses would use their own determined measures of ESG, which can be non-rigorous, comparative and unverified [60].

To illustrate, the rise of fast-fashion giants, such as Shein and H&M, has been accused of promoting so-called eco-friendly collections, when a closer look reveals that neither sourcing nor manufacturing is responsible. These practices are misleading to consumers and degrade the citizens in the general trust of ESG reporting. The narratives can be manipulated without any responsibility, without consistent definitions of such concepts as sustainable, carbon-neutral and ethical sourcing.

Also, the ESG reporting systems, including the GRI, SASB and TCFD, are diverse in their focus and form. Although these frameworks are powerful guidelines, it is not easy to compare the performance of various companies because the regions and sectors do not harmonise. This grey area also poses difficulties for the investors and regulators who have been using ESG data to determine risk and ethical position.

10.9.2 Costs and complexity of transitioning to sustainable practices

The process of changing a traditional business to a sustainable one is labour-intensive. It can frequently entail initial capital investment, operational reorganisation, employee training and supply-chain audit. These transitions can appear costly to the small and medium-sized enterprises (SMEs), particularly in an emerging market [61].

Take the example of the change to organic materials in the production of textiles. Although adopting certified organic cotton instead of traditional cotton would greatly minimise the use of pesticides and environmental degradation, it would require more expensive inputs, reduced production and the establishment of new relationships with suppliers. In the same spirit, making the circular economy model, such as recycling, resale, or repair programmes, necessitates costs in logistics, technology and educating consumers [62].

Delays and budgetary pressures may occur even in large corporations. In 2020, Nike committed to using 100 per cent renewable energy in all its owned and operated facilities by 2025, which is environmentally friendly but requires the infrastructure to be redesigned globally and coordinated with other countries.

Besides, the auditing and certification of ESG is costly. Such certifications as B Corps, Fairtrade, or Cradle to Cradle involve extensive paperwork, constant control and third-party verification, all of which are costly to maintain.

10.9.3 Resistance from traditional stakeholders

Resistance to change is another great barrier; this time, primarily among those stakeholders who are focused on short-term returns, conventional growth models, or who may feel that ESG threatens their profitability. This may be opposed by the investors, board members, suppliers, or internal managers who are doubtful about the economic benefit of sustainability or may fear that it will slow down the production cycles [63].

As an example, in the textile supply chain, several manufacturers are more concerned with cost reduction and speed. These priorities usually take the place of sustainable sourcing, ethical labour practices, or waste reduction objectives. Achieving suppliers switching to less exploitative forms of operation, like underpaid labour or dyes damaging the environment, may face vigorous opposition when it comes to impacting margins.

More so, ESG efforts tend to necessitate cultural change in companies. Transparency, inclusivity and accountability need to be encouraged among leaders on all levels, which is unlike the hierarchical models. Unless ESG is regarded as a significant compliance matter or public-relations strategy, employees and managers might not think much about its objectives.

Lastly, the shareholders with the incentive on quarterly earnings can also compel companies not to invest in long-term sustainability that cannot be directly translated to the bottom line. This is a tension between short-term profits and long-term value creation of ESG, which is a major problem with many publicly traded companies.

10.9.4 Moving forward: overcoming the barriers

To deal with these issues, companies must:

- embrace plausible, internationally accepted ESG standards to achieve transparency and comparability;
- research on ESG education and stakeholder engagement to change the corporate culture and break resistance;

- establish collaborations with suppliers, governments and non-governmental organisations to jointly develop sustainable value chains;
- use digital tools and ESG platforms to efficiently collect and report data and measure impact.

The process of ESG integration is complicated but necessary. As expectations among the consumers increase, investor attention comes with more scrutiny and regulatory requirements; these companies failing to address these challenges risk becoming obsolete. Breaking the barriers is not only a compliance problem but also a strategic necessity of future-proofing the fashion and textile industry.

10.10 Future trends and opportunities

With the shift of sustainability into a fringe issue and then into an international imperative, the future of ESG is being influenced by several transformational trends. These are digital innovations, the changing regulatory environment, investor and consumer activism and the increasing role of the new emerging markets and SMEs. All these trends offer potential to businesses, particularly those in the fashion and textile industries, to have resiliency, innovation and value in the long run.

10.10.1 Digital transformation and ESG

Digital technologies such as artificial intelligence (AI), blockchain and the Internet of Digital Technologies which includes AI, blockchain and Internet of Things are becoming effective ESG implementation tools. Predictive sustainability analytics are becoming one of the applications of AI, including energy consumption, carbon emissions and waste. It may also help the brands keep track of the ethical labour practices by having the suppliers audited in real time and checking the sentiments of the employees [64].

The supply-chain accountability is being changed by blockchain, which has a high level of transparency and traceability. The fashion industry is one of the industries where blockchain, such as Provenance and TrustChain by IBM, is being applied to ensure sustainability in sourcing and fair-trade certification, as well as authenticity of raw materials. This assists the consumers in making better decisions and deters greenwashing.

In the meantime, Internet of Things sensors may be used to monitor the energy consumption of production facilities or to track the amount of water being used in dyeing processes, thus enabling more precise reporting of ESG and real-time corrections.

10.10.2 Regulations and government incentives

Governments and global institutions are becoming more and more stringent in ESG compliance through the use of regulations, tax breaks and procurement policies by governments. The EU's CSRD requires big companies to report on specific measures of sustainability beginning in 2025. Equally, carbon pricing, extended producer responsibility and green finance models are also taking root around the world [65].

These changes in regulations present both threats and opportunities. By taking the initiative to comply with ESG policies, including the TCFD or the SASB, companies may enjoy some first-move benefits, avoid fines and receive the support of the government.

Other countries, such as France and Sweden, are gaining advantages in recycling efforts and circular design through the use of subsidies by textile firms. These are stimulating the development of sustainable materials, dye-less staining and customisable fashion.

10.10.3 Investor and consumer-driven ESG demands

ESG has become the key component of investment and consumerism. ESG is becoming more popular in global investment funds, including BlackRock and Vanguard, as criteria to measure long-term risks and returns. Consequently, those companies that have clear ESG approaches are getting more capital and a higher price [66].

Market expectations are also being reshaped by the consumers (particularly millennials and Generation Z). Brands are required to be authentic, transparent and sustainable by them. This is reflected in the increasing popularity of resale stores such as ThredUp or sustainable brands, such as Reformation and Everlane, that publicly post chain information and measures of environmental impact.

The fashion and textile industries need to adopt customer-focused ESG innovation, such as sustainable packaging, zero-waste designs, or rewarding eco-friendly actions through loyalty programmes.

10.10.4 ESG in emerging markets and SMEs

Although the adoption of ESG has conventionally been a prerogative of big business in the developed world, the next step is to take ESG practices to SMEs and the upcoming economies. These industries embrace huge growth opportunities but may not have the resources or expertise to integrate all ESFs [66].

These gaps have been filled by international organisations and development banks, which are providing technical support, capacity-building initiatives and green loans to Asian, African and Latin American SMEs. Meanwhile, digital tools such as mobile-based ESG scorecards and cloud-based reporting systems assist small businesses in monitoring and reporting on their sustainability performance.

In the case of the fashion industry, where a large portion of the supply chain lies in the emerging economies, the need to empower local suppliers by fair contracts, capacity-building, and ESG literacy is not only a social responsibility but a strategic requirement.

10.11 Call to action for business leaders, investors and policymakers

The global sustainability crisis is growing more severe, and thus, business leaders, investors and policymakers need to act decisively and in a coordinated manner. Company leaders need to go beyond any empty promises and build sustainability into their strategies, practices and culture. This involves the establishment of quantifiable targets, the involvement of employees at all levels, as well as investments in green innovation and ethical supply chains. Investors, on their part, must use their power to ensure that they direct the capital to the corporations that show genuine ESG performances and not only financial profits in the short term. Responsible investment has the capability of speeding up the process of achieving a sustainable economy.

Policymakers are important in the establishment of the regulatory mechanisms and incentives that sustainability in business models requires. Market forces can be adjusted to long-term environmental and social objectives through clear and enforceable policies, which may take the form of carbon pricing, sustainability reporting standards and green subsidies. Cooperation between these groups of stakeholders has ceased to be voluntary: it is compulsory. Through cooperation, they will be able to create a clearer, more responsible and more sustainable global economy. The era of piecemeal activities and procrastination of promises is long gone. Now there is a need to have a strategic and urgent, unified approach to ensure a sustainable and fair future for generations to come.

10.12 Conclusion

This chapter has revealed that the definition of the success of a business is being fundamentally changed. Companies can no longer count their successes in financial terms as the main indicator in an era of environmental crises, high awareness of social issues and stricter regulations and investor control. Rather, the long-term generation of value is becoming increasingly linked to the performance of organisations to integrate CSR, ESG principles and SBMs into their core strategies. The chapter indicates that CSR has developed beyond discretionary philanthropy to a strategic model that defines ethical behaviours, stakeholder relationships and brand credibility. ESG, on its part, has broken these values down into actionable and reportable and investment-relevant ideas that inform decision-making, risk-taking and corporate responsibility. The practical means through which the CSR values and ESG measures are integrated into daily business activities are SBMs, specifically, the models that are based on the principles of the circular economy, shared value creation and the TBL.

The sectoral focus of the chapter, particularly on fashion, textiles and consumer goods, portrays the urgency and the possibility of this change. The sectors that have historically shown to be costly to the environment and society are proving to be more sustainable than previously thought, to encourage innovation, efficiency, resilience and competitive advantage. The case studies of Patagonia, Unilever and Interface Carpets prove the fact that responsible leadership, open governance, and sustainability-based innovation can be at the same level as good financial results and international market relevance.

Simultaneously, the chapter does not deny the fact that the shift towards ESG-oriented and responsible business models is not devoid of challenges. Greenwashing, the absence of global standards, the high cost of transition and opposition by short-term stakeholders are among the problems that remain as barriers to improvement. Nonetheless, there is no reason to disregard the argument of sustainability because they indicate that a more powerful commitment of leadership and unified reporting systems, collaboration with stakeholders, as well as long-term strategic thinking, are necessary.

References

[1] Intergovernmental Panel on Climate Change, 'Climate Change Widespread, Rapid, and Intensifying', (9 August 2021), www.ipcc.ch/2021/08/09/ar6-wg1-20210809-pr/.

[2] Lucas Chancel, Thomas Piketty, Emmanuel Saez and Gabriel Zucman, 'The World Inequality Report 2022 Presents the Most Up-to-Date & Complete Data on Inequality worldwide', (n.d.), https://wir2022.wid.world/executive-summary/.

[3] Earth Overshoot Day, 'How Many Earths? How Many Countries?', (n.d.), https://overshoot.footprintnetwork.org/how-many-earths-or-countries-do-we-need/.

[4] S. Kotsantonis and G. Serafeim, 'Four Things No One Will Tell You about ESG Data', (15 July 2019), https://papers.ssrn.com/abstract=3420297.

[5] Robert G. Eccles and Svetlana Klimenko, 'Shareholders Are Getting Serious about Sustainability', (May–June 2019), https://hbr.org/2019/05/the-investor-revolution.

[6] A. Bilancia, F. Costa and A. P. Staudacher, 'A Holistic Framework to Enhance Sustainable Development and Reporting in Corporate Sustainability', *IFIP Adv Inf Commun Technol*, 728 (2024): 338–51, doi: 10.1007/978-3-031-71622-5_23.

[7] B. Marco-Lajara, P. Zaragoza-Sáez, J. Martínez Falcó and L. A. Millan-Tudela, 'Corporate Social Responsibility: A Narrative Literature Review', in *Frameworks for Sustainable Development Goals to Manage Economic, Social, and Environmental Shocks and Disasters*, IGI Global, (2022), pp. 16–34. doi: 10.4018/978-1-6684–6750–3.ch002.

[8] A. B. Carroll, 'The Pyramid of Corporate Social Responsibility: Toward the Moral Management of Organizational Stakeholders', *Bus Horiz*, 34 (4) (July 1991): 39–48, doi: 10.1016/0007–6813(91)90005-G.
[9] M. Roszkowska-Menkes, 'Porter and Kramer's (2006) "Shared Value"', in S. Idowu, R. Schmidpeter, N. Capaldi, L. Zu, M. Del Baldo and R. Abreu (eds), *Encyclopedia of Sustainable Management*, Cham: Springer, (2021). doi: 10.1007/978-3-030-02006-4_393-1.
[10] Unilever, 'Leading the Next Era of Corporate Sustainability', (n.d.), www.unilever.com/sustainability/.
[11] Cone Communications, '2017 Cone Communications CSR STUDY', (2017), www.cbd.int/doc/case-studies/inc/cs-inc-cone-communications-en.pdf.
[12] P. J. Commer, S. Sci, M. Arslan, D. Abeuova and A. Alqatan, 'Corporate Social Responsibility and Institutional Investors: Evidence from Emerging Markets', *Pakistan Journal of Commerce and Social Sciences*, 15 (1) (2021): 31–57.
[13] A. Stojanović, N. Sofranova, S. Arsić, I. Milošević and I. Mihajlović, 'The Role of Corporate Social Responsibility Activities in Employees' Perception of Brand Reputation and Brand Equity', *Case Studies in Chemical and Environmental Engineering*, 7 (5) (June 2023): 100313, doi: 10.1017/S1062798721000156.
[14] Manyata Rai, 'Corporate Social Responsibility (CSR) in Fashion Supply Chains', Carbon Trail, (n.d.), https://carbontrail.net/blog/corporate-social-responsibility-csr-in-fashion-supply-chains/.
[15] Fashion Revolution, 'People Tree', (n.d.), www.fashionrevolution.org/tag/people-tree/.
[16] SocialTargeter, 'Using Case Studies to Highlight Sustainability Practices in the Fashion Industry', (31 July 2025), www.socialtargeter.com/blogs/using-case-studies-to-highlight-sustainability-practices-in-the-fashion-industry.
[17] J. S. Knudsen and J. Moon, 'Government and Ethical Trade: The Ethical Trading Initiative and Responses to Rana Plaza', *Visible Hands*, (October 2017), pp. 79–102, doi: 10.1017/9781316224908.006.
[18] A. P. Burnstine and R. Ghattas, 'Assessing the Sustainable Circular Fashion Supply Chain as a Model for Achieving Economic Growth in the Global Market', *Sustainability*, 17 (19) (September 2025): 8558, doi: 10.3390/su17198558.
[19] M. N. Khuong, N. K. Truong and T. T. Thanh Hang, 'Stakeholders and Corporate Social Responsibility (CSR) Programme as Key Sustainable Development Strategies to Promote Corporate Reputation: Evidence from Vietnam', *Cogent Business and Management*, 8 (1) (January 2021): 1–21, doi: 10.1080/23311975.2021.1917333.
[20] Auburn University, 'Nike: Managing Ethical Misstep—Sweatshops to Leadership in Employment Practices', (n.d.), https://harbert.auburn.edu/binaries/documents/center-for-ethical-organizational-cultures/cases/nike.pdf.
[21] CSR Jobs, 'Attracting Gen Z: Why Corporate Social Responsibility Matters More than Ever?', (7 December 2025), www.csrjobs.club/blog-article/attracting-gen-z-why-corporate-social-responsibility-matters-more-than-ever/r/recOpMsGlxMhyN1Gx.
[22] KindLink, 'Consumer Brands that Demonstrate Commitment to Sustainability Outperform Those That Don't', (2015), www.kindlink.com/news/materials/consumer-brands-demonstrate-commitment-sustainability-outperform-those-dont.
[23] Retraced, 'What Sustainability Fashion Brands Teach Us about ESG Tactics', (1 May 2025), www.retraced.com/blogs/magazine/what-a-sustainability-fashion-brand-can-teach-us-about-esg-strategies.
[24] Triple Tree, 'ESG: Its Importance and Impact on Textile Industry in 2025', (16 March 2025), www.3-tree.com/blog/esg-its-importance-and-impact-on-textile-industry-in-2025/.
[25] Isadora Ferreira, 'Fashion's Blind Spot: Why Biodiversity Is Critical for a Nature-Positive Future', www.naturemetrics.com/news/why-biodiversity-should-be-on-the-runway.
[26] M. A. Achabou and S. Dekhili, 'Green Production and Consumption of Textiles and Apparel: Importance, Fabrication, Challenges and Future Prospects', *Journal of Open Innovation: Technology, Market, and Complexity*, 10 (2) (June 2024): 100280, doi: 10.1016/j.jbusres.2013.02.011.

[27] resilinc, 'ESG Legislation in the Apparel Supply Chain', https://resilinc.ai/blog/esg-legislation-in-the-apparel-supply-chain/.

[28] Charles J. Corbett, Sriram Narayanan, John Aloysius, Gemma Berenguer, Ebru K. Bish, Margrét V. Bjarnadóttir, et al., 'Diversity, Equity and Inclusion and Operations Management: Critical Linkages and Research Opportunities', *Prod Oper Manag*, 34 (3) (March 2025): 310–30, doi: 10.1177/10591478251318107.

[29] Emily McCoy, 'Fairtrade', (n.d.), Fashion Revolution www.fashionrevolution.org/tag/fairtrade/.

[30] F. Chien, 'The Role of Corporate Governance and Environmental and Social Responsibilities on the Achievement of Sustainable Development Goals in Malaysian Logistic Companies', *Economic Research–Ekonomska Istrazivanja*, 36 (1) (2023): 1610–30, doi: 10.1080/1331677X.2022.2090407.

[31] T. Whelan, U. Atz, T. Van Holt and C. Clark, 'ESG and Financial Performance: Uncovering the Relationship by Aggregating Evidence from 1,000 Plus Studies', NYU Stern Center for Sustainable Business and Rockefeller Asset Management, New York, (2015).

[32] S. Nosratabadi, A. Mosavi, S. Shamshirband, E. K. Zavadskas, A. Rakotonirainy and K. W. Chau, 'Sustainable Business Models: A Review', *Sustainability*, 11 (6) (March 2019): 1663, doi: 10.3390/su11061663.

[33] N. Bocken, 'Sustainable Business Models', in W. Leal Filho (ed.), *Encyclopedia of Sustainability in Higher Education*, Cham: Springer, (2021), pp. 963–975, doi: 10.1007/978-3-319-95867-5_48.

[34] K. E. Rathobei, H. Ranängen and Å. Lindman, 'Stakeholder Integration in Sustainable Business Models to Enhance Value Delivery for a Broader Range of Stakeholders', *Bus Strategy Environ*, 33 (4) (May 2024): 3687–706, doi: 10.1002/bse.3651.

[35] E. Jeronen, 'Sustainability and Sustainable Development', in S. O. Idowu, N. Capaldi, L. Zu and A. D. Gupta (eds), *Encyclopedia of Corporate Social Responsibility*, Berlin, Germany: Springer, (2013), 2370.

[36] Life Cycle Initiative, 'What is Life Cycle Thinking?', (n.d.), www.lifecycleinitiative.org/activities/what-is-life-cycle-thinking/.

[37] S. Schaltegger, E. G. Hansen and F. Lüdeke-Freund, 'Business Models for Sustainability: Origins, Present Research, and Future Avenues', *Organ Environ*, 29 (1) (March 2016): 3–10, doi: 10.1177/1086026615599806.

[38] L. Macchion, A. C. Toscani and A. Vinelli, 'Sustainable Business Models of Small and Medium-Sized Enterprises and the Relationships to Be Established within the Supply Chain to Support These Models', *Corp Soc Responsib Environ Manag*, 30 (2) (March 2023): 563–73, doi: 10.1002/csr.2374.

[39] P. Puntillo, 'Circular Economy Business Models: Towards Achieving Sustainable Development Goals in the Waste Management Sector – Empirical Evidence and Theoretical Implications', *Corp Soc Responsib Environ Manag*, 30 (2) (March 2023): 941–54, doi: 10.1002/csr.2398.

[40] X. Chen, H. A. Memon, Y. Wang, I. Marriam and M. Tebyetekerwa, 'Circular Economy and Sustainability of the Clothing and Textile Industry', *Materials Circular Economy*, 3 (1) (December 2021): 12, doi: 10.1007/s42824-021-00026-2.

[41] M. E. Porter and M. R. Kramer, 'Creating Shared Value: How to Reinvent Capitalism – and Unleash a Wave of Innovation and Growth', *Harvard Business Review*, 89 (1/2) (2011): 62–77.

[42] A. Harmon, 'Triple Bottom Line (Accounting)', Research Starters: Environmental Sciences, EBSCO Information Services, (2022).

[43] C. Coppola, A. Vollero and A. Siano, 'Developing Dynamic Capabilities for the Circular Economy in the Textile and Clothing Industry in Italy: A Natural-Resource-Based View', *Bus Strategy Environ*, 32 (7) (November 2023): 4798–820, doi: 10.1002/bse.3394.

[44] The Natural Step, 'Interface Case Study', (n.d.), www.thenaturalstep.de/about/interface-case-study/.

[45] Interface, '30+ Years of Sustainability: Ray Anderson's Innovative Impact', (n.d.), www.interface.com/US/en-US/sustainability/our-history.html.

[46] T. Brasch and C. Eckert, 'Development of the ESG Pillar Scores and Data Availability: Empirical Evidence from the Insurance Industry', *Journal of Risk and Financial Management*, 18 (8) (August 2025), doi: 10.3390/jrfm18080447.

[47] Global Reporting, 'GRI: Standards', (n.d.), www.globalreporting.org/standards/.
[48] IFRS Foundation, 'Understanding SASB Standards', IFRS, (2026), www.ifrs.org/issued-standards/sasb-standards/understanding-sasb-standards/.
[49] Task Force on Climate-Related Financial Disclosures, 'Recommendations of the Task Force on Climate-related Financial Disclosures', Financial Stability Board, Basel, Switzerland, (June 2017), https://assets.bbhub.io/company/sites/60/2020/10/FINAL-2017-TCFD-Report-11052018.pdf.
[50] Corporate Sustainability Reporting Directive, (n.d.), 'Everything You Need to Know', www.csrdreadiness.com.
[51] T. Maak and N. M. Pless, 'Responsible Leadership in a Stakeholder Society: A Relational Perspective', *Journal of Business Ethics*, 66 (1) (May 2006): 99–115, doi: 10.1007/s10551-006-9047-z.
[52] E. B. Ntiamoah, P. O. Egyiri, M. Kwamega, E. B. Ntiamoah, P. O. Egyiri and M. Kwamega, 'Corporate Social Responsibility Awareness, Firm Commitment and Organizational Performance', *Journal of Human Resource and Sustainability Studies*, 2 (2) (June 2014): 91–101, doi: 10.4236/jhrss.2014.22008.
[53] IFRS, 'About', (n.d.), www.ifrs.org/issued-standards/integrated-reporting/.
[54] Z. Chen, A. Rose, F. Roberts and A. Tucci, 'Regional Supply-Chain Impacts of Mississippi River Fertilizer Shipments Disrupted by Climate Change', *Risk Analysis*, 45 (10) (October 2025): 3090–109, doi: 10.1111/risa.70064.
[55] Y. Gunawan, A. Matahariza and W. K. Putri, 'The Dark Side of Fast Fashion: Examining the Exploitation of Garment Workers in Bangladesh', *Journal Hukum dan Peradilan*, 12 (3) (November 2023): 441, doi: 10.25216/jhp.12.3.2023.441–68.
[56] Jon Hale, 'A Broken Record: Flows for US Sustainable Funds again Reach New Heights', Morningstar, (28 January 2021), www.morningstar.com/sustainable-investing/broken-record-flows-us-sustainable-funds-again-reach-new-heights.
[57] Deloitte, 'Deloitte Global Gen Z and Millennial Survey 2025', (n.d.), www.deloitte.com/global/en/issues/work/genz-millennial-survey.html.
[58] IBM, 'IBM Global Consumer Study: Sustainability Actions Can Speak Louder Than Intent', (13 April 2022), https://newsroom.ibm.com/2022-04-13-IBM-Global-Consumer-Study-Sustainability-Actions-Can-Speak-Louder-Than-Intent.
[59] A. Tuteja, D. Joshi, S. J. Kumar M., A. Prakash, and S. Gaur, 'Sustainability Strategies in Contemporary Business Management: Integrating Environmental, Social, and Governance (ESG) Principles', *Educational Administration: Theory and Practice*, 30 (5) (2024): 7562–8, doi: 10.53555/kuey.v30i5.4211.
[60] W. Chen, Y. Xie and K. He, 'Environmental, Social, and Governance Performance and Corporate Innovation Novelty', *International Journal of Innovation Studies*, 8 (2) (June 2024): 109–31, doi: 10.1016/j.ijis.2024.01.003.
[61] ACCA, 'SMEs: Business Challenges and Strategic Innovation Opportunities', Association of Chartered Certified Accountants, London, (January 2023), www.accaglobal.com/content/dam/ACCA_Global/professional-insights/sme-business-challenges/PI-SME-CHALLENGES-INNOVATION%20v4.pdf.
[62] T. U. Din, N. Zaman, M. N. Uddin, W. Khan, N. Qayum, S. Qayum, S. Iqbal, I. Liaqat, A. Mumtaz, S. Z. Safi, A. Tahir, and M. Arshad, 'The Circular Economy Concept in the Textile Industry', in M. Arshad (ed), *Enzymes in Textile Processing: A Climate Changes Mitigation Approach*, Singapore: Springer, (2025), pp. 331–61. doi: 10.1007/978–981–97–8058-7_13.
[63] S. Doetsch and A. Huchzermeier, 'The Intersection of Supply Chain Resilience and Sustainability from a Theoretical and Practical Perspective: An Integrated Framework', *J Clean Prod*, 484 (3) (December 2024): 144306, doi: 10.1016/j.jclepro.2024.144306.
[64] K. Salame, 'Artificial Intelligence and Sustainability: Innovations in Business and Managerial Practices', *SBS Journal of Applied Business Research*, 1 (2026): 46–59, doi: 10.70301/SBS.MONO.2025.1.3.
[65] Economist Intelligence Unit, 'Sustainability Disclosure Requirements are Tightening', (1 September 2023), www.eiu.com/n/sustainability-disclosure-requirements-are-tightening/.
[66] A. Baratta, A. Cimino, F. Longo, V. Solina and S. Verteramo, 'The Impact of ESG Practices in Industry with a Focus on Carbon Emissions: Insights and Future Perspectives', *Sustainability*, 15 (8) (April 2023): 6685, doi: 10.3390/su15086685.

11 Life-cycle assessment and product environmental footprint in textiles

Raja Muhammad Waseem Ullah Khan, Zeeshan Ahmed, and Ali Raza Shafqat

11.1 Introduction

The textile business is a major contributor to environmental degradation, and it is estimated that the textile industry is contributing about 10 per cent of the total carbon emission in the globe and the consumption rate of the industry is estimated to be consuming about 141 billion cubic metres of water per year and thus there is a pressing case of sustainability measures to reduce the effects of the industry [1]. The importance of sustainability in the textile industry cannot be overstated because the industry generates 92 million tonnes of waste each year, and production increased twice from 2000 to 2015, and the time of wearing garments was reduced by 36 per cent, which proves the transition to overconsumption and the need to develop textile-related products that would be environmentally friendly [2]. Product environmental footprint (PEF) and life-cycle assessment (LCA) are crucial towards assessing environmental performance through the quantification of impacts over the entire life cycle of the raw material, from to end of life to enable brands to pinpoint hotspots and enhance sustainability rates [3]. Such tools are used in informed decision-making, including the mitigation of microfibre pollution, which comprises 9 per cent of the total ocean microplastic pollution annually by textiles and the management of chemical pollution by dyeing processes that impact 20 per cent of clean water in the world. The sustainability objectives, such as the UN Sustainable Development Goals (SDGs), highlight the need to decrease the textile-related greenhouse gas (GHG) emission (2–8 per cent of the total global), and the EU Green Deal is focused on achieving circularity to reduce waste and ensure resiliency in value chains by 2030 [4]. The effects of the textile sector on the habitats and resources require grounding to these targets, which may lead to the global recovery of almost $200 billion towards sustainable production and consumption by 2030 [5].

11.2 Fundamentals of LCA and PEF

LCA, as stipulated by ISO 14040/44, is a way of assessing the environmental effects of a product or a process over its life cycle, i.e., the extraction of the resources to the disposal through various phases of the life cycle, such as goal and scope definition, inventory analysis, impact assessment and interpretation [6] (shown in Figure 11.1). The PEF is a methodology developed by the EU Commission, using the principles of the LCA, intended to be used to calculate and communicate the environmental performance of products in sixteen categories of impacts across raw material extraction, Processing and product manufacturing, logistics and distribution, consumption and product's end-of-life strategy, saving inconsistency in sustainability claims [7]. PEF category rules provide guidelines for each specific sector and define specific benchmark calues, allocation rules, system boundaries, functional unit and data quality requirements. PEF

DOI: 10.4324/9781003638148-14

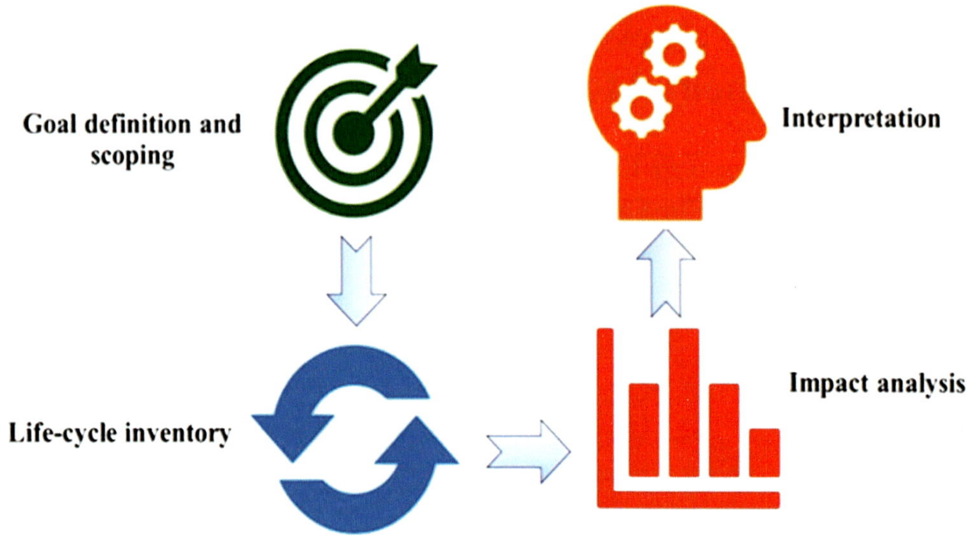

Figure 11.1 Components of LCA.

helps demonstrate access impacts of recycling and waste volarisation, help in crafting product declarations, support circular and sustainable design in textile and fashion industry and help in enabling businesses in global trade rediness. Although LCA is more flexible in its scope and use as a method of making general environmental assessments, PEF is more standardised and granular in its approach, with more specific rules regarding data quality and modelling and thus is appropriate when used in regulatory compliance and comparative purposes of products [8]. In the textile industry, LCA could concentrate on backgrounds that are customisable to manufacturing effects, such as cradle-to-gate, whereas PEF focuses on total life-cycle granularity, such as end-of-life situations to measure circularity [9]. In textile LCA, system boundaries, including between the production of fibres and the wearing and disposal of garments, are used to ensure the full coverage of impacts and the functional unit, which may be a garment or kilogram of fabric, is used as a reference point to measure performance. The aspects permit customised studies in textile, where the boundaries may leave out some phases in gate-to-gate research, whereas PEF must have a wider scope to match the EU regulations concerning the environmental footprint [10].

11.3 Current trends in LCA and PEF in the textile industry

LCA and PEF are moving towards cradle-to-cradle analysis and circular economy models in the textile industry, with the emphasis on closed-loop models that minimise waste production and promote material reuse, which is present in models that are consistent with planetary boundaries and those that make environmental pressures lower [11]. The use of digital algorithms and AI in the modelling of LCA is developing, and it becomes possible to predict the possibility of global warming and energy consumption with the help of machine learning, which does not contradict the results of the experiment and simplifies the process of evaluating textile products [12]. Increasing attention to the aspect of water footprint, microfibre release and chemical load in textiles emphasises their contribution to LCA, where research indicates that synthetic fabrics release more microfibres during washing and printing [13]. There is an increase in the use of environmental product declarations (EPDs) and product category rules (PCRs), which offer standardised guidelines to EPD development in fabrics, ensuring verifiable environmental information depending on the full life-cycle phases [14]. These trends help in the sustainability in that they serve to verify the claims of green and help in compliance; these PCRs are requirements of categories such as woven or knitted textile to improve comparability [14]. All in all, the use of AI tools and circular models is reshaping LCA practices and encouraging the adoption of responsible practices and the elimination of social and environmental risks in the global textile value chain [15].

11.4 Regulatory and market requirements

Due to the increasing environmental concerns, regulatory frameworks and legislation, the demand for sustainable textiles has been increasing during the past decade and is projected to grow even more [16]. The field of sustainable textiles has seen a profound transformation, mostly due to standardised and scientific evaluations related to involvement [17]. The PEF and LCA are the most important tools in the context of the current and upcoming legislation [18]. These standards ensure that textile and fashion products follow and demonstrate coherence between supply chain governance, environmental claims and product designs. A sketch of regulatory requirements is provided in Figure 11.2.

11.4.1 EU Green Deal and PEF regulation updates

The EU Green Deal is a comprehensive set of strategies related to climate change and resource efficiency, and it helps the textile and fashion sector, as one of the most important and top-priority sectors [18]. Within the EU Green Deal, the PEF methodology helps in implementing a wide approach of environmental performance measurement strategies across various categories,

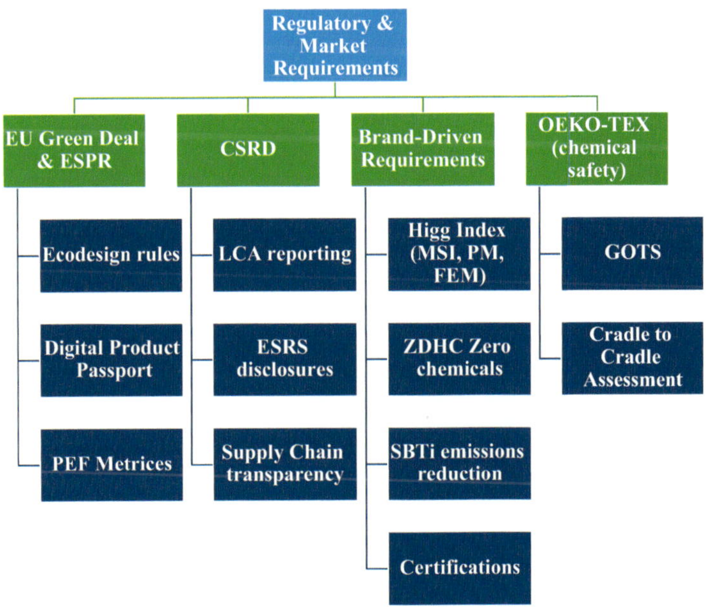

Figure 11.2 A sketch of regulatory requirements in the fashion and textile sector.

including water consumption, resource depletion, climate change toxicity, etc. [19]. The Ecodesign for Sustainable Products Regulation (ESPR) requires the apparel and fashion companies and brands to implement the following [20]:

- demonstrate compliance with eco-design criteria, including product longevity, fibre identification, disassembly and microfibre release;
- implement digital product passports that must contain the PEF details related to the use of chemicals and materials used during manufacturing, the recyclability and repairability of the product and the product's longevity.

It also binds the brands to show transparent information to customers, market-surveillance authorities and waste-management committees. This information should be based on the complete life cycle of the product.

The European Green Deal, PEF and ESPR altogether provide a consistent environmental policy for establishing a clear picture of reporting expectations, providing manufacturers with clear design expectations and reducing the supply-chain fragmentation [21].

11.4.2 Corporate Sustainability Reporting Directive implications

The Corporate Sustainability Reporting Directive (CSDR) binds the brands and companies to disclose the data related to the environment in accordance with European Sustainability Reporting Standards [22]. The CSDR requires companies operating in the European Union (EU) and companies based outside the EU that have significant revenue being generated from the EU to disclose and report this data according to European Sustainability Reporting Standards (ESRS) [23]. The key requirements of the CSRD are demonstrated visually in Figure 11.3. The PEF and LCA are utilised to measure, maintain and report this data. The major outputs of these assessments include the consumption of water during product manufacturing and the wastewater released, the impacts of raw materials on environment, the solid waste produced during product manufacturing, the resources utilised, the toxicity level of the released chemicals and the overall chemical footprints of the product during utility life and GHG emissions including Scope 1, Scope 2 and Scope 3 level emissions [24, 25].

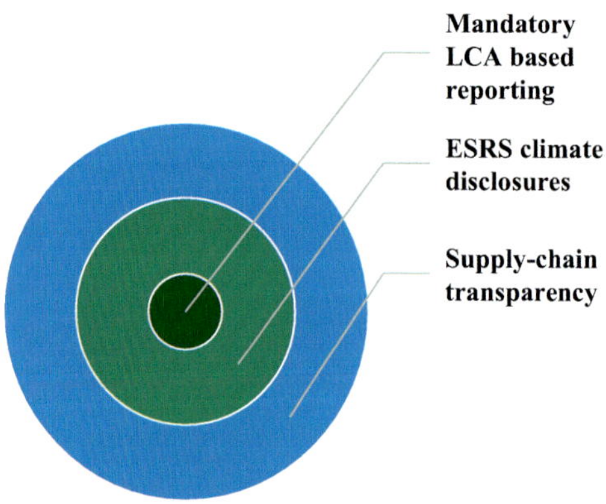

Figure 11.3 Key requirements of the Corporate Sustainability Reporting Directive.

The CSDR formalises the reporting of environment-related parameters, it requires the brands and companies to demonstrate that all the claims of sustainability they are pledging to are based on evidence, are verifiable and quantifiable [26].

11.5 Brand-driven sustainability requirements (e.g., Higg Index, Sustainable Apparel Coalition (SAC), Zero Discharge of Hazardous Chemicals (ZDHC)

Large corporations, big companies and leading brands are now requesting the suppliers and their collaborators to provide complete data related to sustainability as a prime condition for procurement; furthermore, to properly manage, measure and report the data, LCA is being implemented so that the data is deemed authentic and reliable [27]. Some of these initiatives and systems are discussed below.

11.5.1 ZDHC road map to zero

The ZDHC road map guides suppliers to refrain from using hazardous chemicals in the complete supply chain of textile and fashion products. It links the management of chemicals to LCA impacts as the primary category [28].

11.5.2 Higg Index (SAC)

Higg Product Module (PM) and Higg Material Sustainability Index help in the assessment of facility and material impacts using LCA datasets [29].

11.5.3 The Science-Based Target Initiative

The Science-Based Target Initiative (SBTi) guides companies to reduce emissions and provides them with targets according to the life cycle, and it specifies targets for each sector separately, providing clear paths towards decarbonisation. The SBTi provides specific guidelines for the fashion and textile industry, as well as leading towards a specific decarbonisation pathway for this sector [30].

Nowadays, brands are pushing their suppliers to employ these tools so that they can be held accountable; meanwhile, the brands can make data-driven procurement decisions so that they can align themselves well with the objectives of global climate preservation [31].

11.6 Role of certifications (e.g., Oeko-Tex, Global Organic Textile Standard) in LCA/PEF context

Several certifications help standardise the environmental performance of textile and fashion products, often providing third-party validation services. These certification bodies use LCA-based approaches to evaluate the impacts of various stages of the product life cycle on the environment. Some of these examples are discussed below.

11.6.1 Global Organic Textile Standard

Global Organic Textile Standard (GOTS) requires the use of environmentally sustainable processing of fashion and textile products along with the use of organic fibres, and it aligns with the PEF, resource utilisation indicators [32].

11.6.2 Cradle-to-cradle certification

This certification employs a rigorous framework focused on circularity, and it analyses the product circularity, evaluating the health of materials, the use of renewable energy, social fairness and water stewardship during the life cycle of the product [33].

11.6.3 Oeko-Tex standard 100

This standard provides our solution to screen the use of harmful substances and link it with the LCA related to the chemical impact of the product [34].

These standards and certifications help the brands and manufacturers to align themselves with the ever-changing regulatory and legislative framework and deliver according to the market expectations, meanwhile embedding a standardised life-cycle-based criteria throughout the fashion supply chain uniformly.

11.7 Application of LCA/PEF in textile products

Through the applications of PEF and LCA as tools, stakeholders across the textile product supply chain can understand the consequences of textile processing techniques, product design-related decisions and the end-of-life strategies on the environment. These tools provide detailed assessments for each kind of textile product on which it is implemented, and they help in the identification of hotspots and illuminate opportunities to reduce these impacts.

11.7.1 Examples of the application of LCA and PEF

The PEF and LCA can be applied to a single textile and fashion product, for example, a cotton T-shirt can be subjected to these assessments. For example, the cotton comes from the agriculture sector and has a great impact on the environment. The impact of cotton production is far-reaching, and fertilisers and pesticides, along with water for irrigation. It also requires energy for conversion from fibre into yarn and then into knitted fabric and then needs chemicals and water again during the dying, printing and other manufacturing processes [35]. The LCA can be applied to the cotton T-shirt to study in detail the environmental impacts caused by the production of the T-shirt or to study how replacing the regular cotton with organic cotton can reduce the chemical or the whole environmental footprint of the T-shirt. It can also help understand how organic cotton can reduce the toxic chemicals, but it does not reduce the water consumption without the use of regenerative agriculture. LCA can also be applied to products like denim jeans, which have huge environmental impacts incurred during the bleaching and finishing process and multiple washes, the release of pumice stone and potassium permanganate and require high energy input [36]. The LCA can be implemented to study how new technologies like ozone washing, laser finishing and waterless dyeing can reduce the impacts of denim on the environment [37]. Another example can be LCA of activewear made of polyester, the polyester fibre comes from petrochemical origin and cause resource depletion and climate change although it requires less water than then cotton provides longevity into the textile and fashion products but it contributes to pollution and microplastic release during the utility life in form of washing, recycled polyester on the other hand helps reduce the reliance on virgin feedstock has reduced GHG emissions but LCA can be implemented to study its impact on environment including if the use of recycled polyester can help mitigate the environmental concerns related to microplastic pollution [38].

11.7.2 Impact hotspots

These strategies help identify the hotspots that should be focused on and studied in depth for the products to have insights related to the environmental concerns. For example, it can help make informed choices between raw materials to be used for product manufacturing, as raw material choice can help mitigate approximately 70 per cent of all the impacts of water use, energy and chemical inputs. Implementing these tools can also help in studying the impact of each process during the manufacturing of a product; for example, it can help understand how each process is designed and how resource-intensive and chemically complex it is. With increased understanding, the impact of that specific process can be controlled or mitigated [39].

Not only the processing, but also the utility life of the product can be extended by applying these tools [40]. For example, we can study how washing and laundering the textile product impacts the environment in various contexts, including the release of microplastics, consumption of water and utilisation of energy. The end of lives of textile products can also be studied in great detail using these LCA approaches most textiles are even landfilled or incinerated within five years of their production which prevents their circular life cycle, implementation of LCA can provide deeper insights on how much waste is generated by the product and how would it impact the environment if this product is recycled or upcycle and brought back into loop or circular economy [41]. Some hotspots in fashion and textiles are provided in Figure 11.4.

11.7.3 Strategies for impact reduction based on LCA/PEF insights

The application of these tools and strategies can help in reducing the environmental impact of these products. For example, they can provide deeper insights on how the use of bio-based regenerative or recycled fibres can help reduce the environmental impact of the products. Furthermore, they can also help adopt to low impact dyeing and finishing processes or utilising the enzymatic CO_2 based processes they can also help study how take back systems and the extended producer responsibility (EPR) policies can help reduce the impacts by providing deeper insights and measurements of various variables it can promote sustainability not only among designers or manufacturers but can also promote enhanced consumer behaviour towards extending the use of their government or reduce water consumption during washing [42].

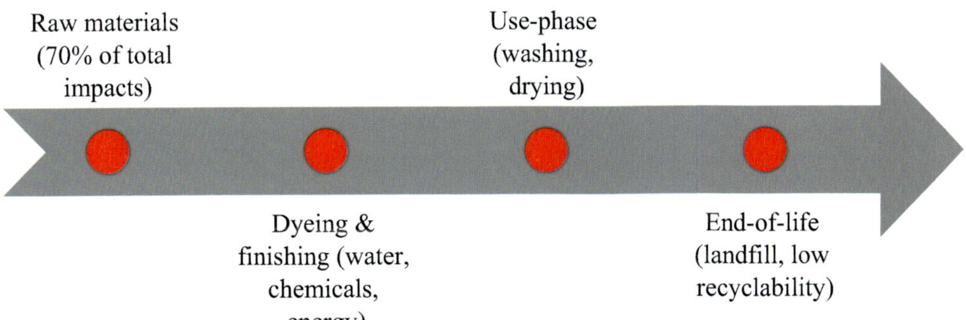

Figure 11.4 Schematics of hotspots in the fashion and textile supply chain.

11.8 Challenges and limitations

Although the implementation of PEF and LCA helps provide in detailed analysis of the environmental impacts of a product, several limitations exist towards global scalability and complete reliability of these tools, as shown in Figure 11.5.

11.8.1 Data availability and quality issues

The textile supply chain is spread beyond borders and is often distributed throughout the globe. These applications consist of multiple businesses collaborating, due to which it is very difficult to obtain accurate data among various suppliers, producers, manufacturers and brands, many of which are insufficient in monitoring systems. Therefore, the implementation of these tools requires significant dependence on datasets from third-party or generic databases available online. Furthermore, most of these sectors are informal, due to which limited data is available on a small scale. Furthermore, the geographical distinct locations impose difficulty in having correct data on chemical utilisation and energy intensity of the processes [43].

11.8.2 Methodological inconsistencies and regional differences

The number of different choices between various methods, including the system boundaries, functional units and the allocation rules, incorporates the variations in the results by default. For example, the PEF uses fixed impact categories, whereas LCA provides a flexible and

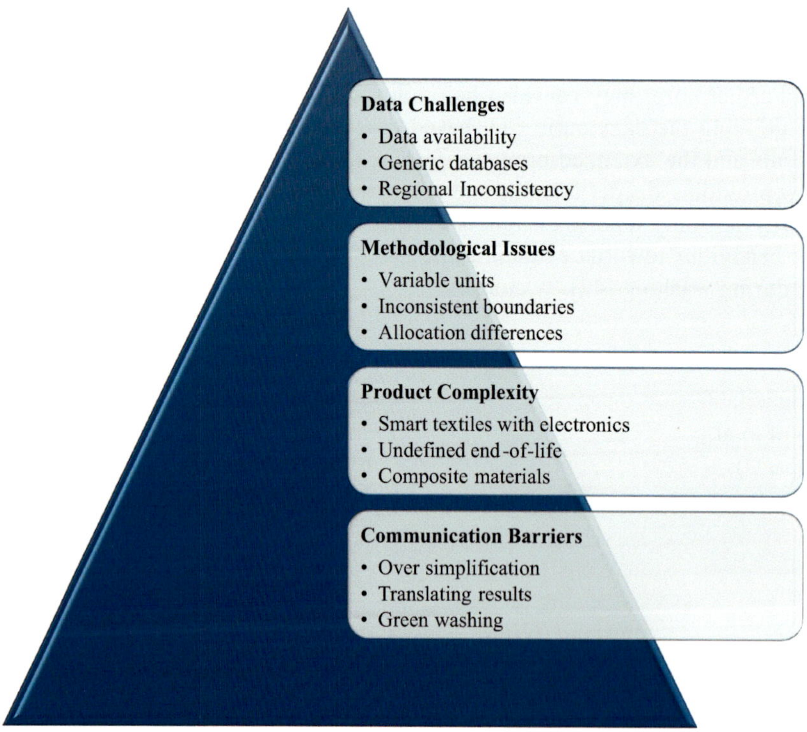

Figure 11.5 The current challenges of LCA and PEF analysis.

inconsistent selection of the impact type, due to which inconsistency between these is caused. Furthermore, the difference is due to the different priorities of various regions, including energy waste management or water scarcity can significantly influence the results of the analysis [44].

11.8.3 Complexity in multifunctional textile products (e.g., smart textiles)

Sometimes the fashion and textile products offer multi-functionalities. For example, the technical textiles or smart textiles that have embedded electronics within them impose huge difficulties for implementation of LCA or PEF these products impose difficulties in determining the right boundaries, modelling across different interactive functionalities of the product evaluating the energy consumption during product manufacturing or product use and ultimately assessing the disposal impacts of the products especially the electronic components. These difficulties can also be imposed due to the use of multiple materials in these products [45].

11.8.4 Limitations in communicating results to non-experts (designers, consumers)

The LCA and PEF analysis can be difficult to be understood by broad stakeholders and sometimes difficult terminologies impose understanding limitations, to mitigate these limitations these analysis should be translated into digestible information for end consumers as well as designers but it should be done with great care and caution because this oversimplification of data may confuse impact categories, for example GHG emission or wastewater emission. This oversimplification can sometimes hide methodical and processes assumptions and sometimes may greenwash the product or cause selective reporting [46].

11.9 Future directions

The implementation of LCA and PEF techniques in the textile and fashion industry is deeply interlinked with circularity matrices, social sustainability, digital transformation and global legislation and frameworks. Some examples of key research areas of the fashion and textile industry's future are discussed in Figure 11.6.

11.9.1 Alignment of global textile LCA databases

Initiatives like textile exchange, LCA repositories, PEF and the Higg data set can help in improving the data accuracy, harmonise the variation arising due to distinct geographical locations and enhance the cross-stakeholder and cross-industrial comparison of data [47].

11.9.2 Blockchain and traceability integration in LCA tools

Digital tools, like blockchain and digital product passports, can help understand the real-time environmental impacts of the product, the automation of data retrieval from the manufacturing sites and secure proper verification of chemical utilisation during the process, or the origin of raw material [48].

11.9.3 Inclusion of social and circularity indicators in LCA and PEF

It is forecast that the LCA and PEF frameworks and strategies will somehow evolve to include the circularity matrices like repairability, reuse, recyclability and recovery rate from the waste

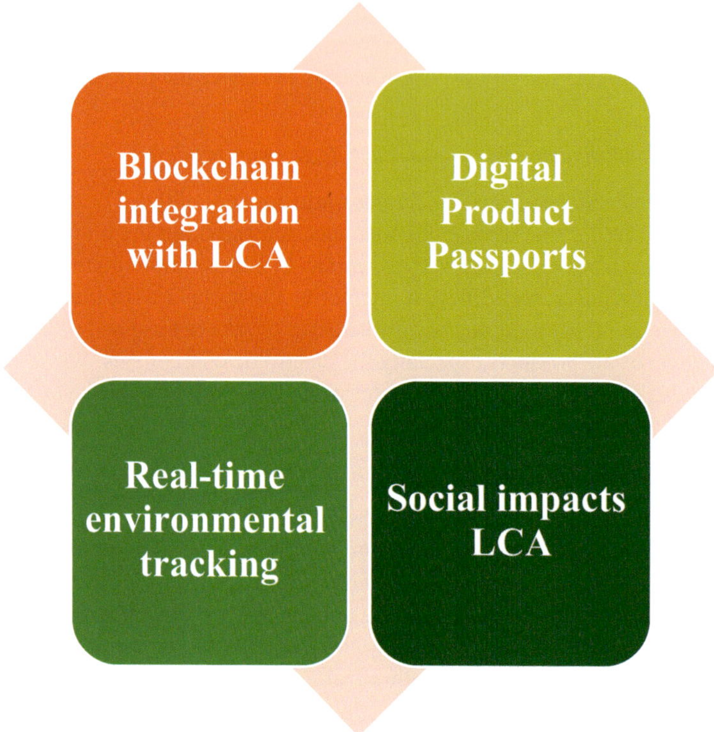

Figure 11.6 Key research areas for future for fashion and textile sector.

material, and the social indicators, including occupational health, community well-being and labour rights etc. [49], to have better integration with the social aspects of sustainability, along with environmental impacts.

11.9.4 Role in supporting global sustainability frameworks

PEF and LCA are both very essential tools for making the global commitments possible; these implementations would have a direct impact on global SDGs, including SDG 12, resource responsible consumption, as the G6 related as water and SDG 13, related to climate action. [50]. It would help align with the Paris Agreement on carbon-neutral transformation of production and help achieve circular economy policies implemented by different nations, including the EU, Japan and China, and many other Global South countries that are integrating various strategies to study these impacts based on LCA [51, 52]. By embedding these strategies into supply chain design innovation, governance and public policy, the fashion and textile sector can be transitioned into a more climate-aligned align transparent and circular sector [48, 51, 53].

11.10 Conclusion

Overall, this chapter has enlightened the heightened importance of LCA and PEF in the development of sustainability in the textile sector, both in the quantification of environmental effects throughout product life cycles and in the development of regulatory and innovation directions.

Combining these approaches, the stakeholders can solve such crucial issues as high carbon emissions, lack of water and chemical pollution, and employ such trends as AI-enhanced modelling and the principles of the circular economy to streamline the processes and decrease hotspots in the raw material, manufacturing and end-of-life stages. The regulatory frameworks established by the EU Green Deal, CSRD and ESPR, as well as brand programmes like the Higg Index as well and certifications such as GOTS, strengthen the necessity of standardised, verifiable data, which can increase transparency and responsibility across the global supply chains. Despite the flaws in terms of data availability, methodological disparities and the issues of multifunctional products, the instruments support a data-driven methodology that supports the eco-design, resource efficiency and consumer awareness. In the future, data standardisation, use of blockchain as a traceability tool, and adding more indicators to include social and circularity will be critical again in the facilitation of frameworks such as the SDGs and the Paris Agreement. Finally, the wide usage of LCA and PEF can make the textile industry an example of environmental responsibility to serve economic gain through the application of sustainable technologies and open up the prospects of a climate-neutral future.

References

[1] N. Cardona, 'Environmental Impact of Fast Fashion Statistics', (2025), www.uniformmarket.com/statistics/fast-fashion-statistics.

[2] United Nations Environment Programme, 'Unsustainable Fashion and Textiles in Focus for International Day of Zero Waste 2025', (27 March 2025), www.unep.org/news-and-stories/press-release/unsustainable-fashion-and-textiles-focus-international-day-zero.

[3] B. Zamani, 'Life Cycle Assessments for Textile and Fashion Products', (6 July 2023), www.carbonfact.com/blog/knowledge/lca-fashion-textile.

[4] Geneva Environment Network, 'Environmental Sustainability in the Fashion Industry', (24 April 2026), www.genevaenvironmentnetwork.org/resources/updates/sustainable-fashion/.

[5] Neil Balchin, 'How to Make Textile and Garment Production More Sustainable', (2023), www.weforum.org/stories/2023/11/textile-garment-industry-sustainability-policy-measures/.

[6] ISO, 'Environmental Management – Life Cycle Assessment – Requirements and Guidelines', (2006), www.iso.org/standard/38498.html.

[7] European Commission, 'Product Environmental Footprint Method Understanding Key Concepts and How to Implement Them', (n.d.), https://green-forum.ec.europa.eu/green-business/environmental-footprint-methods/pef-method_en.

[8] Clara, 'LCA, EPD, and PEF: What Are the Differences?', Hedgehog, (7 August 2023), www.hhc.earth/knowledge-base/articles/lca-epd-and-pef-what-are-the-differences.

[9] Makersite, 'Product Environmental Footprint vs. Life Cycle Assessment', (n.d.), https://makersite.io/insights/pef-vs-lca/.

[10] P. Gholkar, 'Everything You Need to Know about Functional Units in LCA', (11 August, 2025), https://ecochain.com/blog/functional-unit-in-lca/.

[11] N. Islam, M. R. Hall, C. MacMillan and S. Gordon, 'Life Cycle Assessment and the Emergence of Product Circularity: A Review with Context to the Textile Supply Chain in Australia', *Environ Impact Assess Rev*, 117 (March 2026): 108198, doi: 10.1016/j.eiar.2025.108198.

[12] I. de Paula Salgado, F. Conrad, C. Signorini, E. Günther, S. Ihlenfeldt and V. Mechtcherine, 'Integrating Life Cycle Assessment (LCA) and Machine Learning for Sustainable Designs: A Case Study on Protective Layers Made of Mineral-Bonded Fiber-Reinforced Composites', *Int J Life Cycle Assess*, 30 (2025): 2403–22, doi: 10.1007/s11367-025-02454-7.

[13] Sinem Hazal Akyildiz, Silvia Fiore, Martina Bruno, Hande Sezgin, Ipek Yalcin-Enis, Bahattin Yalcin and Rossana Bellopede, 'Release of Microplastic Fibers from Synthetic Textiles during Household Washing', *Environmental Pollution*, 357 (September 2024): 1–15, doi: 10.1016/j.envpol.2024.124455.

[14] EPD, 'New PCR on Fabrics', (6 September 2022), www.environdec.com/news/new-pcr-on-fabrics.
[15] UL Solutions, 'What Are Product Category Rules (PCRs)?', (n.d.), www.ul.com/resources/product-category-rules-pcrs.
[16] M. T. Rahaman and M. S. H. Khan, 'Green Merchandising of Textiles and Apparel in a Circular Economy: Recent Trends, Framework, Challenges and Future Prospects towards Sustainability', *Journal of Open Innovation: Technology, Market, and Complexity*, 11 (1) (March 2025): 100457, doi: 10.1016/J.JOITMC.2024.100457.
[17] E. Faghih, Z. Saki and M. Moore, 'A Systematic Literature Review: AI-Enabled Textile Waste Sorting', *Sustainability*, 17 (10) (May 2025): 4264.
[18] Directorate-General for Research and Innovation, 'A Framework for an Accessible Environmental Footprint Database to Simplify Life Cycle Assessments: Research and Innovation', (31 January 2025), https://research-and-innovation.ec.europa.eu/news/all-research-and-innovation-news/framework-accessible-environmental-footprint-database-simplify-life-cycle-assessments-2025-01-31_en.
[19] European Commission, 'Environmental Footprint Methods', (n.d.), https://green-forum.ec.europa.eu/green-business/environmental-footprint-methods_en.
[20] European Commission, 'Ecodesign for Sustainable Products Regulation', (n.d.), https://commission.europa.eu/energy-climate-change-environment/standards-tools-and-labels/products-labelling-rules-and-requirements/ecodesign-sustainable-products-regulation_en.
[21] European Commission, 'The European Green Deal', (n.d.), https://commission.europa.eu/strategy-and-policy/priorities-2019-2024/european-green-deal_en.
[22] European Commission, 'Corporate Sustainability Reporting', (9 December 2025), https://finance.ec.europa.eu/capital-markets-union-and-financial-markets/company-reporting-and-auditing/company-reporting/corporate-sustainability-reporting_en.
[23] H. Oskarsdottir, T. O. Sigurjonsson and R. S. Steinthorsson, 'The European Union's Corporate Sustainability Reporting Directive: Impact on Corporate Boards', *International Journal of Law and Management*, (2025), https://doi.org/10.1108/IJLMA-03-2025-0106.
[24] B. Moutik, J. Summerscales, J. Graham-Jones and R. Pemberton, 'Life Cycle Assessment Research Trends and Implications: A Bibliometric Analysis', *Sustainability*, 15 (18) (September 2023): 13408, doi: 10.3390/SU151813408.
[25] G. J. A. Aguiar, L. R. Almeida, B. S. Fernandes, S. Gavazza, G. L. Silva and S. Machado Santos, 'Use of Life Cycle Assessment as a Tool to Evaluate the Environmental Impacts of Textile Effluents: A Systematic Review', *Environ Sci Pollut Res Int*, 30 (31) (July 2023): 76455–70, doi: 10.1007/S11356-023-27785-6.
[26] E. Rivo-López, M. Villanueva-Villar, C. Pardo-López and M. Cortés-García, 'Environmental Sustainability Reporting: A Systematic and Bibliometric Review of Two Decades of Research', *Future Business Journal*, 11 (1) (July 2025): 165, doi: 10.1186/S43093-025-00580-0.
[27] E. Davis, 'How Businesses Can Effectively Respond to Sustainability Data Requests in the Supply Chain', (18 April 2024), www.elliottdavis.com/insights/how-businesses-can-effectively-respond-to-sustainability-data-requests-in-the-supply-chain.
[28] Roadmap to Zero, 'Homepage', www.roadmaptozero.com.
[29] International Leather Maker, 'Sustainable Apparel Coalition Completes First Higg Index Reviews', (n.d.), https://internationalleathermaker.com/sustainable-apparel-coalition-completes-first-higg-index-reviews/.
[30] Science Based Targets, 'Ambitious Corporate Climate Action: Science Based Targets Initiative', (n.d.), https://sciencebasedtargets.org/.
[31] J. K. Runtuk, P. K. Ng and S. Y. Ooi, 'Challenges and Solutions in Working with Green Suppliers: Perspective from a Manufacturing Industry', *Sustainability*, 16 (20) (October 2024): 8744, doi: 10.3390/SU16208744.
[32] Global Organic Textile Standard, 'The Standard', (n.d.), https://global-standard.org/the-standard.
[33] Cradle to Cradle, 'Introducing C2C Certified® Circularity', (14 October 2024), https://c2ccertified.org/articles/introducing-c2c-certified-circularity.
[34] OEKO-TEX, 'Oeko-Tex® Standard 100: The Original Safety Standard', (n.d.), www.oeko-tex.com/en/our-standards/oeko-tex-standard-100/.

[35] Merna Aly, 'The Environmental Impact of Cotton: Conventional vs. Organic – EmVide', (22 October 2024), https://emvide.com/the-environmental-impact-of-cotton-conventional-vs-organic/.
[36] Fashion Dive, 'Nearly a Quarter of Denim Finishing Contains Hazardous Chemicals: Report', (21 April 2025), www.fashiondive.com/news/jeanologia-denim-sustainability-report-finishing/745890/.
[37] A. P. Periyasamy and S. Periyasami, 'Critical Review on Sustainability in Denim: A Step toward Sustainable Production and Consumption of Denim', *ACS Omega*, 8 (5) (February 2023): 4472, doi: 10.1021/ACSOMEGA.2C06374.
[38] C. Palacios-Mateo, Y. van der Meer and G. Seide, 'Analysis of the Polyester Clothing Value Chain to Identify Key Intervention Points for Sustainability', *Environ Sci Eur*, 33 (1) (December 2021): 2, doi: 10.1186/S12302-020-00447-X.
[39] A. Pacana, D. Siwiec, L. Bednárová and J. Petrovský, 'Improving the Process of Product Design in a Phase of Life Cycle Assessment (LCA)', *Processes*, 11 (9) (August 2023): 2579, doi: 10.3390/PR11092579.
[40] M. Hussain and A. R. Shafqat, 'Sustainability and Materials', in *Engineering Materials*, (2025), pp. 257–74.
[41] M. W. Iqbal, M. B. Ramzan, H. Manzoor and S. M. Qureshi, 'Value Extraction from End-of-Life Textile Products in Pakistan', *Recycling*, 10 (3) (May 2025): 101, doi: 10.3390/RECYCLING10030101.
[42] P. Gabrielli, M. Gazzani and M. Mazzotti, 'The Role of Carbon Capture and Utilization, Carbon Capture and Storage, and Biomass to Enable a Net-Zero-CO_2 Emissions Chemical Industry', *Ind Eng Chem Res*, 59 (15) (April 2020): 7033–45, doi: 10.1021/ACS.IECR.9B06579.
[43] T. K. Agrawal and R. Pal, 'Traceability in Textile and Clothing Supply Chains: Classifying Implementation Factors and Information Sets via Delphi Study', *Sustainability*, 11 (6) (March 2019): 1698, doi: 10.3390/SU11061698.
[44] O. J. Robinson, A. Tewkesbury, S. Kemp and I. D. Williams, 'Towards a Universal Carbon Footprint Standard: A Case Study of Carbon Management at Universities', *J Clean Prod*, 172 (January 2018): 4435–55, doi: 10.1016/J.JCLEPRO.2017.02.147.
[45] A. R. Shafqat and A. Saccavini, 'Eco-design of Textiles', in S. Rubab Batool, S. Ahmad, Y. Nawab and M. Hussain (eds), *Eco-design of Textiles*, (2023), pp. 39–58, Springer, doi: 10.1007/978-3-031-49479-6_3.
[46] E. Kaynak, I. S. Piri and O. Das, 'Revisiting the Basics of Life Cycle Assessment and Lifecycle Thinking', *Sustainability*, 17 (16) (August 2025): 7444, doi: 10.3390/SU17167444.
[47] Textile Exchange, 'Ensuring Integrity in the Use of Life Cycle Assessment Data', (7 August 2025), https://textileexchange.org/knowledge-center/reports/ensuring-integrity-in-the-use-of-life-cycle-assessment-data/.
[48] M. J. Kim, C. H. Han, K. J. Park, J. S. Moon and J. Um, 'A Blockchain-Based Digital Product Passport System Providing a Federated Learning Environment for Collaboration between Recycling Centers and Manufacturers to Enable Recycling Automation', *Sustainability*, 17 (6) (March 2025): 2679, doi: 10.3390/SU17062679.
[49] G. Batool, M. Bashir, S. A. Baig and A. R. Shafqat, 'Ecotourism Taking Modernity to the Deprived Regions of the World', in M. Abrar, M. Asim Nawaz, F. Kiran and F. Asmi (eds), *Supporting Environmental Stability through Ecotourism*, IGI Global, (2024), pp. 1–15. doi: 10.4018/979-8-3693-1030-4.ch008.
[50] A. R. Shafqat, M. Hussain, Y. Nawab, M. Ashraf, S. Ahmad and G. Batool, 'Circularity in Materials: A Review on Polymer Composites Made from Agriculture and Textile Waste', *Int J Polym Sci*, 1 (2023): 1–21, doi: 10.1155/2023/5872605.
[51] United Nations Framework Convention on Climate Change, 'Circular Economy Crucial for Paris Climate Goals', (22 January 2019), https://unfccc.int/news/circular-economy-crucial-for-paris-climate-goals.
[52] United Nations, 'The Paris Agreement', (4 November 2016), www.un.org/en/climatechange/paris-agreement.
[53] M. Zhang, R. Liu and H. Sun, 'Achieving Carbon-Neutral Economies through Circular Economy, Digitalization, and Energy Transition', *Sci Rep*, 15 (1) (December 2025): 13779, doi: 10.1038/S41598-025-97810-W.

12 Eco-labelling, certifications and traceability in textiles

Amna Siddique, Shahid Rasul, and Jawad Naeem

12.1 Background

With increase in population of world at alarming rate, there is a drastic increase in demand for food, energy and resources [1–3]. Due to global warming, environmental pollution have become focus of interest for industrial manufacturers and government authorities [4, 5].

Clothing and the fashion industry are major contributors to causing substantial pollution to water and generation of solid wastages [6]. Textile wet processing industry uses hazardous chemicals and produce toxic pollutants, emit greenhouse gases (GHGs) and consume huge quantity of water [7–11]. This approach is not sustainable. As a result, it is imperative to develop products that have minimum carbon footprints [3, 6, 9–13]. Sustainability concerns in the textile industry are rising because of the overuse of natural resources and poor working conditions [5, 14]. Environment conscious consumers, government institutes and NGOs along with all forms of media are playing pivotal role in pressurising the clothing manufactures and fashion industry to adapt sustainable practices by altering their traditional manufacturing procedures [15–17].

The integration of advanced textile machines and technology highlighted the start of considerable changes in industry of textiles at the end of the eighteenth century [18]. From the beginning of the twentieth century, a lot of manufacturing industries acquired essence of development. For improving the quality of life and easy access to consumer goods, a series of studies on synthetic and chemical textile substrates were carried out [19]. In first half of previous century, the scale of production of textiles was small and its contribution to polluting environment was insignificant. Furthermore, the nature of chemicals was not hazardous. However, the fast fashion industry has significantly transformed customers' shopping nature apart from increment in population [18, 20]. In second half of the twentieth century and recent years, production and utility of synthetic fibre was substantially enhanced [3].

Fast fashion is related to design and production of cloth in a very short period i.e., in almost fourteen days, they are available in stores. Due to low labour cost of underdeveloped countries, retailers are gaining benefits. However, poor working conditions for operators are creating lot of challenges. As the cost of manufacturing of clothes decreases, the profit margin of retail organisations enhances significantly along with increase in requirement of production at swift rate [21, 22]. This creates difficulties of textile organisations which utilise cotton and manual labour for process of bleaching and dyeing. Both these processes require time. Increase in productivity of these processes requires exploitation of work force and natural raw materials [23].

Environmental complications ascended from utility of pesticides for crops of cotton, chemicals in textile processing along with management of pre-consumer wastages [22]. Madsen et al. [25] mentioned release of number of GHGs, ingesting of natural raw materials, wastages, contamination of air and water along with mutilation of biodiversity and its negative influence on

DOI: 10.4324/9781003638148-15

textile organisations. The requirement of water and oil is essential for growth of cotton. A great amount of energy resources are consumed during washing and drying of clothes [26] and as per Dahllof's investigation [27], a huge quantity of water (almost 90 per cent) is needed for raw material growth and processing.

Societal problems are connected to poor working conditions and production provisions, especially in underdeveloped nations. This inculcated less payment, prolonged working hours and forced child labour along with racism and harassment problems at place of work. [22]. According to Madsen et al. [25], workers' rights along with health and safety issues must also be included in societal problems created by the increasing demand of fast fashion. The ignorance of retailers about working conditions in manufacturing organisations is also a big problem especially when the production facilities are disseminated [22]. The protection of worker and their rights, mitigation of poverty, health care of community near textile industry along with its socio-economic influence are highlighted as significant factors affecting textile industry [25]. In underdeveloped countries having less grip on laws and regulations, these above-mentioned issues take place in production and growth phase especially when call of fair trade want to lecture these problems [25]. Furthermore, the issue of life and safety of labour are huge concerns due to requirement of pesticides during production and growth of cotton.

In underdeveloped countries, challenges of noise and dust due to utility of equipment in production of cotton requires a lot of improvement [28]. On the other hand, mitigation of poverty due to provisions of jobs from textile industry have positive influence on society in underdeveloped countries where literacy rate is very low. However, there is a training requirement in this aspect which most of workers are unable to attain [25].

12.2 Corporate social responsibility

According to Todeschini et al. [29], corporate social responsibility (CSR) is a movement that enhances sustainability for corporate business. It was noted that three factors are pivotal portions of CSR:

1. No sweatshop conditions i.e., working environment is just and equitable.
2. Fair wages and work environment is health.
3. Sourcing from local suppliers in production facilities close to consumers resulted in reduction in the emission of harmful gases during transportation of materials.

Complicated supply chain of fashion business makes CSR more pertinent due to dissemination of huge number of suppliers all over the world. The outsourced production of fast fashion is due to cheap labour in underdeveloped countries. Due to less implementation of labour laws, the fashion brands are forcing textile industry of underdeveloped countries to alter hazardous working conditions and encouraging these organisations to strictly compliance of the labour laws and standards. In consequence, the CSR of leading fashion brands are trying to convert the negative effect of textile industry to positive affect. Monitoring of various stakeholders, implementation of labour laws and sustainable practices, fair trade and third-party audits are few examples of CSR [29].

12.3 Sustainable consumption and awareness of consumers about fashion industry

Niinimäki [30] states that sustainable consumption in terms of less purchasing of garments which are particularly significant to end consumer due to greater quality of standards and enhanced life cycle. The emphasis of sustainable consumption is on frequent wearing of garments and these garments should not be discarded due to minor damage. All those activities like ethical and sustainable consumption which can enhance the life cycle of garment must be given great importance to support ethical and make awareness in consumer to discourage fast fashion [20, 31].

12.3.1 Slow-fashion campaign

The slow-fashion campaign is one of the instances of consumer awareness towards sustainable consumption. Buying a smaller number of garments for restricted period emphasises end user purchasing less in a more cautious and conscious way. In the present era, the movement of sustainable fashion is altering the behaviour of wearer to hunt for various alternative options instead of fast fashion. The combined effect of slow consumption and the slow-fashion campaign have deep influence on business models of fashion industry. As a consequence, the fast-fashion brands like Zara, H&M and C&A have altered their business model towards circular economy through sustainable practices for capturing environment conscious consumers [29].

12.3.2 Green fashion and ethical fashion

It is associated to slow fashion campaign and often used as substitute of ethical or slow fashion [32]. Fashion summit in 2019 defined eco-fashion as superior quality and long-lasting without having adverse impact on surrounding climate.

According to Joergens [34], highlighted difficulty in illustrating green fashion due to lack of availability of industry standards at that time. However, according to Joergens, green fashion inculcates clothing fulfilling trends of fashion and manufactured under fair working conditions [34].

According to Ethical Fashion Forum 2016, ten principles of eco fashion were established:

1. It is opposite to fast fashion and its adverse effects.
2. Industry must have appropriate working climate.
3. Operators must be given salary on timely basis.
4. Harmful chemicals must be banned.
5. Eco-friendly materials must be encouraged [35].
6. The utility of water in textile industry must be reduced.
7. Materials must be recycled.
8. Energy must be used in efficient way.
9. Sustainability must be given utmost importance.
10. More awareness must be given to consumers about ethical practices in fashion [24]

12.3.3 Exterior factors impacting decisions of consumers

- Impact of price has major impact on consumers while purchasing garments.
- Problems related to institution, society and culture and their effect on consumer behaviour [36].
- Desire of consumers to reduce their effort in sorting.
- Limitation of time for buying clothes with little amount of information keeping in view costing and ethical practices.
- Increase in cost of product must be ethically justified [37].

- Shifting of consumers towards less expensive garments due to costly nature of eco-friendly products.
- Quality of eco-friendly products might be problem for customers.

12.3.4 Organisational factors

These factors ensure the availability of eco-friendly products which are restricted in access. In most cases, these products cannot be purchased in markets, and consumers are not willing to purchase them from other dedicated areas. Two challenges for the consumers in purchase of ethical products are: remembering their brand names and what kind of alternates are available in category of ethical brands [24, 37].

12.3.5 Intrinsic factors

These factors are enthusiasm, awareness and that influence ethical purchase decisions are motivation, knowledge, behaviour and self-control and moral values [36]. Factors of demographics can also be included in this class. The inspiration to vary purchase attitude of consumer influence ethical purchase [37]. In study mentioned by Kollmuss and Ageyman [36], it was noted that the motivation of consumer can be overpowered by easiness and less expensive shopping. Apart from that lack of information, experience of shopping also impact on purchase of green products [24].

12.4 Eco-label

The idea of eco-label was established in the global community for preserving quality of environment in sustainable manner [38, 40]. Eco-label is undertaking by the company indicating that its product undergo certification procedure by a third party to access its quality of conformance to eco-friendly performance during its entire life cycle [40, 43]. Eco-labelling serves as tool for informing consumers that product of company is eco-friendly i.e., it fulfils all the requirements of sustainability [49]. The certification is assurance for consumer that product they are purchasing is eco-friendly [50, 51].

One of the oldest eco-label logos is Blue Angel. It was endowed by Germany almost forty years ago and it is functional today [54]. Some of other eco-friendly label devices are Eco mark from India and Japan, Ecolabel from EU and Environmental Choice from Canada and Sweden [40, 43]. All over the world, the International Organization for Standardization (ISO) has dispensed several global standards serving as reference materials for eco-label actions [51, 55–60]. ISO categorises labels into three types: Type I, Type II and Type III.

12.4.1 Type I

Issued by a third party, which can be government, private or non-commercial entity [56, 61]. They are discretionary, multi-spectral [62] and usually employed for public purchases. Analysis of life cycle of product is performed on basis of pass and failure. EU flower is an example of Type I eco-label. ISO standard 14024:2018 deals with Type I label.

12.4.2 Type II

It is a declaration from manufacturing organisation along with importing and distributing firms about environmental aspects. This indicated that announcement is self-claimed and partial as it is one sided [56, 62]. Claims made on packaging of product is an example of Type II label. ISO 14021:2016 standard is related to Type II label.

12.4.3 Type III

Constitutes quantified information of product dependent on influence of life cycle [56]. It can be associated with distinct groups of products [56, 63]. The evidence is exhibited in the form reports concerning sustainability evaluation reports [64]. ISO 14025:2006 standards deal with Type III eco-labels.

Previous research highlighted that eco-labelling delivers mutual benefits for regulating bodies, government institutes and broader society. Eco-labels serve as marketing tool by assisting the organisations to accomplish sustainable goals by evading less productive practices. This will enable the customers to purchase environment friendly products in more conscious way [17].

Furthermore, textiles manufactured through conventional methods might have remnant chemicals and uncomfortable skin absorbing dyes. This might become the cause of cancer and allergies in patient [9]. To overcome these adverse issues, restrictions must be made on utility of typical chemicals and dyes [65, 66]. Furthermore, rules and regulation for health of worker and protection of environment must be consolidated [3].

It is pertinent that textile substrates do not evacuate hazardous chemicals during their usage. The influence of maintenance on environment is far more than that of its manufacturing [67, 68]. A huge quantity of water is consumed during washing of fabric [33]. Due to this reason, there is dire need of those textile products which require cost-effective washing with less quantity of washing [3].

Textile substrates must be developed in eco-friendly manner so that they can be utilised for longer periods. Their longevity can be enhanced in three ways: design for durability, design for disassembly and design for durability [69, 70]. It is imperative to use high-quality materials and seams that avoid tearing for longer periods. Recent fashion trends must be considered through cautious intake of design elements to avert the textile product from rejection in shorter period. Longevity use and reusage of textile clothing must be encouraged. In the end, the dismantling of textile item must be simple followed by repair and recycling [3].

12.5 Classification of eco-friendly textile fibres

Eco-friendly fibres can be natural or human-made, but their production process must have a low effect on surrounding climate as compared to traditional fibres [19]. The lives of consumers and farmers are at less risk due to these eco-friendly fibres as various pesticides and fertilisers are eradicated during their production [71]. Other hazardous impacts can also be decreased further by substituting harmful pesticides through bio pesticides due to their less hazardous impact on surrounding climate [72]. Insecticides can also be replaced with integrated pest management [73]. Classification of eco-friendly textile fibres is shown in Figure 12.1.

Owing to the rapid increase in human population, the fashion business is a critical element of the world economy. The apparel business in Europe is worth €1.1 trillion and is responsible for the employment of 300 million people [12]. Modern industry has been reformed into sustainable approach due to enhanced cognisance of end users about ethical production of textile goods. This has resulted in reduction in the utility of virgin cotton and raw ingredients. It was mentioned by Grand View Research [74], that eco-fibre market might reach US$77.17 billion in 2030. The government and other regulation authorities are making efforts in modification of laws. In this way, industry can be persuaded to obey against innumerable limitations [3].

Figure 12.1 Classification of eco-textile fibres.

The primal organic textile standards are: The Organic Content Standard (OCS) and the Global Organic Textile Standard (GOTS). The OCS is a non-mandatory standard that accomplishes norms for independent (third-party) certification of organic contribution and evidence trail. The main objective of OCS is to increase productivity of organic agriculture. In 2020, number of OCS certified organic places increases from 6,181 to 8,680 [75].

12.6 Eco-label certification in textile sector

The acquisition of eco-labels in clothing guarantees that garments conform to standards related to social, environmental and ecological concerns. It is noted from available data, there are almost 100 different labels for certification of textile products and more than 8,000 textile companies have got certifications [18, 76]. The fair global eco-labelling system invigorate the markets, but the local eco-labelling system might put limitations on entrance to market and diminishes worldwide competition [3]. The most commonly used eco-labels in textile industry are:

- **Brazilian Association of Technical Standards (ABNT) eco-label**: This type of eco-label is related to life cycle of product and is discretionary approach towards certification, product labels and eco-friendly execution services in Brazil.
- ***Bluesign***: This standard will assist manufacturing of textiles in sustainable approach to decrease the influence of textile business on surrounding way.
- ***CarbonFree***: This certification ensures reduction in carbon footprints of products and increases in awareness of customers about emissions of product.
- ***Cradle-to-Cradle***: By employing cradle-to-cradle certified products, a company might depict its association with eco-friendly design.
- ***Degree of Green***: It accesses the products and delivers explicit details of assessment. In this way, the extent of green grading programme helps the consumers.
- **EU Ecolabel**: Created the optional EU Ecolabel or EU Flower initiative in year 1992.
- **Eco-Leaf**: This label accesses the life cycle of eco-friendly products and serves as declaration for this eco-friendly product.

- **Fairtrade**: It is a methodology of creditable business practices which classifies the necessities of customers. Furthermore, it enables agriculturists and labours in developed countries to have better lifestyle a secure future.
- ***GOTS***: This standard is used for organic fibres. It inculcates autonomous certifications of entire textile supply chain along with eco-friendly and social standards.
- ***Green* products standard**: It assists customers to have improved understanding of increasing amount of environmentally friendly items in market.
- **Institution for Market Ecology**: The global association for inspection of items, certification and quality management of eco-friendly items.
- ***Korean* Eco-label**: The Korea Eco-Products Institute performs several tasks associated with benefits of environmentally friendly items.
- **Label STEP**: It is granted to handicraft carpet businesses who obey fair-trade procedures like avoiding misuse of forced child labour and reasonable pricing.
- **Migros ECO**: It is guaranteed that no ingredient is probable to exacerbate skin allergies and must be mentioned on the label of item.
- ***Nordic* Eco-label *or* 'Swan'**: It determines that item is environmentally friendly selection. The 'Swan' is symbol in Scandinavian countries for sixty-five item groups.
- **Oeko-Tex Standard 100**: It is a worldwide testing and certification body for raw materials of textiles, intermittent products and completed items for all phases of production.
- ***Processed Chlorine Free***: It reviews the ecological impact of production procedures and requires chain of evidence for all raw materials.
- ***Singapore Green Label Scheme*** *(SGLS)*: Its main objective is to assist the identification of eco-friendly products which fulfil established eco-standards in Singapore.
- **Terra Cycle**: Performs waste collection activities in ten different countries. In these countries, post-consumer non-recyclable garbage are taken and converted into new items and raw materials.
- ***USDA* certified bio-based**: Developed by the US Farm Bill in 2002. The main aim was to encourage utility of bio-based items [3].

12.7 Traceability in textile supply chains

Traceability is defined by ISO 9001:2015 as the ability to identify and trace the history, distribution, location and application of products, parts, materials and services. Within the textile and clothing sector, a robust traceability system can provide visibility into suppliers, production methods, logistics and sustainability metrics [77].

The supply chain of textile and apparel industry is extremely complex and is distributed worldwide, due to which there are several gaps in information and data reliability issues. The absence of reliable information has generally resulted in severe consequences, including unsafe working conditions, fake products and problems inspection and liability [78]. These underlying weaknesses are the major concern for governments and consumers, emphasising the importance of better tracking and traceability [14].

12.7.1 The need for traceability in textile supply chains

12.7.1.1 Information asymmetry and counterfeits

A lack of transparency allows fake products to enter the markets, resulting in major losses. Between 2007 and 2012, fake products resulted in an annual revenue loss to the textile and clothing industry of about €26.3 billion, beside government tax losses and job reductions [79].

12.7.1.2 Consumer and government pressure

The consumers' consciousness towards sustainability, traceability and labour rights has increased over time. They question about the origin, stuff and ecological impact of purchased items [80]. Governments have also responded by strengthening controls on product safety, labour practices and chemical usage in textile processing. These external pressures drive strong motivation towards adopting the traceability systems.

12.7.1.3 Complexity of global supply chains

The textile supply chain covers raw material, spinning, weaving, dyeing, finishing, apparel manufacturing and marketing [81]. The production is usually done in developing countries due to lower labour costs, while retail markets concentrate in developed countries. This geographical division makes information sharing more difficult thus gaps in traceability and accountability [82].

12.7.2 Concept and benefits of traceability

Traceability refers to the ability of monitoring a product journey forward and backward through the supply chain. According to Moe [83], it is the ability to track a product batch and its complete history across all stages of production, transportation and distribution. Within textiles, traceability covers raw materials, processing conditions, logistics and compliance certifications.

The benefits of implementing traceability systems includes transparency and accountability across supply-chain actors, fraud prevention and counterfeit control, quality assurance and efficient product recalls, regulatory compliance, particularly regarding conflict free materials and chemical safety and sustainability monitoring, including carbon footprints and labour practices.

Moreover, research suggests that consumers are willing to pay a premium price for traceable products, equating traceability with certification and authenticity [78].

12.7.3 Certification systems and standards

Certification offers a controlled approach to ensure both traceability and accountability within the textile and clothing sector. Most traceability certification systems follow four key steps: (1) establishing product tracking, (2) defining clear standards, (3) auditing supply chains and (4) issuing certification. Several prominent frameworks guide this process. The GOTS emphasises environmentally and socially responsible practices, while Fair Trade Certification prioritises equitable labor conditions. In addition, the 2017 OECD guidelines highlight the importance of traceability in addressing challenges such as counterfeiting and limited supply-chain visibility [84].

Such systems not only foster consumer trust but also enable brands to identify inefficiencies and environmental risks across their operations. However, despite these advantages, traceability in the textile and clothing sector remains underdeveloped. Several barriers contribute to this challenge. Technological limitations persist, such as affordable and specialised traceability tools are still limited. A lack of consensus among stakeholders on what information should be shared and how further complicates the progress. Unlike food and pharmaceuticals, the textile industry also faces an absence of strong regulatory mandates to implement traceability. Finally, concerns over confidentiality discourage companies from disclosing too much information, as this may compromise their competitive advantage [81].

12.7.4 Frameworks for traceability implementation

Frameworks can support efficient adoption of traceability across the textile and clothing supply chain. Kumar et al. [78] proposed a generic data model based on business-integration tools such as requirement planning and enterprise modelling. Their framework emphasises traceability within organisations and across organisations, using shared standards to ensure systems can work together.

A Delphi-based methodology has also been used to identify critical factors influencing traceability adoption, highlighting the importance of stakeholder perception, key information sets and consensus building [85, 86].

12.7.5 Way forward for industry traceability

The textile industry can strengthen traceability through multiple strategic approaches. The adoption of digital technologies such as blockchain, radio frequency identification (RFID) and cloud-based platforms enable real-time tracking while safeguarding data integrity. Collaboration across brands, suppliers and regulators is essential for establishing industry wide standards, while consumer demand for greater transparency can further accelerate adoption. Government support through policies and subsidies also plays a key role in encouraging investment in traceability. In addition, ongoing research and development in areas like low-cost sensors, smart labelling and AI-driven monitoring systems hold the potential to transform supply-chain visibility. Ultimately, creating resilient and transparent textile and clothing supply chains requires a careful balance between transparency and confidentiality, supported by frameworks that integrate sustainability, regulatory compliance and consumer trust [87].

Ultimately, building resilient and transparent textile and clothing supply chains requires balancing transparency with confidentiality and adopting frameworks that integrate sustainability, regulatory compliance and consumer trust.

References

[1] K. Kalkanis, D. Alexakis, E. Kyriakis, K. Kiskira, J. Lorenzo-Llanes, C. S. Themelis, and N. J. Psomopoulos, 'Transforming Waste to Wealth: Achieving Circular Economy', *Circular Economy and Sustainability*, 2 (2022): 1541–59.

[2] K. Hatzilyberis, P. Tsakanika, L. A. Lymperopoulou, T. Georgiou, K. Kiskira, K. M. O. Tsopelas, F. Ochsenkühn and M. Petropoulou, 'Design of an Advanced Hydrometallurgy Process for the Intensified and Optimized Industrial Recovery of Scandium from Bauxite Residue', *Chemical Engineering and Processing – Process Intensification*, 155 (2020): 108015.

[3] S. Plakantonaki, M. Stergiou, G. Panagiotatos, K. Kiskira and G. Priniotakis, 'A Review of Sustainability Standards and Ecolabeling in the Textile Industry', *Sustainability*, 15 (15) (2023): 1–18. doi: 10.3390/su151511589.

[4] E. Muslim and D. R. Indriani, 'Analisis Pengaruh Eco-Label terhadap Kesadaran Konsumen untuk Membeli Green Product', *Jurnal Manajemen Teknologi*, 13 (1) (2014): 86–100.

[5] J. Naeem and A. Siddique, 'Sustainable Consumption and Production Trends of Textiles and Apparel Industry in Pakistan', in Sadhna, R. Kumar, H. Memon and S. Greeshma (eds), *Consumption and Production in the Textile and Garment Industry: SDGs and Textiles*, Singapore: Springer, (2024), pp. 17–32. doi: 10.1007/978–981–97–6577-5_2.

[6] A. Vercalsteren, M. Nicolau and E. Lafond, 'Textiles and the Environment in a Circular Economy', *Eionet Report*, ETC/WMGE, (2019).

[7] A. Siddique, A. Nazir, T. Hussain, S. Ahmad and S. Abid, 'Foam Dyeing: A Review of Its Potential to Reshape Sustainable Textile Processing', *Textile Research Journal*, (2025) (online first). doi: 10.1177/00405175251324669.

[8] S. Ahmad, M. Ashraf, S. Abid, M. Jabbar, F. Shafiq and A. Siddique, 'Recent Developments in Laser Fading of Denim: A Critical Review', *Journal of Natural Fibers*, (2022): 1–11. doi: 10.1080/15440478.2022.2029793.

[9] C. Sivaramakrishnan, 'Pollution in Textile Industry', *Colourage*, 16 (2009): 66–8.

[10] B. Huang, J. Zhao, Y. Geng, Y. Tian and P. Jiang, 'Energy-Related GHG Emissions of the Textile Industry in China', *Resources, Conservation and Recycling*, 119 (2017): 69–77.

[11] WWF, 'Cotton Facts', (2022), www.worldwildlife.org/industries/cotton.

[12] A. ten Wolde and P. Korneeva, 'Circular Fashion Advocacy: A Strategy towards a Circular Fashion Industry in Europe', Ecopreneur.eu, (2019).

[13] S. Plakantonaki, M. Stergiou, G. Panagiotatos, K. Kiskira and G. Priniotakis, 'Regenerated Cellulosic Fibers from Agricultural Waste', *AIP Conference Proceedings*, 2430 (2022): 080006.

[14] V. Kumar, T. K. Agrawal, L. Wang and Y. Chen, 'Contribution of Traceability towards Attaining Sustainability in the Textile Sector', *Textiles and Clothing Sustainability*, 3 (1) (2017): 5. doi: 10.1186/s40689-017-0027-8.

[15] K. Niinimäki and L. Hassi, 'Emerging Design Strategies in Sustainable Production and Consumption of Textiles and Clothing', *Journal of Cleaner Production*, 19 (2011): 1876–83.

[16] United Nations Environment Programme, 'Green Economy', UNEP, 2013. www.unep.org/explore-topics/green-economy.

[17] C. E. Henninger, 'Traceability: The New Eco-Label in the Slow-Fashion Industry? Consumer Perceptions and Micro-Organisations Responses', *Sustainability*, 7 (5) (2015): 6011–32. doi: 10.3390/su7056011.

[18] R. A. K. Choudhury, 'Development of Eco-Labels for Sustainable Textiles', in S. S. Muthu (ed.), *Roadmap to Sustainable Textile and Clothing: Regulatory Aspects and Sustainability Standards of Textiles and the Clothing*, Cham: Springer, (2015), pp. 137–74.

[19] J. Periyasamy and A. P. Militky, 'Sustainability in Regenerated Textile Fibers', in S. S. Muthu and M. A. Gardetti (eds), *Sustainability in the Textile and Apparel Industries*, Cham: Springer, (2020), pp. 63–95.

[20] J. Naeem and A. Siddique, 'Textile and Fashion Industry of Mozambique', in *Textile and Fashion Industry of African Countries*, Singapore: Springer, (2023), pp. 157–73. doi: 10.1007/978-81-99-1320-6_8.

[21] K. Fletcher, 'Slow Fashion: An Invitation for Systems Change', *Journal of Design, Creative Processes and the Fashion Industry*, 2 (2) (2010): 259–65.

[22] A. Kozlowski, M. Bardecki and C. Searcy, 'Social and Eco-labelling of Textile and Clothing Goods as Means of Communication and Product Differentiation', *Fibres and Textiles in Eastern Europe*, 19 (4) (2011): 20–6.

[23] K. Fletcher, 'Slow Fashion', *The Ecologist*, (June 2007), https://theecologist.org/2007/jun/01/slow-fashion.

[24] J. Madsen, B. Hartlin, S. Perumalpillai, S. Selby and S. Aumônier, *Mapping of Evidence on Sustainable Development Impacts that Occur in Life Cycles of Clothing: A Report to the Department for Environment, Food and Rural Affairs*, London: Defra, (2007).

[25] American Fiber Manufacturers Association, *Resource and Environmental Profile Analysis of a Manufactured Apparel Product: Life Cycle Analysis (LCA) – Woman's Knit Polyester Blouse*, Washington, DC: American Fiber Manufacturers Association, (1993).

[26] L. Dahllöf, 'Life Cycle Assessment (LCA) Applied in the Textile Sector: The Usefulness, Limitations and Methodological Problems – A Literature Review', Environmental Systems Analysis, Chalmers University of Technology, Göteborg, (2003).

[27] K. Kooistra and A. Termorshuizen, 'The Sustainability of Cotton: Consequences for Man and Environment', Biological Farming Systems Group Report, Wageningen University, (2006).

[28] B. V. Todeschini, M. N. Cortimiglia, D. Callegaro-de-Menezes and A. Ghezzi, 'Innovative and Sustainable Business Models in the Fashion Industry: Entrepreneurial Drivers, Opportunities, and Challenges', *Business Horizons*, 60 (2017): 759–70.

[29] K. Niinimäki, *Sustainable Fashion: New Approaches*, Helsinki: Aalto ARTS Books, (2013).

[30] K. Niinimäki, 'Sustainable Consumer Satisfaction in the Context of Clothing', in C. Vezzoli, C. Kohtala and C. Srinivasan (eds), *Product-Service System Design for Sustainability*, Sheffield: Greenleaf, (2014), pp. 218–37.

[31] L. Carey and M. Cervellon, 'Ethical Fashion Dimensions: Pictorial and Auditory Depictions through Three Cultural Perspectives', *Journal of Fashion Marketing and Management*, 18 (4) (2014): 483–506.

[32] C. Joergens, 'Ethical Fashion: Myth or Future Trend?', *Journal of Fashion Marketing and Management: An International Journal*, 10 (3) (2006): 360–71.

[33] Ethical Fashion Forum, 'The Source: Industry Platform for Ethical Fashion', (2016), www.ethicalfashionforum.com.

[34] A. Kollmuss and J. Agyeman, 'Mind the Gap: Why Do People Act Environmentally and What Are the Barriers to Pro-environmental Behaviour?', *Environmental Education Research*, 8 (3) (2002): 239–60.

[35] M. Wiederhold and L. F. Martinez, 'Ethical Consumer Behaviour in Germany: The Attitude–Behaviour Gap in the Green Apparel Industry', *International Journal of Consumer Studies*, 42 (2018): 419–29.

[36] A. Carlson and C. Palmer, 'A Qualitative Meta-synthesis of the Benefits of Eco-labeling in Developing Countries', *Ecological Economics*, 127 (2016): 129–45.

[37] H. Uchida, Y. Onozaka, T. Morita and S. Managi, 'Demand for Ecolabeled Seafood in the Japanese Market: A Conjoint Analysis of the Impact of Information and Interaction with Other Labels', *Food Policy*, 44 (2014): 68–76. doi: 10.1016/j.foodpol.2013.10.002.

[38] D. Brécard, 'Consumer Misperception of Eco-Labels, Green Market Structure and Welfare', *Journal of Regulatory Economics*, 51 (3) (2017): 340–64.

[39] F. Isharyadi and K. Ediyanto, 'Analysis of Eco-label Implementation and Green Products in Indonesia', *Open Engineering*, 11 (1) (2021): 556–69.

[40] F. Isharyadi, A. Surachman, S. Suryani and C. Sitompul, 'Analysis of Eco-label Certification Implementation on Eco-friendly Products in Indonesia', *IOP Conference Series: Earth and Environmental Science*, 1108 (1) (2022). doi: 10.1088/1755-1315/1108/1/012002.

[41] F. G. Topolansky Barbe, M. M. Gonzalez-Triay and A. Hensel, 'Eco-Labels in Germany', *Journal of Customer Behaviour*, 12 (4) (2013): 341–59. doi: 10.1362/147539213X13875568505868.

[42] M. Nadanyiova and J. Kliestikova, 'Eco-labels: A Tool for Green Marketing or Just a Blind Mirror for Consumers?', *Electronic Green Journal*, 1 (42) (2019).

[43] European Commission, 'Study on Different Types of Environmental Labelling (ISO Type II and Type III): Executive Summary', European Commission, 2011, https://ec.europa.eu/environment/ecolabel.

[44] Lembaga Ekolabel Indonesia, 'Sertifikasi LEI', Lembaga Ekolabel Indonesia, 2022, www.lei.or.id.

[45] Global Ecolabelling Network, 'Life Cycle-Based Ecolabels and Global Ecolabelling Network Descriptions', (2022), www.globalecolabelling.net.

[46] Ministry of Environment and Forestry of Indonesia, 'Center for Facilitating the Implementation of Environmental and Forestry Instrument Standards', Database Ekolabel, (2022), https://ekolabel.menlhk.go.id.

[47] H. U. de Haes and G. R. de Snoo, 'Eco-labelling of Agricultural Food Products', in U. Sonesson, J. Berlin and F. Ziegler (eds), *Environmental Assessment and Management in the Food Industry*, Cambridge: Woodhead Publishing, (2010), pp. 374–97.

[48] European Commission, 'Green Public Procurement and the European Ecolabel', (2011), https://ec.europa.eu/environment/gpp.

[49] Global Ecolabelling Network, 'Trade as an Environmental Policy Tool? Environment as a Trade Policy Tool?', (2003), www.globalecolabelling.net.

[50] E. Pedersen and P. Neergaard, 'Caveat Emptor: Let the Buyer Beware! Environmental Labelling and the Limitations of "Green" Consumerism', *Business Strategy and the Environment*, 15 (2006): 15–29.

[51] Global Ecolabelling Network, 'GEN Position on the ISO 14024 Guidance Standard', (2003), www.globalecolabelling.net.

[52] A. Patra and S. R. K. Pariti, 'Restricted Substances for Textiles', *Textile Progress*, 54 (2022): 1–101.

[53] B. Dulo, B. De Graef, V. K. Mubiana, R. Blust, G. Verschraegen and K. De Wael, 'Evaluating the Potential of Natural Dyes from Nutshell Wastes: Sustainable Colouration and Functional Finishing of Wool Fabric', *Sustainable Materials and Technologies*, 34 (2022): 518–31.

[54] N. Lant, A. S. Hayward, M. Peththawadu, K. Sheridan and J. R. Dean, 'Microfiber Release from Real Soiled Consumer Laundry and the Impact of Fabric Care Products and Washing Conditions', *PLOS One*, 15 (2020): e0233332.

[55] C. Yun, S. Patwary, M. L. LeHew and J. Kim, 'Sustainable Care of Textile Products and Its Environmental Impact: Tumble-Drying and Ironing Processes', *Fibers and Polymers*, 18 (2017): 590–6.

[56] L. Ibanez, 'Ecolabels: Are They Environmental-Friendly?', in G. Marciano and G. B. Ramello (eds), *Encyclopedia of Law and Economics*, New York: Springer, (2016), pp. 1–9.

[57] D. G. K. Dissanayake and D. Weerasinghe, 'Towards Circular Economy in Fashion: Review of Strategies, Barriers and Enablers', *Circular Economy and Sustainability*, 2 (2021): 25–45.

[58] T. Durham and T. Mizik, 'Comparative Economics of Conventional, Organic, and Alternative Agricultural Production Systems', *Economies*, 9 (2021): 64–84.

[59] D. Srijita, 'Biopesticides: An Ecofriendly Approach for Pest Control', *World Journal of Pharmacy and Pharmaceutical Sciences*, 4 (2015): 250–65.

[60] Beyond Pesticides, 'Defining a Strong IPM or EPM Program', (n.d.), www.beyondpesticides.org.

[61] Grand View Research, 'Textile Market Size, Share and Trends Analysis Report', (2021), www.grandviewresearch.com.

[62] L. R. Pepper and L. Truscott, 'Preferred Fiber and Materials Market Report', Textile Exchange, (2021), https://textileexchange.org.

[63] Ecolabel Index, 'All Ecolabels', (2022), www.ecolabelindex.com.

[64] International Organization for Standardization, *ISO 9000:2015: Quality Management Systems – Fundamentals and Vocabulary*, Geneva: ISO, (2015), www.iso.org/standard/45481.html.

[65] T. K. Agrawal and R. Pal, 'Traceability in Textile and Clothing Supply Chains: Classifying Implementation Factors and Information Sets via Delphi Study', *Sustainability*, 11 (6) (2019). doi: 10.3390/su11061698.

[66] N. Wajsman, C. A. Burgos and C. Davies, 'The Economic Cost of IPR Infringement in the Clothing, Footwear and Accessories Sector', Office for Harmonization in the Internal Market (OHIM), Alicante: OHIM, (2015).

[67] K. Fletcher, *Sustainable Fashion and Textiles: Design Journeys*, 2nd edn, Abingdon and London: Routledge, (2012). doi: 10.4324/9781849772778.

[68] V. Kumar, C. Hallqvist and D. Ekwall, 'Developing a Framework for Traceability Implementation in the Textile Supply Chain', *Systems*, 5 (2) (2017). doi: 10.3390/systems5020033.

[69] G. Gereffi and S. Frederick, 'The Global Apparel Value Chain, Trade and the Crisis: Challenges and Opportunities for Developing Countries', World Bank Policy Research Working Paper No. 5281, Washington, DC: World Bank, (2010).

[70] T. Moe, 'Perspectives on Traceability in Food Manufacture', *Trends in Food Science and Technology*, 9 (5) (1998): 211–14. doi: 10.1016/S0924–2244(98)00037-5.

[71] C.-C. Hsu and B. A. Sandford, 'The Delphi Technique: Making Sense of Consensus', *Practical Assessment, Research and Evaluation*, 12 (10) (2007).

[72] J. Kembro, D. Näslund and J. Olhager, 'Information Sharing across Multiple Supply Chain Tiers: A Delphi Study on Antecedents', *International Journal of Production Economics*, 193 (2017): 77–86. doi: 10.1016/j.ijpe.2017.06.032.

[73] C. F. Mason, 'The Economics of Eco-Labeling: Theory and Empirical Implications', *International Review of Environmental and Resource Economics*, 6 (4) (2013): 341–72. doi: 10.1561/101.00000054.

[74] S. Saberi, M. Kouhizadeh, J. Sarkis and L. Shen, 'Blockchain Technology and Its Relationships to Sustainable Supply Chain Management', *International Journal of Production Research*, 57 (7) (2019): 2117–35. doi: 10.1080/00207543.2018.1533261.

Section 4

Future trends in sustainable design for textiles and fashion

13 Future trends in sustainable design for textiles and fashion

Amna Warda and Syed Zameer Ul Hassan

13.1 Introduction

In the current era where the market is shaped by consumer awareness, climate urgencies and increasing regulatory requirements, the textile and fashion industry has reached a point of inflection, which means what we wear has never been this personal, environmental, or political. After being driven by the trends and consumption cycle, fashion is now reimagined as a tool for regeneration, innovation, and social responsibility [1, 2]. The textile and fashion industry has long been criticised for its high consumption of resources, pollution and waste, which has pushed it towards ethical and eco-friendly alternatives. This shift depends on the profound rethinking of raw materials and the production process. Recycled and bio-based textiles, for example, mycelium leather, seaweed-derived fibres and Tencel, are becoming industrial standards rapidly, which are replacing cotton-like resource-intensive conventional crops. Innovative digital designs and waterless dyeing technologies are minimising pollution and waste drastically, while circular economy models are redefining the product's life cycle by emphasising its reuse, recycling and repair [3, 4]. In 2025, with growing demands for ethical sourcing, transparency and durable, timeless fashion articles that challenge the fast-fashion pattern paradigm, consumer behaviour is also evolving [5, 6]. Meanwhile, to make sustainability a legal as well as moral obligation, worldwide regulatory frameworks are prompting brands to adopt fully circular and carbon-neutral systems [6]. The textile and fashion sector has long been criticised for its high consumption of resources, contribution to waste and pollution, and push towards ethical and eco-friendly alternatives [7].

13.2 Eco trends in sustainable design

The textile sector is the world's largest and least sustainable sector. So, the textile industry leaves a serious environmental impact through its production journey, i.e., from cotton farming to dyeing fabrics and to manufacturing garments. The textile industry generates almost 20 per cent of water waste and 10 per cent of carbon emissions according to World Bank calculations [8]. As environmental concerns intensify and consumer awareness also grows, the fashion and textile industries are focusing on sustainability more than ever before. In 2025, sustainability is no longer a niche but has become a central pillar in shaping materials, production and business models. Sustainability in fast fashion and fast-moving textiles drives them towards eco-friendly and sustainable operational materials (Table 13.1) [8].

Globally, governments also promote eco-labelling and sustainability standards, and this encourages 'eco-friendliness' in textiles. A product must be responsible at all stages of its production, i.e., manufacturing, utilisation and maintenance, to be considered the product a 'green'

DOI: 10.4324/9781003638148-17

Table 13.1 Key sustainable materials in textile production.

Material	*Environmental benefits*	*Challenges*	*Example brands*
Organic cotton	Reduced water and pesticide usage	Higher cost of production	Patagonia, Stella McCartney
Recycled polyester	Reduces plastic waste, lowers energy use	Lower durability compared to virgin polyester	Adidas, H&M
Hemp	Requires little water, pesticide-free	Limited consumer awareness, higher costs	Patagonia, Levi's
Tencel (Lyocell)	Biodegradable, uses less water	Expensive production, less widely available	H&M, Eileen Fisher
Bamboo fibre	Biodegradable, low environmental impact	Chemical processing can be harmful	Boody, Caribola
Recycled cotton	Reduces textile waste, lowers water use	Fibre quality and colour variation	Reformation, Levi's

product. These products that are safe for humans and the environment are given specific certification from an independent organisation [9, 10].

13.3 Eco-friendly fibres

Eco-friendly fibres can be natural or synthetic, but their production and processing should have a lower environmental impact compared to conventional fibres [11]. They pose fewer risks to people, especially farmers, when synthetic pesticides and chemical fertilisers are reduced or eliminated during production [12]. Instead of synthetic chemical pesticides, biopesticides can be used, as they are significantly less harmful to the environment and human health [13]. Many organic farms use integrated pest management (IPM) instead of chemical pesticides. Overall, managing pests according to current standards and using improved techniques can lead to a reduction in pesticide use. The use of chemical pesticides is linked to harmful effects on human health and the environment [14]. Several organic fibres, including cotton, wool, silk, recycled cotton, naturally coloured cotton, lyocell, corn-based and soy-based fibre, recycled polyester and others, have recently become popular in the textile sector as they have a positive impact on the environment (see Figure 13.1). As a result of human development and an escalating population, the world is producing more fibres and consuming the same. In recent years, sustainable clothing has experienced a revolution, which was fuelled by the increasing number of ethical awareness concerning the manufactured materials. Such a transformation not only offers another source of economic advantage but also offers new hiring grounds in the market and minimises the utilisation of virgin resources and raw materials. Such trends will also be instrumental in ensuring that the organic fibre industry is advanced in the forecast period [16]. The use of chemicals and pesticides to enhance crop yields is gradually gaining momentum, thus causing massive destruction to the environment. This has, in turn, forced governments and other regulating bodies to impose a lot of restrictions [12].

These regulations lead to the production and disposal of synthetic agricultural fibres. As a result, alternative products, such as organic cotton, hemp and other fabric types sensitive to the environment, have become promising. The market opportunity is expected to grow during the period covered by the forecast because the demand for clothes and environmentally friendly fabrics is growing. This is expected to grow in the major economies like India, China, Japan and Australia in the future [16].

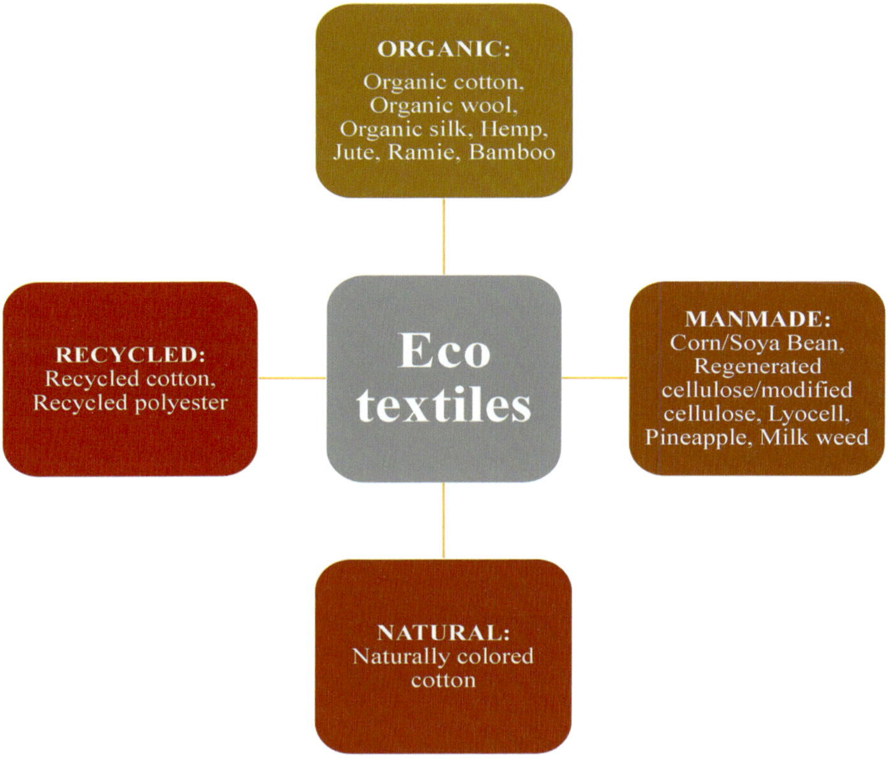

Figure 13.1 Classification of the most common eco-friendly fibres [15].

13.4 Eco-friendly product design and manufacturing

The usage, proper handling and maintenance of the product should always be considered. When creating a product, designers must prioritise environmental sustainability. The choice of raw materials directly impacts the environment. Some components of clothing, such as coatings and zippers, are not eco-friendly. Despite having high ambitions, a manufacturer dedicated to environmentally responsible production will consider replacing such components with environmentally acceptable and eco-friendly alternatives. Environmentally friendly and sustainable materials can also be achieved using traditional textile production methods [16, 17]. Chemicals and dyes that are volatile or absorbed into the skin can, in some cases, cause allergic reactions or even cancer. To prevent these negative effects and in response to facts and growing environmental awareness, there are increasingly strict regulations and limitations on the use of certain chemicals and dyes.

Textile products mustn't release volatile organic compounds or other hazardous substances [18]. The use of compounds or particularly hazardous chemicals, such as heavy metals, during use and maintenance can cause more environmental impact than the manufacturing process itself. For example, items like worn clothing typically require excessive amounts of water for regular washing. Consequently, the textile products must be designed in such a way that they do not require much washing and care, and preferably they should use less water. Additionally, it is easier and less damaging to the environment to wash clothes in low-temperature water. The regular use, maintenance and final use and use of a product should all be taken into account.

Designers could also be significant when it comes to the product creation process, as they can make the product environmentally acceptable [20].

The textile products must be long-lasting because eco-design will help the user to use the products longer. The notion is founded on the three main design principles: design to disassemble, design to enduring style and design to durability. Another important consideration is fashion sensitivity, which considers the preferences of customers and the ever-changing fashion trends. Increasing the value of timeless items encourages their long-term use, extended wear and potential for reuse. At last, how textile products are assembled also affects how easily they can be disassembled for repair, maintenance, reuse of components, or recycling (see Figure 13.2) [19]. With the reuse of components and recycling in mind, many fashion houses and retailers are now considering ways to make their products more 'green' (environmentally friendly).

13.5 Eco-labelling

'Eco textiles' is the term that refers to such materials that are made from natural fibres, regenerated fibres and organically dyed materials that consume less water, chemicals and non-harmful chemicals [21, 22]. That's why eco-labelling is introduced to help consumers identify eco-friendly products. Eco-labelling can complement traditional policy tools such as taxes and subsidies to promote more beneficial and sustainable economic development. It helps to introduce new environmentally friendly products, technologies and production systems. It can be used by companies to attract the perceptions and preferences of consumers towards green products and by governments as a powerful way of internalising negative externalities by imposing taxes [23].

Eco-labelling does not involve spending much budget. The government should not spend heavily; rather, it should provide transparency in the markets and proper regulation of the certification systems. It can also ensure that the excessive use of labels and environmental declarations is curtailed, as it regularly complicates the way to discern what indeed is the difference between products, a factor that brings about confusion and diminishes the trustworthiness of the information being relayed. The eco-declarations do not always meet the reliable, accurate and clear legal requirements. They are sometimes used intentionally to deceive consumers, leading to unfair positioning of products and services and even companies as a whole.

This is also referred to as greenwashing and has been a big challenge in the realisation of sustainable consumption patterns [23]. Moreover, in order to facilitate environmental development and innovations and the creation of green improvements, the government ought to train businesses and the people generally to push the demand for green products at all levels of the income bracket. It requires the government to undertake strict certification initiatives by eco-labelling as a way of curbing the confusion in regard to the environmental assertion by firms and also to counter the increasing scepticism and market support for green products [22].

A survey conducted on the awareness of eco-labelling in Russia revealed that the Russian consumers know the theoretical aspects involved in eco-labelling. Figure 13.3 illustrates the most popular eco-labels in the Russian market. Nevertheless, when making purchasing choices, they listen to eco-labels with little regard and usually fail to examine and notice the most prevalent eco-label marks. Thus, it is possible to conclude that, due to the awareness customers raise, the demand for the product, especially the demand for greener goods, has also increased. The more people are ready to spend on such products, the more it affects consumer behaviour. Applications should be created for smartphones that would help buyers ensure that the ISO standards

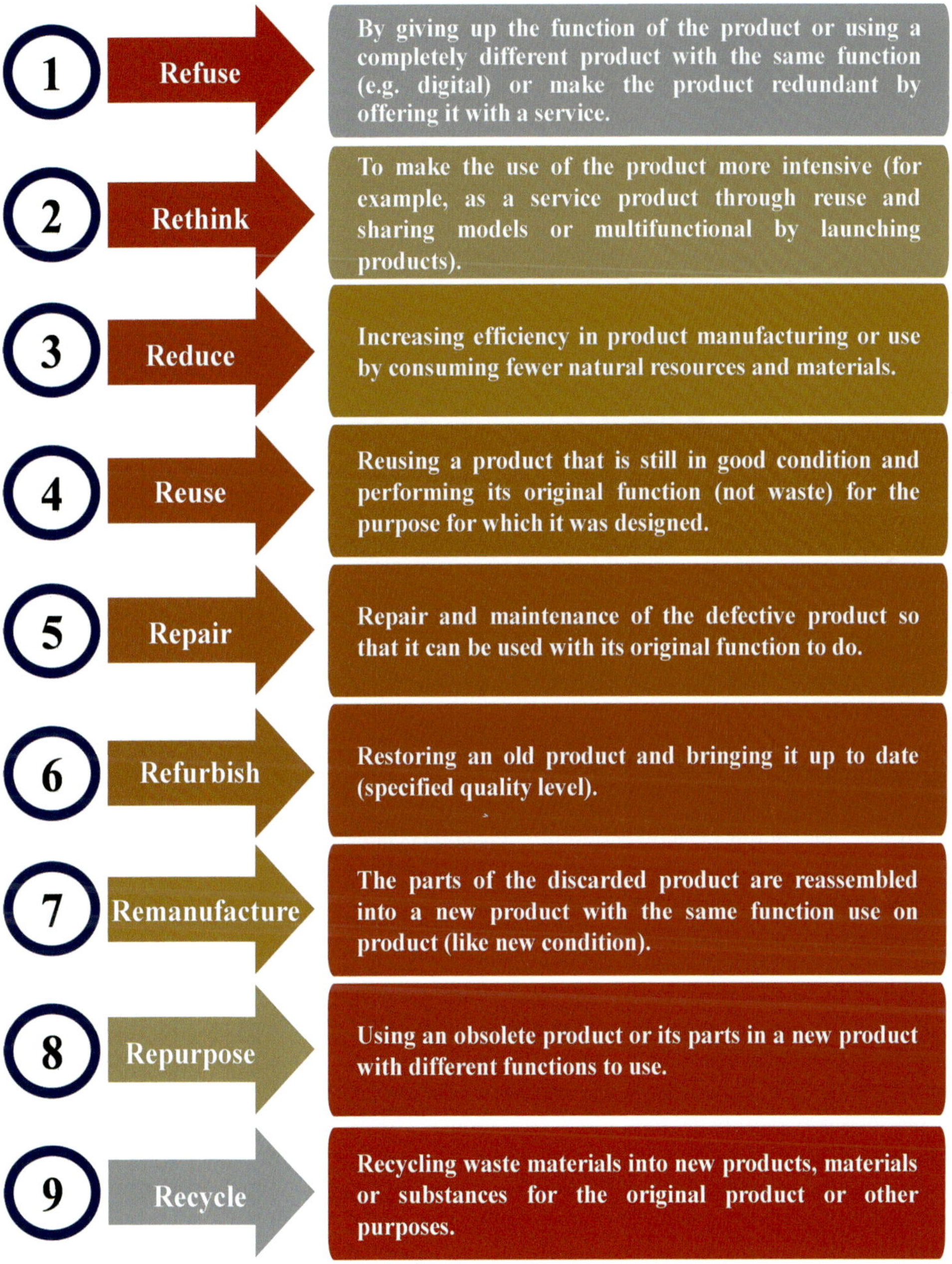

Figure 13.2 The 9R framework in a circular economy [20].

Figure 13.3 The most popular eco-labels on the Russian market [23].

are met when purchasing the product by scanning the eco-labels on the product directly [24]. Such a strategy may be considered a rather cost-efficient and simple method to proceed, increasing consumer awareness of the concept of eco-labelling and decreasing the chances of greenwash [23].

13.6 Smart textiles and sustainability

The area of smart textiles is largely questionable and is concerned with sustainability. The future of smart clothes cannot be denied, and one does not need a lot of imagination to comprehend where such doubts arise. The textile industry, as well as the microelectronics industry, has not had a good history of sustainability. Besides, the collaboration of electronics as well as textile materials, particularly at a large scale, is a new challenge towards the attainment of sustainability. Adding electronics to the current textiles to make their usage smart complicates and makes it harder to process and manage them at the end of their life [25, 26]. The combination of chemical compounds in textiles and the incorporation of electronics creates new and unfamiliar sustainability challenges. As an environmental consideration, on top of the material implication, there is one more issue regarding the total life of the product. This is regardless of electronic and textile products, as they tend to have short service lives and relatively low reuse or recycling rates [25].

The intermittent nature of use of these two kinds of products, though they have a short life cycle, also enhances the concern for the environment. There is a sense of urgency to develop a holistic design of products in order to promote sustainability in smart textiles. This is achieved by adopting materials that are friendly to the environment, developing manufacturing processes that are environmentally friendly and providing effective end-of-life solutions. In order to accommodate the creation of the next-generation smart and sustainable textiles, such products can be recycled into high-value products, or they can be degraded safely in landfills without damaging the environment [27]. The future of the waste issue can be solved by the circular and environmentally friendly design of smart textiles only [28].

Designers would be advised to embrace a multidimensional approach and take into consideration different aspects during the overall design and production process. As an illustration, in the design stage, there should be a process of working on the awareness and perception of smart textiles to the consumer. During the production stage, the selection of materials is quite important in enhancing the sustainability of smart textiles. More emphasis on the use of major materials that are biodegradable, reusable and recyclable can play a significant role in designing smart textiles that are environmentally friendly [27]. Manufacturers use such materials to ensure the services they offer support the circular economy, as well as decrease waste. The materials are recyclable, which can be reused in the textile manufacturing process or biodegrade and do not harm the environment. Smart textiles that are developed with biodegradable polymers could play a great role in promoting sustainable production procedures [28, 29]. With these environmentally friendly materials, modern textiles that not only meet functional and technical demands but also reduce the environmental impact can be manufactured. This strategy is correlated with the increasing consumer desire towards sustainable goods and is aimed at creating a solid background of environmental innovation in the sphere of smart textiles. In addition, implementing more environmentally friendly manufacturing systems, including inkjet printing, 3D printing, knitting and using renewable energy sources, may contribute greatly to the eco-design of smart textiles [30]. Also, designers need to focus on the creation of detachable electronic elements to make their separation simple when dismantling the device or recycling it [31]. The strategy not only encourages sustainability but also makes material reuse and recycling occur more efficiently and with less effect on the environment [32].

13.7 Fast fashion and sustainability

Fast fashion is a phrase that is employed to denote a design in the fashion sector that is speed-oriented in terms of quick and cheap manufacturing. Globalisation has created an opportunity whereby fast-fashion companies are able to obtain raw materials, produce and deliver the goods to customers in every corner of the world. This has made the production and consumption of clothes increase significantly, with considerable changes in the global fashion industry. Fast fashion has a large negative impact on the environment, and every stage of the life cycle of a garment can cause some environmental and workplace risks [33]. The most widespread synthetic fibre, polyester, which is highly used in fashion garments, is a product of petroleum, and it can take hundreds of years to decompose. The production growth in the fashion industry has almost doubled the production of synthetic fibres. Technical Textiles reports that synthetic fibre manufacturing and more so of polyester, has greatly increased over the past fifteen years, coupled with a steep growth in demand. Production of cotton in large quantities at low prices is the biggest cause of fashion globalisation, but it is detrimental to the environment because it uses a lot of pesticides during production [34]. Fast fashion companies usually manufacture high quantities of clothing that are not developed based on the demand of consumers and are disposed of in a short period of time, adding to the textile industry waste, with the majority being put into landfills. In contrast to food waste, textile waste cannot be composted, both natural and synthetic waste, since it goes through several chemical processes during the manufacturing process [35]. The chemicals may leak onto groundwater in a landfill or into the atmosphere through the burning of the clothing. Furthermore, synthetic fibres also pose the same environmental issues because they are mostly petroleum-based and they may take hundreds or even thousands of years to decompose. The connection between fast fashion and sustainability is contradictory by its nature, as it tends to contribute to biased decisions made by consumers, which is typically referred to as the present over future or the other way around [34, 36].

Even though fast-fashion companies are not completely sustainable yet, it can be stated that much can be done. To illustrate, a significant number of fast-fashion companies are still manufacturing in large quantities, and this is something that promotes overconsumption. Some of the steps that fast-fashion companies have implemented to lower their environmental impact are through their introduction to 'Join Life' programmes, which are geared towards the utilisation of sustainable resources, the reduction of resources, and the enhancement of the welfare of their workers. Zara is a Spanish multinational retailer in the clothing sector and a secondary of the Inditex Group. Amancio Ortega founded it in 1975 with the vision of providing high-fashion clothes at affordable prices in Arteixo, Galicia, Spain [37]. Zara has established a reputation for a fast-fashion business model that allows the brand to introduce new designs into the market at a very quick pace, using a vertically integrated supply chain. This enables Zara to control the whole production process, including design, manufacturing, to retail and is in control, ensuring speed, efficiency and responsiveness to the changing fashion trends. Its brand aims to make 50 per cent of recycled or sustainably sourced materials its aim by 2025. Moreover, it is working to enhance the traceability and sustainability of its supply chain, ensuring greater transparency and environmental responsibility throughout its operations. Zara is utilising energy-efficient lighting, HVAC systems and recycled materials in the construction of its stores [38]. The brand has launched a water-conservation programme and partnered with renewable energy providers to promote the utilisation of clean energy. These efforts were also recognised by many different certifications. Zara's recycling programme collects cardboard, used clothes, plastic and packaging material from its stores. As there are many fast-fashion brands that are working on sustainable design, but buyers should also be responsible while making choices. There are many ways to support sustainability by choosing better quality and buying fewer products, by taking care of the products and by focusing on second-hand things in search of unique fashion. So, there are many simple guides for following sustainability:

- **Cradle to cradle**: The product is designed and manufactured with multiple life-cycle possibilities so that it can be recycled or composted [39].
- **Reuse and redesign**: Because of the vast availability of textiles and garment waste, reusing designs is becoming a popular concept in fashion, even among influential designers and millennial users. Brands can capitalise on this trend by creating capsule collections, launched in limited editions. The uniqueness of this type of garment significantly reduces the risk of becoming outdated or unfashionable, making it more appealing to conscious buyers [39, 40].
- **Partnership for co-creation and recycling**: Recycling encompasses both downcycling and upcycling and during this process materials may either lose or enhance their qualities and value, depending on the design and treatments applied. Therefore, strategic alliances and collaborations with experts and professionals are essential for the brands to ensure effective material transformation and innovation within this process [37].
- **Change in business models**: The fashion industry should adopt innovative business models to advance sustainability because of increasing environmental and social concerns. These transformative approaches not only aim to protect the environment but also represent a form of social responsibility and improve people's life quality, enabling new avenues for economic growth and profitability [34].

Resource utilisation and waste generation can be significantly minimised by a strategic shift, for example, transitioning from a product-centric model to a service-based model. Simultaneously, the integration of eco-friendly practices throughout the value chain is significant for developing a more responsible and resilient fashion and textile industry [37].

13.8 Waste management

Looking at the current state of waste generation in the fashion industry, some critical questions are being raised related to the traditional linear economy model that has been followed for a long time. There is an increasing demand for a shift from this linear model based on take, make and dispose, towards the circular economy model (Figure 13.4). The linear model typically consists of the process, which involves material extraction, manufacturing, distribution, consumption and disposal. This model follows a flow of resources mainly leading to extra waste generation and environmental degradation [38].

In contrast, the circular economy model, rooted in sustainability, aims to close the loop of product life cycles. The essence of the circular economy model is to reduce wastage and maximise resource use through reusing, repairing, refurbishing, re-manufacturing and recycling of materials and products. Instead of terminating at disposal, resources are continually returned to the system, thus minimising the extraction of virgin material and decreasing the ecological footprint. The circular economy is formed on the basis of three principles: reduce, recycle and reuse [41]. The second major source of pollution is reported to be the textile and clothing industry, which contributes to nearly 5 per cent of the total global waste. The textile wastes are broadly classified into two major groupings, about production and manufacturing: pre-consumer and post-consumer waste. The classification can be used to develop specific waste management and recycling policies in the textile and fashion industry [42, 43].

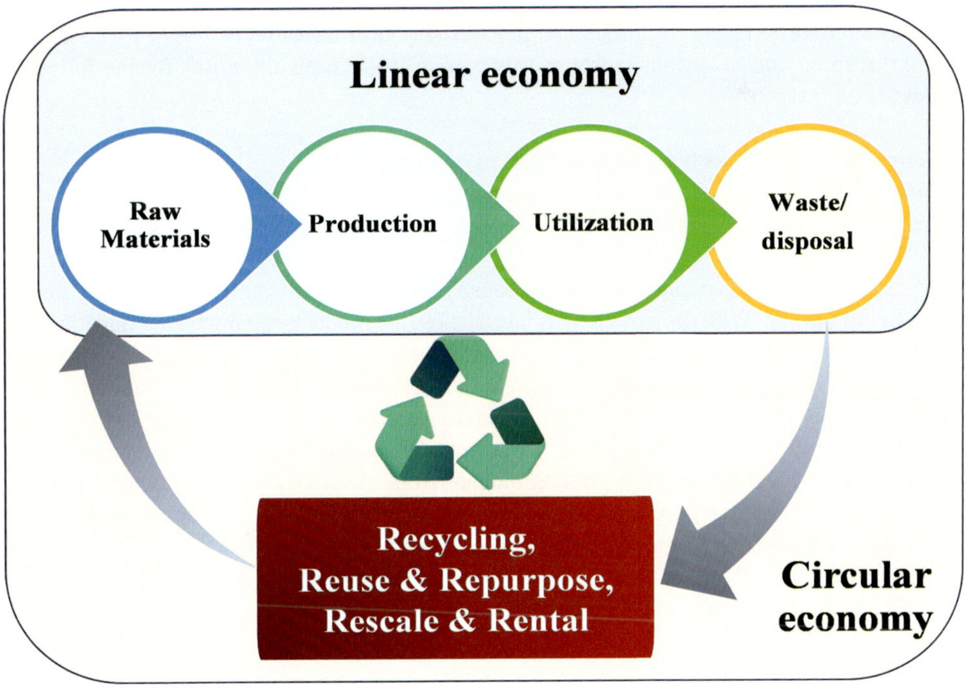

Figure 13.4 Transforming from a linear to a circular economy.

13.8.1 Pre-consumer

Pre-consumer waste, also known as production waste, includes the leftover materials produced during the manufacturing process. This includes fibre scraps, consumer textile waste, and offcuts that are generated during the garment development process. The percentage of pre-consumer textile waste during the garment development process is based on many factors, like:

- the complexity of the garment design;
- the characteristics of the fabric (e.g., elasticity, texture, print orientation);
- the efficiency of pattern-making and cutting techniques.

Effectively managing pre-consumer waste is crucial for improving resource efficiency and reducing the environmental footprint of the textile industry. The sources of pre-consumer waste within the garment manufacturing supply chain include:

- production faults, such as stitching errors, fabric defects, or sizing issues;
- production waste: including leftover fabric scraps, trimmings and offcuts generated during cutting and sewing processes;
- overproduction: when excess stock is created that never reaches the market;
- quality control rejects: items that fail to meet quality standards during various stages of production [44].

13.8.2 Post-consumer waste

Post-consumer waste refers to fabrics and apparel that have already been used by consumers and are discarded due to various reasons. This type of waste typically arises for the following reasons:

- **Wear**: The clothing has been worn out due to regular use.
- **Damage**: The item develops a defect (e.g., tearing, fading).
- **Trend**: The garment becomes outdated or unfashionable due to changing fashion trends.

Post-consumer waste is generated by the end consumer and is significantly influenced by the dynamics of fast fashion. With the paradigm of fast fashion, the products are frequently manufactured using low-quality and less durable materials. Consequently, the quality of the products is reduced, and the fabric's life is also diminished, resulting in its faster disposal following the slightest use [45, 46]. The handling of textile wastes is a very imperative process in ensuring that the circular economy is effective. In order to achieve proper waste management, many plans and technologies have been ventured into and put in place. Among the numerous effective measures, some of them are enhancing the recycling infrastructure, sustainable consumption, extended producer responsibility, as well as the introduction of modern technologies and already-implemented recycling strategies [47, 48]. Moreover, such alternatives as second-hand clothes and the development of large-scale recycling centres can not only assist in the minimisation of waste but also provide more jobs [43].

13.5.3 Zero-waste management

The idea of zero waste has developed as a result of different cultural, economic and technological influences in different historical eras. The simplest types of clothing were made in the early history of mankind by stitching leather fragments with plant fibres, by primitive methods of draping and fastening to satisfy minimum requirements in the sphere of protection. They may be viewed as one of the first specimens of a zero-waste clothing design [19, 34]. Nevertheless, other more developed civilisations such as Ancient Egypt, Greece and Rome invented comparatively more sophisticated techniques in converting fabrics into clothes. These methods were planned to utilise materialised material and generate a minimum or no trash, which expressed an early awareness of resource efficiency. In these designs, where the entire cloth is converted into covering, there is no pre-consumer waste generated, and they are also valuable as a source of reference to make contemporary zero-waste clothes [49].

This style is currently used in most countries and cultures, both as an everyday style or, at most, a traditional or ceremonial style. However, in Asia and Africa, it is complex. The pre-consumer zero-waste fashion design consists of designing clothes with a concentration on the fabric efficiency based on the principle of sustainability. This process contrasts the conventional pattern-making techniques, which do not focus on experimental usage in the pattern development process, re-patterning and adaptable designing, such that a single product can be converted into various design variations. In this area, the concept of geometric alignment is often employed in order to have the maximum harmony of materials. Nevertheless, outside the field of geometry, 3D modelling and computer-aided design (CAD)/computer-aided manufacturing (CAM) software have good potential in the advancement of novel techniques and improving design procedures to implement more efficient and sustainable garment manufacturing [30, 50].

Zero waste can be categorised into the following key areas:

- **Modular and transformable design**: Creating garments that can be easily adjusted, reconfigured, or adapted for multiple uses, extending their life cycle and reducing waste.
- **Material use and efficiency**: Maximising fabric utilisation and minimising leftover scraps through thoughtful design and planning.
- **Pattern and cutting techniques**: Employing innovative pattern-making and cutting methods that eliminate or significantly reduce fabric waste during the production process.

These categories collectively support the goal of sustainable fashion by optimising resources and minimising environmental impact throughout the design and manufacturing stages [49].

13.6 Digitalisation and AI in sustainable design

In the textile industry, digital technologies play a significant role in enforcing the rules for sustainable production and the circular economy. They encourage the use of renewable sources and materials. The industry can implement a faster supply-chain cycle through the use of digital technologies, including online design exhibitions and the selling of products. These technologies make it more efficient and more environmentally friendly and capable of adaptation to fashion. Digital transformation of the textile industry involves the implementation of innovative technologies such as the Internet of Things, blockchain, AI and big data analytics in the traditional business manufacturing process and supply chain. These tools improve efficiency, transparency and sustainability by helping to monitor in real-time, anticipate maintenance, share information in a secure manner and make decisions throughout the entire value chain and source of raw materials, production, distribution and communication with customers. Textile and fashion industry digital technologies also include CAD and CAM, digital printing and 3D printing. These technologies facilitate the design and manufacturing process, minimise the wastage of materials and permit a high level of personalisation as well as accuracy. In addition, there are several significant benefits in the development of the 3D printing trend, which include:

- on-demand production, reducing overstock and waste;
- rapid prototyping enables faster design iterations and innovation;
- complex and lightweight structures that are difficult to achieve with traditional methods;
- customisation at scale, allowing unique designs tailored to individual consumers.

This shift towards digital and additive manufacturing is significantly reshaping the future of sustainable and efficient textile production [30].

AI has emerged as a transformative force, reshaping the fashion industry's approach to sustainability. It plays a critical role in the design, production and marketing of sustainable products, while also influencing consumer behaviour towards more eco-conscious choices. It enables brands to:

- design smarter by analysing trends, consumer preferences and sustainability metrics;
- streamline production to reduce waste, optimise material use and improve efficiency;
- enhance marketing strategies through personalised recommendations and targeted campaigns focused on eco-conscious consumers.

The concept of AI is changing the creation of sustainable materials. Machine-learning models can recreate the material properties and, therefore, shorten the process of innovation and commercialisation of eco-friendly fabrics like lab-grown leather, mycelium-based textiles and recycled polyester. Biomimicry has been used to understand how organisms such as spider silk are produced to create biodegradable materials that provide high-performance and limited environmental impacts. This is a combination of technology and nature that is introducing a new age of green fabric innovation. The AI is being used to improve the textile practice, specifically, in dyeing, which is consuming a lot of water and emitting toxic chemicals, with reduced consumption.

Non-toxic alternatives to dyes can be recognised with the help of AI, which contributes to the minimisation of environmental impact and reduce the total energy usage in the textile industry. AI applications, including Generative Adversarial Networks, can help designers digitally design clothes, enabling them to choose and optimise styles on the computer and then print them to be made. This not only saves on material waste but also accelerates the design process and helps in sustainable practices [44].

The use of AI by designers is becoming a central aspect in the human–computer interaction community, since AI systems have deep learning platforms capable of simulating the more traditional manual design processes. These systems can help the designers with repetitive tasks, create through data-driven insights and provide quicker, more original and sustainable design solutions. In the past, a designer was left alone with the design process. Modern textile design is based today on collaboration and communication between designers. This has led to the importance of critical thinking and problem-solving skills in a team. In order to work more collaboratively, Google Ventures came up with Design Sprints in which small groups of specialists collaborate within a limited time. Such sprints bring together experts to identify issues and develop new solutions, which is a more effective and proactive way of solving problems than the traditional ones [24].

13.7 Consumer behaviour and ethical design

Environmental consumerism, sometimes called green purchasing, includes the selection and utilisation of eco-friendly products. Ethical consumerism shows consumers who choose products deliberately to support brands and products aligned with social and environmental protection supporting aspects. This includes the sustainable practices, for example, equitable labour conditions, environmentally responsible production and a transparent supply chain. The enhanced awareness about the harmful effects of traditional fashion practices normally shifts consumer behaviour. Ethical consumerism is developed by those factors that influence buying. The theory of planned behaviour indicates four main elements: ethical identity, social norms, attitude and perceived control over actions, with trust that plays a key moderating role [51]. Personal and social norms have a great influence on ethical purchases, while moral identity places a lot on the intentions of consumers. Among the younger consumers, sustainability moved beyond consumer preferences, with second-hand fashion and vintage, attaining significant popularity, influencing and shifting preferences towards distinctive and sustainable clothing. This resurgence is fuelled by economic influences and shifting preferences towards sustainable and distinctive clothing. The sellers increasingly moved to the uniqueness of meaningful narratives, bound to the craftsmanship of vintage garments [17]. There are some other factors that influence consumer behaviour, e.g., emotions, individual motivation, perceptions, decision-making and attitude [17]. According to a survey, ethical consumerism is moving beyond sustainability into the age of regenerative models that aggressively look for improvement, where companies target to restore the ecosystem and make a positive social change. Although the adaptation and awareness of these models are growing, the challenges related to cost, measurement and policy support are still there [52]. A study on Chinese and Spanish consumers depicted that sustainability and social responsibility are attaining prominence in the fashion industry, where consumers favour ethical businesses naturally. These businesses develop positive social, economic and environmental outputs along with reducing the negative impact of textile products. Institutions, manufacturers and retailers promoting ethical values must acknowledge that the unique cultural traits of every region influence shaping the consumer perception, attitudes and intentions. Thus, adopting a localised approach to foster dialogue and enhance awareness about ethical fashion may be more effective than depending completely on a global approach based on globalisation [51].

13.8 Challenges of sustainable design

In recycling, there are big challenges that behave like barriers, including technological limitations, material complexity, limited recycling-oriented design and some gaps in consumer awareness. For the advanced materials, all these factors, including some other factors, affect the development of effective recycling. Mitigating these challenges requires a collaborative effort between the stakeholders, for example, manufacturers, consumers and policymakers, for the sustainable practice and improvement of recycling technologies [32]. Another big challenge, according to a survey, is that organic and renewable materials are costly compared to conventional fabrics. Most of the material that is referred to as sustainable is difficult to find in the local market. Most are available online, while consumers are frustrated over the colour and quality of the material. Because of this, many consumers had to choose the dyeing option to achieve the required colours. There must be an option for swatch testing in the online market. On the other hand, through the dyeing process, achieving the required colour is also difficult, because sometimes it needs multiple dyeing sessions. A lot of consumers cannot find green material easily [53]. Handcrafted items are also becoming less in demand, as it takes a lot of effort and time, i.e., hand spinning, hand weaving, natural dyeing, printing, embroidery and embellishment etc. It is a major part of the rural community, as it is a major source of income for them. Due to a lack of demand for handicrafts, unemployment rises and leads locals to leave their villages. Waste management, especially zero-waste material, is also a big challenge for the textile and fashion industry [22].

13.9 Conclusion

The textile and fashion industries are joined at a specific point where sustainability is no longer a trend but a requirement for future needs. In this chapter, multidirectional approaches to sustainable designs from the adoption of eco-friendly raw materials, circular approaches for different production, to the integration of digital technology and ethical consumer behaviours have been explored. The recent transfer of the industry towards the regenerative low-impact raw materials, for example, organic cotton, bio-based textiles and recycled polyester elaborated its potential to minimise the overall environmental footprint. Furthermore, the AI-driven production, innovations in digital design and smart textiles provide promising pathways to reduce waste generation and increase efficiency. However, some other significant issues still exist, for example, fast fashion continues to contribute to excessive utilisation of raw materials and textile waste generation, which highlights the requirement for systematic change in consumer behaviour as well as in the business model. Some other factors also impact the overall progress, including a lack of standardised recycling infrastructure, persistent greenwashing practices and high cost for sustainable materials. To solve these obstacles, it is important that multilateral efforts by policymakers, manufacturers, designers, and consumers are made to create transparency, encourage sustainable policies and educate about ethical fashion. Finally, sustainable textile design lies in the future, but it will have to be holistic, balancing the innovative with the responsible, the economically viable with the eco-friendly, and the short-term trends with the long-term sustainability. The fashion industry can become a global sustainability leader through the adoption of circularity, the use of technology and the focus on making products ethically instead of being a significant polluter. It will be a complicated process, yet with further dedication and imagination, a genuinely sustainable fashion ecosystem is a possibility.

References

[1] S. Islam and G. Bhat, 'Environmentally-Friendly Thermal and Acoustic Insulation Materials from Recycled Textiles', *J Environ Manage*, 251 (December 2019): 1–21.

[2] T. Vadicherla, D. Saravanan, M. M. Ram and K. Suganya, *Textiles and Clothing Sustainability*, (2017).

[3] A. Khan, 'Cutting Down on Textile Waste: Top Tech Innovations in 2025 & Beyond', (2025), www.3-tree.com/blog/cutting-down-on-textile-waste-top-tech-innovations-in-2025-beyond/.

[4] browzwear, 'Sustainable Fashion Trends to Watch Out for in 2025', (21 April 2025), https://browzwear.com/blog/sustainable-fashion-trends.

[5] Ethik, 'Sustainable Fashion Trends in 2025 for the Conscious Consumer', (2025), https://ethikonline.com/blogs/featured-articles/sustainable-fashion-trends-2025.

[6] A. Borbely, 'Insight of the Day: 10 Sustainable Fashion Trends to Watch in 2025', (2025), www.insighttrendsworld.com/post/insight-of-the-day-10-sustainable-fashion-trends-to-watch-in-2025.

[7] M. T. Rahaman and M. S. Hossain Khan, 'Biomaterials for Manufacturing Environmentally Sustainable Textiles and Apparel: Sources, Applications, Challenges, Enablers and Future Directions', *Int J Environ Sci Technol*, 22 (10) (June 2025): 9655–710, doi: 10.1007/s13762-025-06481-7.

[8] A. A. Jessica, 'Sustainable Merchandising Practices in the Textile Sector', (March 2025).

[9] B. Jena, B. P. Das, A. Khandual, S. Sahu and L. Behera, 'Ecofriendly Processing of Textiles', *Mater Today Proc*, 2 (4–5) (2015): 1776–91, doi: 10.1016/j.matpr.2015.07.020.

[10] M. Koszewska, 'Life Cycle Assessment and the Environmental and Social Labels in the Textile and Clothing Industry', in S. Senthilkannan Muthu (ed.), *Handbook of Life Cycle Assessment (LCA) of Textiles and Clothing*, Woodhead Publishing, (2016), pp. 325–44. doi: 10.1016/B978-0-08-100169-1.00015-0.

[11] S. Senthilkannan, M. Miguel and A. Gardetti, *Sustainable Textiles: Production, Processing, Manufacturing & Chemistry Sustainability in the Textile and Apparel Industries Sourcing Synthetic and Novel Alternative Raw Materials*, Springer, (2020).

[12] T. C. Durham and T. Mizik, 'Comparative Economics of Conventional, Organic, and Alternative Agricultural Production Systems', *Economies*, 9 (2) (2021): 1–22, doi: 10.3390/economies9020064.

[13] G. M. W. Lengai and J. W. Muthomi, 'Biopesticides and Their Role in Sustainable Agricultural Production', *J Biosci Med*, 6 (6) (2018): 7–41, doi: 10.4236/jbm.2018.66002.

[14] Beyond Pesticides, 'Defining a Strong IPM or EPM Program', (n.d.), www.beyondpesticides.org/resources/safety-source-on-pesticide-providers/what-is-integrated-pest-management.

[15] S. Plakantonaki, K. Kiskira, N. Zacharopoulos, I. Chronis, F. Coelho, A. Togiani, K. Kalkanis and G. Priniotakis, 'A Review of Sustainability Standards and Ecolabeling in the Textile Industry', *Sustainability*, 15 (15) (July 2023): 1–18, doi: 10.3390/su151511589.

[16] P. Examiner and M. Bernshteyn, 'Biodegradable Textile Yarn and Textile Made from Recycled Materials', US 10,858,512 B1, (2020).

[17] J. Krywalski-Santiago, 'Navigating Sustainable Transformation in the Fashion Industry: The Role of Circular Economy and Ethical Consumer Behavior', *J Intercult Manag*, 16 (3) (September 2024): 5–36, doi: 10.2478/joim-2024–0009.

[18] G. Saxena, R. Kishor, S. Zainith and R. N. Bharagava, 'Environmental Contamination, Toxicity Profile and Bioremediation Technologies for Treatment and Detoxification of Textile Effluent', in G. Saxena, V. Kumar and M.P. Shah (eds), *Bioremediation for Environmental Sustainability*, Elsevier, (November 2021), pp. 415–34. doi: 10.1016/B978-0-12–820524-2.00017-1.

[19] D. Çınar, 'Zero Waste Supply Chain: Conceptual Framework, Application Areas, and Future Perspectives', *Uluslararası Batı Karadeniz Sos. ve Beşeri Bilim. Derg.*, 8 (2) (December 2024): 262–79, doi: 10.46452/baksoder.1572222.

[20] S. Muñoz, M. R. Hosseini and R. H. Crawford, 'Towards a Holistic Assessment of Circular Economy Strategies: The 9R Circularity Index', *Sustain Prod Consum*, 47 (June 2024): 400–12, doi: 10.1016/j.spc.2024.04.015.

[21] M. Umair and R. M. W. U. Khan, 'Fibers for Sports Textiles', in S. Ahmad, A. Rasheed and Y. Nawab (eds), *Fibers for Technical Textiles*, Springer, (2020), pp. 93–115. doi: 10.1007/978-3-030-49224-3_5.

[22] M. Jain, 'Challenges for Sustainability in Textile Craft and Fashion Design', *Int J Appl Home Sci*, 5 (2) (2018): 489–96, http://scientificresearchjournal.com/wp-content/uploads/2018/01/Home-Science-Vol-5_A-489-96-Full-Paper.pdf.

[23] S. Ratner, K. Gomonov, S. Revinova and I. Lazanyuk, 'Ecolabeling as a Policy Instrument for More Sustainable Development: The Evidence of Supply and Demand Interactions from Russia', *Sustainability*, 13 (17) (August 2021): 1–21, doi: 10.3390/su13179581.

[24] D. Jung and S. Suh, 'Enhancing Soft Skills through Generative AI in Sustainable Fashion Textile Design Education', *Sustainability*, 16 (16) (2024), doi: 10.3390/su16166973.

[25] B. Piribauer and A. Bartl, 'Textile Recycling Processes: State of the Art and Current Developments – A Mini Review', *Waste Manag. Res. J. a Sustain. Circ. Econ.*, 37 (2) (February 2019): 112–19, doi: 10.1177/0734242X18819277.

[26] M. E. Rashid, M. R. Khan, R. U. Haque and M. Hasanuzzaman, 'Challenges of Textile Waste Composite Products and Its Prospects of Recycling', *J. Mater. Cycles Waste Manag.*, 25 (3) (May 2023): 1267–87, doi: 10.1007/s10163-023-01614-x.

[27] D. Bhatia, A. Sharma and U. Malhotra, 'Recycled Fibers: An Overview', *Int J Fiber Text Res*, 4 (4) (2014): 77–82.

[28] Aleksandra Ivanoska-Dacikj, Lutz Walter, Karin Eufinger, Ariadna Detrell, Yesim Oguz Gouillart, Enrico Venturini, Raul Fangueiro, Daniela Zavec, Paulo Cadeia, Vincent Nierstrasz, Bruno Mougin, 'Smart Textiles: Paving the Way to Sustainability', *Maced J Chem Chem Eng*, 43 (1) (2024): 149–64, doi: 10.20450/mjcce.2024.2821.

[29] A. Martikkala, B. Mayanti, P. Helo, A. Lobov and I. F. Ituarte, 'Smart Textile Waste Collection System: Dynamic Route Optimization with IoT', *J Environ Manage*, 335 (March 2023): 117548, doi: 10.1016/j.jenvman.2023.117548.

[30] M. Glogar, S. Petrak and M. Mahnić Naglić, 'Digital Technologies in the Sustainable Design and Development of Textiles and Clothing: A Literature Review', *Sustainability*, 17 (4) (2025): 1–26, doi: 10.3390/su17041371.

[31] A. Noor, M. A. Saeed, T. Ullah, Z. Uddin and R. M. W. Ullah Khan, 'A Review of Artificial Intelligence Applications in Apparel Industry', *J Text Inst*, 113 (3) (2022): 505–14, doi: 10.1080/00405000.2021.1880088.

[32] M. G. Tadesse, M. T. Abate, J. Felix Lübben and M. Rauch, 'Recycling and Sustainable Design for Smart Textiles: A Review', *Adv Sustain Syst*, 9 (8) (2025): 1–25, April 2025, doi: 10.1002/adsu.202401072.

[33] G. Rebitzer, T. Ekvall, R. Frischknecht, D. Hunkeler, G. Norris, T. Rydberg, W.-P. Schmidt, S. Suh, B.P. Weidema, and D.W. Pennington, 'Life Cycle Assessment', *Environ Int*, 30 (5) (July 2004): 701–20, doi: 10.1016/j.envint.2003.11.005.

[34] I. Zdonek, M. Podgórska and B. Hysa, 'The Sustainable Fashion Value Proposition of Companies Identifying with the Zero Waste Movement', *Sustainability*, 17 (3) (January 2025): 887, doi: 10.3390/su17030887.

[35] A. Joy, J. F. Sherry, A. Venkatesh, J. Wang and R. Chan, 'Fast Fashion, Sustainability, and the Ethical Appeal of Luxury Brands', *Fash. Theory*, 16 (3) (September 2012): 273–95, doi: 10.2752/175174112X13340749707123.

[36] N. Tokatli, 'Global Sourcing: Insights from the Global Clothing Industry – The Case of Zara, a Fast Fashion Retailer', *J Econ Geogr*, 8 (1) (August 2007): 21–38, doi: 10.1093/jeg/lbm035.

[37] P. Smita Pathak and M. G. Hegde, 'Fast Fashion and Sustainability: A Paradoxical Relationship', *SISDM-2023 Sustainable Fashion & Design – II FAST*, (2024), no. March, pp. 184–94.

[38] Fashion ABC, 'Zara: A Spanish Fast Fashion Retailer Brand', (n.d.), www.fashionabc.org/wiki/zara/.

[39] R. M. W. U. Khan and K. Shaker, 'Life Cycle Assessment of Textile Products', in S. R. Batool, S. Ahmad, Y. Nawab and M. Hussain (eds), *Circularity in Textiles*, Cham: Springer Nature, (2023), pp. 159–76. doi: 10.1007/978-3-031-49479-6_6.

[40] R. M. W. U. Khan, A. F. M. Seyam, M. Hussain and Y. Nawab, 'Valorizing Denim and Polyethylene Sheet Waste to Fiber-Reinforced Composites', *Polym Compos*, 46 (13): 12091–111, doi: 10.1002/pc.29730.

[41] S. Kushwaha and C. Swami, 'Upcycling of Leather Waste to Create Upcycled Products and Accessories', *Int J Home Sci*, 2 (2) (2016): 187–92.

[42] Kim and Joo Hyun, 'A Study of High Value-Added Upcycled Handbag Designs for the Dubai Luxury Fashion Market', *Indian J Tradit Knowl*, 14 (1) (2014): 173–88.

[43] A. K. Das, M. F. Hossain, B. U. Khan, M. M. Rahman, M. A. Z. Asad and M. Akter, 'Circular Economy: A Sustainable Model for Waste Reduction and Wealth Creation in the Textile Supply Chain', *SPE Polym*, 6 (1) (2025): 1–34, doi: 10.1002/pls2.10171.

[44] K. Rana, S. P. Banerjee, V. Agarwal and P. Chadha, 'Examining the Role of Artificial Intelligence in Influencing Consumer Preferences and Purchase Intentions toward Green Fashion', *Metall Mater Eng*, 31 (4) (April 2025): 982–92, doi: 10.63278/1542.

[45] S. Cuc and M. Vidovic, 'Environmental Sustainability through Clothing Recycling', *Oper Supply Chain Manag An Int J*, 4 (2) (December 2014): 108–15, doi: 10.31387/oscm0100064.

[46] A. S. Sigaard and K. Laitala, 'Natural and Sustainable? Consumers' Textile Fiber Preferences', *Fibers*, 11 (2) (January 2023): 1–23, doi: 10.3390/fib11020012.

[47] J. P. Juanga-Labayen, I. V. Labayen and Q. Yuan, 'A Review on Textile Recycling Practices and Challenges', *Textiles*, 2 (1) (2022): 174–88, doi: 10.3390/textiles2010010.

[48] M. A. Bukhari, R. Carrasco-Gallego and E. Ponce-Cueto, 'Developing a National Programme for Textiles and Clothing Recovery', *Waste Manag Res J a Sustain Circ Econ*, 36 (4) (April 2018): 321–31, doi: 10.1177/0734242X18759190.

[49] H. Y. Gözene and H. H. MetliOğlu, 'Zero Waste Approach in Sustainable Fashion Design: Designer Perspective for Pre-Consumer Waste Management', *Tekst ve Muhendis*, 32 (137) (2025): 79–93, doi: 10.7216/teksmuh.1471793.

[50] B. D. Caddock and K. E. Evans, 'Microporous Materials with Negative Poisson's Ratios. I. Microstructure and Mechanical Properties', *J Phys D Appl Phys*, 22 (12) (December 1989): 1877–82, doi: 10.1088/0022–3727/22/12/012.

[51] A. Pérez, J. Collado-Agudo and M. T. Liu, 'A Cross-cultural Comparison of Consumer Intention to Purchase Ethical Fashion', *J Sustain Mark*,6 (1) (2025): 1–21, doi: 10.51300/JSM-2025–143.

[52] J. A. Antonyraj, 'Ethical Consumerism and the Rise of Regenerative Business Models: Beyond Sustainability towards Net-Positive Impact', *International Journal of Scientific Research and Engineering Development*, 8 (3) (2025): 203–7.

[53] H. J. Gam and J. Banning, 'Addressing Sustainable Apparel Design Challenges with Problem-Based Learning', *Cloth Text Res J*, 29 (3) (2011): 202–15, doi: 10.1177/0887302X11414874.

14 Conclusion

Ali Raza Shafqat and Yasir Nawab

14.1 Introduction

This is the concluding chapter of the book, and it brings together the framework's insights and case studies discussed throughout the book with a single purpose to move the knowledge of sustainability into action. Throughout the book, we have tried to weave the social, economic, environmental and cultural dimensions of sustainability into a questioning narrative that demonstrates how the education design business and policy together can help shape a more sustainable future of the fashion and textile sector. The mission of this book has been clear to equip the designers, educators, brands, policymakers and ultimately the readers of this book with the strategies, tools and conceptual foundations required to transform the business from the linear model of economy towards the sustainable and circular ecosystem based on the equity cultural expectations and regeneration of materials. Now, as the readers have been armed with the strong foundations of concept and practical case studies, we should transition towards a compelling call to action and highlight the responsibilities of various stakeholders and discuss the opportunities in realising this systematic transition. Figure 14.1 demonstrates how systems thinking observed sustainability design thinking and concept

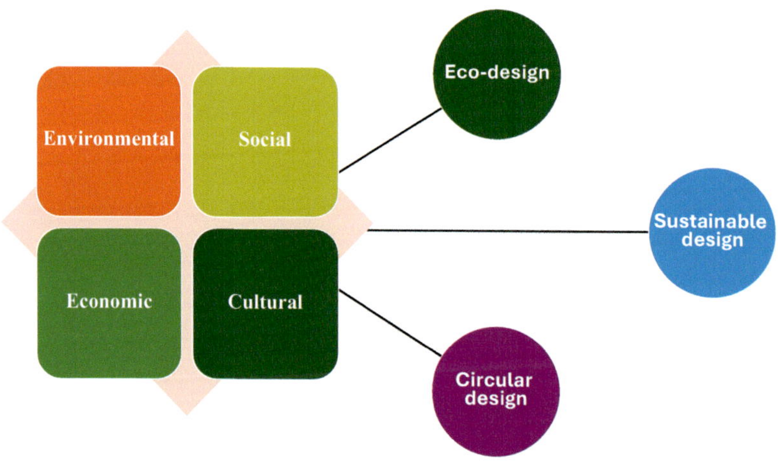

Figure 14.1 Sustainability design thinking and concept in terms of systems thinking.

DOI: 10.4324/9781003638148-18

14.2 Recap of key insights, takeaways and conceptual foundations

We studied the evolution of design paradigms by tracing the journey of eco design and green design principles through sustainable design History and moved towards formal material loops and models like cradle to cradle. We then expanded the scope of sustainability beyond mere environmental protection and included the cultural, social and economic factors. It should be kept in mind that sustainable design is not only about creating a natural ecosystem but also balances the regeneration of material with the aesthetic values, community well-being and profitability.

The discussion on sustainable materials covered key aspects of recycled and regenerative agricultural inputs, along with the merits and demerits of regular and organic fibres. We then studied innovations in detail, including low-impact finishes, waterless dyeing and finishing, along with ZHC-aligned chemical interventions. After that, the design for disassembly and circular economic systems and business models were explored, keeping in view the use of circular strategies, the application of modular designs, fibre-to-fibre recycling systems, upcycling, take-back programmes and other techniques to enable closing the supply chain circular loops. Figure 14.2 shows life-cycle-based highlights of sustainability. The frameworks were then discussed, including the application of blockchain digital product passports and certifications like OEKO-TEX, Global Organic Textile Standard, Higg Materials Sustainability Index (MSI), Product Environmental Footprint (PEF), Material Circularity Indicator (MCI), etc., that help provide transparency and verification across the infrastructure and hold the supply chains accountable. After that, we discussed the current and emerging regulatory landscape and Ecodesign for Sustainable Products Regulation (ESPR), Anti-Gaspillage pour une Économie Circulaire, California's Global Warming Potential (GWP) Act and, Extended Producer Responsibility (EPR), etc., that are promoting the concepts of sustainability into practice.

14.3 Stakeholder roles in sustainable design

The implementation of sustainable design involves multiple stakeholders, ranging from educators, designers, businesses, brands and policymakers [1]. Figure 14.3 shows the important roles of all stakeholders in achieving sustainable design.

The educators can help shape the capacity of their students, who are future professionals in the field. These institutions can equip them with theory, practice, knowledge of interdisciplinary concepts and applied learning, which can ultimately foster innovation in this field [2].

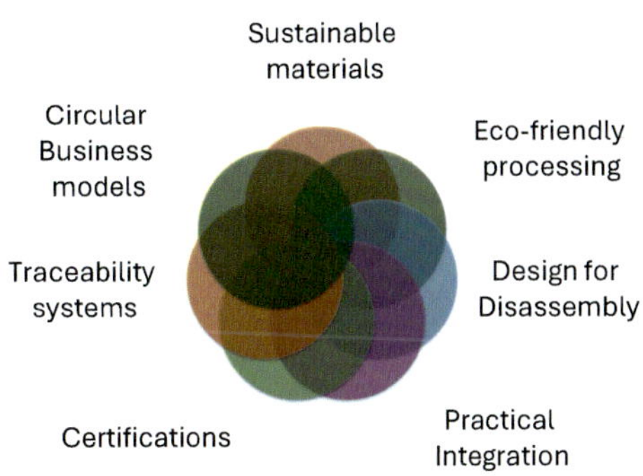

Figure 14.2 Life-cycle-based highlights of sustainability.

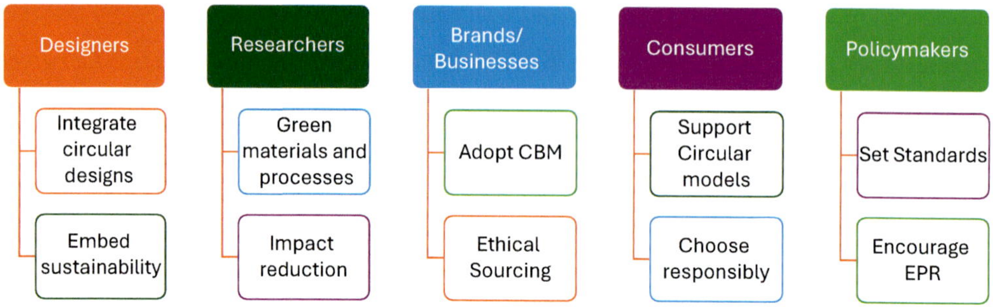

Figure 14.3 Stakeholders' roles in achieving sustainable design.

Businesses and brands can bear the responsibility for value retention and create value from the end-of-life phase of their products. They can also embed the circular economy by investing in supply-chain innovations and upholding circular business models like resale, repair and rental services, etc. [3].

The designers are really in charge of product design; they can help achieve sustainability by selecting the raw materials and processes that have the lowest impact on the environment, and then, at the end of the life of their products, they can also engage across the complete supply chain and interdisciplinary networks during each phase of their product's life [4]. Their intervention could be very helpful in achieving sustainability in the fashion and textile industry.

Policymakers can also be very impactful; they can enable sustainability by enforcing frameworks, support the skills and innovations in the field and binding the businesses and brands in extended responsibilities.

14.4 Call to action for key stakeholders

This section provides practical recommendations for all stakeholders involved in the implementation of sustainability in the fashion and textile industry; this section highlights their importance, outlines their roles and responsibilities to drive this change.

14.4.1 The role of designers

The designers can embed sustainability into the products right from the initial concept of the product [5], and they can incorporate life-cycle thinking along with circular design strategies like the use of mono material, modularity and other concepts, even before initiating production within their design briefs [6]. The early-stage assessments and incorporation of tools like Higg MSI, life-cycle assessment (LCA), MCI, etc., to make the process choices or use benchmark materials can help achieve sustainability right from the very beginning. The collaboration of designers with recyclers, policymakers and material scientists can help achieve more sustainable design, along with keeping up with emerging regulations [7, 8]. The important roles of designers for ensuring sustainability in fashion and textile sector are highlighted in Figure 14.4.

The designers can use the tools for upcycling discs, probably, and modularity. For example, they can integrate their tools like CLO 3D and Product Lifecycle Management (PLM) to enhance their product designs and incorporate the recycling outcomes and disassembly at the design stage. By using proper tools, they can establish systems and specify the easily removable components like panels, frames and zippers, as demonstrated by Yikes Zip, which is easily

Figure 14.4 Roles of designers for ensuring sustainability in fashion and textile sector.

detachable and recyclable. The designers should utilise the tools and platforms like Browzwear, CLO 3D, etc., to reduce waste in their design processes. Designers can also partner up with the recycling systems on various platforms that help reclaim materials like recovered textile systems and Renewcell, etc. They can also collaborate with upcycling and repairing forums to show that their designs are supportive of longer value retention and ultimately have the support of community engagement [9, 10].

14.4.2 The role of businesses and brands

Businesses and brands can redefine their values beyond the mere profits; they can help transition from having a single-time profit by selling the product and can offer the product as a service by sharing subscription and rental modes to decouple the material throughput from the revenue. So, in this way, they can generate income and profits throughout the complete life cycle of the product rather than just getting it once. Furthermore, they can also get incentives from the implementation of sustainability via eco-modulated fees or carbon taxes [11, 12]. Few important roles of brands and businesses for ensuring sustainability in fashion and textile sector are highlighted in Figure 14.5.

Businesses can adopt circular business models and launch or scale the take-back programmes like those offered by The North Face or Patagonia, and they can also introduce the refurbishment and repair of these models. If these businesses cannot do these models on their own, they can collaborate with startups and small and medium enterprises to introduce material innovation in their products and can also diversify their own portfolio, which can also help them minimise their own reliance on virgin raw material [13, 14]. The brands and businesses can use MSI, BRM, Higg Index and Open LCA to measure their environmental impact and have better reportability. The businesses and manufacturers can use True Trace and Textile Genesis to incorporate material verification mechanisms and supply chain transparency [15].

These businesses and brands can invest in like capacity and innovation, by offering the grants and technical education to the suppliers they can uplift their own materials supply and can transition to low environmental impact processes and have fully functional digital traceability systems for their own good, Furthermore they can also invest in green materials development

Figure 14.5 Roles of brands and businesses for ensuring sustainability in fashion and textile sector.

and recycling infrastructure to strengthen their pledge for upstream circularity especially for blended material products.

14.4.3 The role of educational institutions

Educational institutions and universities should embed sustainability and circularity as core elements in their curriculum. They should integrate interdisciplinary courses that combine environmental science, design and business principles. They can train the students with hands-on training on upcycling LCA, designing for disassembly and real-world solutions towards a more sustainable fashion and textile industry [16]. The students must be facilitated to form team-based capstones involving students from different fields and backgrounds, like supply chain, engineering and design. They should be trained in an industrial setup in the co-design studios, incubators, innovation labs and prototypes of circular solutions according to the needs of industrial stakeholders [17]. The important roles of educational institutions for incorporate sustainability in fashion and textile Scholars are highlighted in Figure 14.6.

Figure 14.6 Roles of educational institutions for incorporate sustainability in fashion and textile Scholars

Figure 14.7 Roles of policymakers to incorporate sustainability in fashion and textile sector.

The students should be supported for their sustainability projects and should be provided with seed funding and necessary mentorship so that the student-led circular fashion initiatives can be supported. The impacts of their projects should be disseminated via exhibitions, conferences and other networking events to bring their initiatives into broader visibility. The educators can use IDEO circular labs and CIRCO track, etc., as toolkits for integrating sustainability in their programmes for inspiring circular thinking in professionals of tomorrow.

14.4.4 The role of legislation and policymakers

Policymakers should develop robust and inclusive policies, facilitate access to green innovative solutions and strengthen sustainability-related skills. The government should enforce extended producer responsibilities and apply eco-modulated products to reach recycling targets [18]. They should introduce criteria for minimal design for procurement or trade of sustainable materials and products related to the fashion and textile industry. They should implement strategies to utilise taxes, incentives and loans to support the small industries adopting these circular business model technologies and certifications. They should be provided with the proper finances to make this happen [19]. The important roles of policymakers to incorporate sustainability in fashion and textile sector are highlighted in Figure 14.7.

The government should help in building private-public partnerships for shared recycling infrastructure and municipal waste collection systems. They should also partner up with educational institutions and industries to provide training that supports circular economy skills by embedding sustainability and circularity as core criteria in professional and vocational training programmes [20]. Policymakers can employ the United Nations Environment Programme fashion policy tracking departments or learning systems to have insights into fashion-related best regulatory practices around the globe.

14.5 The future for sustainable fashion and textiles

The application of sustainability and the circular economy in the fashion and textile industry have far-reaching consequences, along with several opportunities [21]. Figure 14.8 presents some hotspots for future perspectives in fashion and textile sector from perspective of sustainability. This section helps with the identification of the upcoming trends that would shape the sector with a good new wave of innovations, some of which are discussed below:

By introducing rental and refurbishment business models, fashion textile products can normalise non-ownership styles in routine, which helps not only in reducing the waste but also enables consumers to have a wide variety of products or have their products useful for a longer period of time [22].

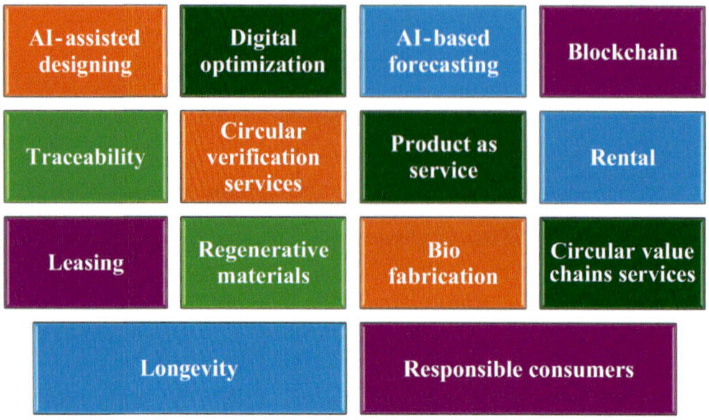

Figure 14.8 Hotspots for future perspectives in fashion and textile sector from perspective of sustainability.

The consumer mindset shift from consumption towards a meaningful engagement with the product by attaching the cultural identity or longevity to the garment, which can be very central and crucial for the cultural scalability of sustainable products [23]. Mainstreaming the global value chains by ensuring that the circular economy practices are followed equitably both in the Global North and the Global South, which requires massive building infrastructure investment and governance alike in both parts of the world [24].

Artificial intelligence, digital twins and blockchain all can help in establishing a new generation of design systems that use digital material passports, an artificial intelligence-optimised supply chain, enhanced transparency, use of predictive and forecasting analytics for circular alignment and customisation of products [25]. Use of bio-fabricated and regenerative materials has introduced a lot of innovations, such as the development of leather in a lab, the development of fibres from algae, banana, pineapple leaves or other unconventional sources, and leather from banana leaves, etc., always suggest opportunities to reduce the use of virgin raw materials [26].

We have arrived at the concluding remarks of this work. It is an important and vital movement where urgency aligns with possibility. The scale of resource depletion, labour treatment and environmental damage is huge, but it also provides the fashion and textile sector with a lot of opportunities for equity regeneration and resilience. Throughout this book, the concepts of circular and sustainable designs have been presented as ethical and creative imperatives, using which educators, designers, businesses and policymakers can build more and more systems that support life rather than exploiting the natural ecosystem. It should be of a size that a meaningful transformation from linear to circular economy and a sustainable textile and fashion sector can happen. It requires long-term commitment, adaptive leadership and a collective action of all stakeholders. The collective action and voice of all stakeholders will determine the trajectories and the potential of the future of the fashion and textile industry. This chapter emphasised that collective integration of sustainability across the whole supply chain should be implemented, and designers, engineers and all other stakeholders should be empowered to implement their concept of the end of life of their designed products using circular thinking strategies. The brands and the businesses should reshape their revenue-generating models by investing in upstream innovation and services; the educational curriculum should be reformed to enable sustainable and circular literacy while producing future professionals. The public policy should be enforced in such a way that it not only penalises but also acts as a catalyst towards circular transition.

References

[1] M. Hussain and A. R. Shafqat, 'Sustainability and Materials', in K. Shaker and Y. Nawab (eds), *Engineering Materials*, Springer, (2025), pp. 257–74, doi: 10.1007/978-3-031-72263-9_11.

[2] O. M. Ventista and C. Brown, 'Teachers' Professional Learning and Its Impact on Students' Learning Outcomes: Findings from a Systematic Review', *Social Sciences & Humanities Open*, 8 (1) (January 2023): 100565, doi: 10.1016/J.SSAHO.2023.100565.

[3] D. Stabler, H. Hakala, T. Huikkola and A. L. Mention, 'Aligning Servitization and Circularity: The Role of Institutional Confluence in Sustainable Business Models', *J Clean Prod*, 462 (July 2024): 142666, doi: 10.1016/J.JCLEPRO.2024.142666.

[4] A. T. Mengistu, M. Dieste, R. Panizzolo and S. Biazzo, 'Sustainable Product Design Factors: A Comprehensive Analysis', *J Clean Prod*, 463 (July 2024): 142260, doi: 10.1016/J.JCLEPRO.2024.142260.

[5] A. R. Shafqat and A. Saccavini, 'Eco-design of Textiles', in *Eco-design of Textiles*, (2023), pp. 39–58, doi: 10.1007/978-3-031-49479-6_3.

[6] S. Val and M. P. Lambán, 'Enhancing Sustainability with LCA: A Comparative Analysis of Design and Manufacturing Processes', *Processes*, 13 (1) (January 2025): 195, doi: 10.3390/PR13010195.

[7] S. S. Muthu (ed.), *Roadmap to Sustainable Textiles and Clothing*, Singapore: Springer, (2015), doi: 10.1007/978–81–287-164-0.

[8] R. Strippoli and T. Gallucci, 'Life Cycle Assessment for Sustainable Business Management', *Top Italian Scientists*, 2 (3) (2025), doi: 10.62684/SBXL2822.

[9] A. Morandi and C. Tonelli, 'The Impact of 3D Technology on Sustainability in Fashion Luxury Product Design: Evidences from an Exploratory Analysis', MS thesis, Politecnico di Milano, Milan, Italy, (2024).

[10] Browzwear, 'How to Integrate Fashion PLM with 3D Clothing Design Software', (20 September 2024), https://browzwear.com/blog/integrate-plm-with-3d-software.

[11] Angus McOuat, Dymfke Kuijpers, Patricio Ibáñez, and Ryan Drassinower, 'A Turning Point for Private Label Brands', McKinsey & Company, (4 November 2024), www.mckinsey.com/industries/retail/our-insights/a-turning-point-for-private-brands-how-retailers-can-seize-the-opportunity.

[12] M. Laukkanen and N. Tura, 'Sustainable Value Propositions and Customer Perceived Value: Clothing Library Case', *J Clean Prod*, 378 (December 2022): 134321, doi: 10.1016/J.JCLEPRO.2022.134321.

[13] P. A. Albinsson, H. R. Chaudhuri, A. Das, S. Ray and L. Nayak, 'Marketing Sustainable Fashion: Trends and Future Directions', *Sustainability*, 15 (7) (April 2023): 6202, doi: 10.3390/SU15076202.

[14] G. M. Michel, M. Feori, M. L. Damhorst, Y. A. Lee and L. S. Niehm, 'Stories We Wear: Promoting Sustainability Practices with the Case of Patagonia', *Fam. Consum. Sci. Res. J.*, 48 (2) (2019): 165–80, doi: 10.1111/FCSR.12340.

[15] N. M. P. Bocken, 'Circular Business Model Innovation: New Avenues and Game Changers', *Business Model Innovation: Game Changers and Contemporary Issues* (January 2024): 193–225.

[16] A. G. Abo-Khalil, 'Integrating Sustainability into Higher Education Challenges and Opportunities for Universities Worldwide', *Heliyon*, 10 (9) (May 2024): e29946, doi: 10.1016/J.HELIYON.2024.E29946.

[17] F. Alhassani, M. R. Saleem and J. Messner, 'Integrating Sustainability in Engineering: A Global Review', *Sustainability*, 17 (15) (July 2025): 6930, doi: 10.3390/SU17156930.

[18] O. Pricopoaia, A. Lupaşc and I. O. Mihai, 'Implications of Innovative Strategies for Sustainable Entrepreneurship: Solutions to Combat Climate Change', *Sustainability*, 16 (22) (November 2024): 9742, doi: 10.3390/SU16229742.

[19] R. Agrawal, S. Agrawal, A. Samadhiya, A. Kumar, S. Luthra and V. Jain, 'Adoption of Green Finance and Green Innovation for Achieving Circularity: An Exploratory Review and Future Directions', *Geoscience Frontiers*, 15 (4) (July 2024): 101669, doi: 10.1016/J.GSF.2023.101669.

[20] G. Batool, M. Bashir, S. A. Baig and A. R. Shafqat, 'Ecotourism Taking Modernity to the Deprived Regions of the World', in M. Abrar, M. Asim Nawaz, F. Kiran and F. Asmi (eds), *Supporting Environmental Stability Through Ecotourism*, IGI Global, (2024), pp. 143–57, doi: 10.4018/979-8-3693-1030-4.ch008.

[21] A. R. Shafqat, M. Hussain, Y. Nawab, M. Ashraf, S. Ahmad and G. Batool, 'Circularity in Materials: A Review on Polymer Composites Made from Agriculture and Textile Waste', *Int. J. Polym. Sci.*, (2023): 5872605, doi: 10.1155/2023/5872605.

[22] N. Islam, M. T. Khawar, M. Hussain and M. B. Ramzan, 'Sustainable Business Model: Feasibility Study for a Clothing Library to Promote Responsible Consumption', *Journal of Engineered Fibers and Fabrics*, 20 (January 2025): 1–15, doi: 10.1177/15589250251346274.

[23] P. Petänen, H. Tuovila and P. Heikkilä, 'Strategic Marketing of Sustainable Fashion: Exploring Approaches and Contradictions in the Positioning of Fashion Rental', *Cleaner Production Letters*, 7 (December 2024): 100075, doi: 10.1016/J.CLPL.2024.100075.

[24] A. P. M. Velenturf and P. Purnell, 'Principles for a Sustainable Circular Economy', *Sustain Prod Consum*, 27 (July 2021): 1437–57, doi: 10.1016/J.SPC.2021.02.018.

[25] I. Markou, D. Sinnott and K. Thomas, 'Current Methodologies of Creating Material Passports: A Systematic Literature Review', *Case Studies in Construction Materials*, 22 (July 2025): e04267, doi: 10.1016/J.CSCM.2025.E04267.

[26] S. Duangsuwan, P. Junkong, P. Phinyocheep, S. Thanawan and T. Amornsakchai, 'Development of Green Leather Alternative from Natural Rubber and Pineapple Leaf Fiber', *Sustainability*, 15 (21) (October 2023): 15400, doi: 10.3390/SU152115400.

Index

For Product Safety Concerns and Information please contact our EU representative GPSR@taylorandfrancis.com Taylor & Francis Verlag GmbH, Kaufingerstraße 24, 80331 München, Germany

Batch number: 10722134

Printed by Printforce, the Netherlands